Climate Obstruction across Europe

CLIMATE OBSTRUCTION ACROSS EUROPE

Edited by
Robert J. Brulle
J. Timmons Roberts
Miranda C. Spencer

OXFORD
UNIVERSITY PRESS

OXFORD
UNIVERSITY PRESS

Oxford University Press is a department of the University of Oxford. It furthers
the University's objective of excellence in research, scholarship, and education
by publishing worldwide. Oxford is a registered trade mark of Oxford University
Press in the UK and certain other countries.

Published in the United States of America by Oxford University Press
198 Madison Avenue, New York, NY 10016, United States of America.

© Oxford University Press 2024

Some rights reserved. No part of this publication may be reproduced, stored in
a retrieval system, or transmitted, in any form or by any means, for commercial purposes,
without the prior permission in writing of Oxford University Press, or as expressly
permitted by law, by licence or under terms agreed with the appropriate
reprographics rights organization.

This is an open access publication, available online and distributed under the terms of a
Creative Commons Attribution – Non Commercial – No Derivatives 4.0
International licence (CC BY-NC-ND 4.0), a copy of which is available at
http://creativecommons.org/licenses/by-nc-nd/4.0/.

You must not circulate this work in any other form
and you must impose this same condition on any acquirer.

Library of Congress Cataloging-in-Publication Data
Names: Brulle, Robert J., editor. | Roberts, J. Timmons, editor. | Spencer, Miranda C., editor.
Title: Climate obstruction across Europe / [edited by] Robert J. Brulle, J. Timmons Roberts,
Miranda C. Spencer.
Description: New York, NY : Oxford University Press, [2024] |
Includes bibliographical references and index.
Identifiers: LCCN 2024002171 (print) | LCCN 2024002172 (ebook) |
ISBN 9780197762042 (hardback) | ISBN 9780197762059 (paperback) |
ISBN 9780197762073 (epub) | ISBN 9780197762080
Subjects: LCSH: Climatic changes—Government policy—Europe. |
Climatic change mitigation—Europe—Prevention.
Classification: LCC QC903.2.E85 C855 2024 (print) | LCC QC903.2.E85 (ebook) |
DDC 363.7/0561094—dc23/eng/20240310
LC record available at https://lccn.loc.gov/2024002171
LC ebook record available at https://lccn.loc.gov/2024002172

DOI: 10.1093/oso/9780197762042.001.0001

Paperback printed by Marquis Book Printing, Canada
Hardback printed by Bridgeport National Bindery, Inc., United States of America

CONTENTS

List of Figures   vii
List of Tables   ix
List of Contributors   xi

1. Introduction: The First Portrait of Climate Obstruction across Europe   1
   Robert J. Brulle and J. Timmons Roberts

2. Climate Obstruction in the United Kingdom: Charting the Resistance to Climate Action   26
   Freddie Daley, Peter Newell, Ruth McKie, and James Painter

3. Climate Obstruction in Scotland: The Politics of Oil and Gas   57
   William Dinan, Victoria Esteves, and Steven Harkins

4. Climate Obstruction in Ireland: The Contested Transformation of an Agricultural Economy   80
   Orla Kelly, Brenda McNally, and Jennie C. Stephens

5. Climate Obstruction in Sweden: The Green Welfare State—Both Progressive and Obstructionist   109
   Kjell Vowles, Kristoffer Ekberg, and Martin Hultman

6. Climate Obstruction in Germany: Hidden in Plain Sight?   136
   Achim Brunnengräber, Moritz Neujeffski, and Dieter Plehwe

7. Climate Obstruction in the Netherlands: Strategic and Systemic Obstruction of Dutch Climate Policies (1980–Present)   162
   Martijn Duineveld, Guus Dix, Gertjan Plets, and Vatan Hüzeir

8. Climate Obstruction in Poland: A Governmental–Industrial Complex   186
   Kacper Szulecki, Tomas Maltby, and Julia Szulecka

9. Climate Obstruction in Russia: Surviving a Resource-Dependent Economy, an Authoritarian Regime, and a Disappearing Civil Society   214
   Marianna Poberezhskaya and Ellie Martus

10. Climate Obstruction in the Czech Republic: Winning by Default   243
    Milan Hrubeš and Ondřej Císař

11. Climate Obstruction in Italy: From Outright Denial to Widespread Climate Delay   268
    Marco Grasso, Stella Levantesi, and Serena Beqja

12. Climate Obstruction in Spain: From Boycotting the Expansion of Renewable Energy to Blocking Compassion Toward Animals   294
    Jose A. Moreno and Núria Almiron

13. Climate Obstruction in the European Union: Business Coalitions and the Technocracy of Delay   320
    Dieter Plehwe, Moritz Neujeffski, and Tobias Haas

14. Conclusion: Ten Lessons about Climate Obstruction in Europe   347
    J. Timmons Roberts and Robert J. Brulle

Index   365

FIGURES

1.1 A visualization of the networks of relationships between agents and organizations seeking to obstruct climate action and the methods they use to do so (money and/or influence).   7

1.2 The rectangles represent the total quantity of greenhouse gases (GHGs) for each of the eleven European countries studied and for the remaining twenty-seven European Union nations.   16

2.1 Total greenhouse gas (GHG) emissions (in MMT CO_2e) and percentage change in emissions in the United Kingdom between 1990 and 2021, inclusive.   28

2.2 A timeline of significant obstruction activities in the United Kingdom, 1988–2022. The upper part of the figure shows key legislative changes or events; the lower section provides examples of obstruction activities and the actors and institutions involved.   33

3.1 Total greenhouse gas (GHG) emissions (in MMT CO_2e) and percentage change in emissions in Scotland between 1990 and 2021, inclusive.   59

4.1 Total greenhouse gas (GHG) emissions (in MMT CO_2e) and percentage change in emissions in the Republic of Ireland between 1990 and 2021, inclusive.   83

5.1 Total greenhouse gas (GHG) emissions (in MMT CO_2e) and percentage change in emissions in Sweden between 1990 and 2021, inclusive.   112

6.1 Total greenhouse gas (GHG) emissions (in MMT CO_2e) and percentage change in emissions in Germany between 1990 and 2021, inclusive.   141

6.2 The development and achievement of greenhouse gas (GHG) emissions (in MMT CO_2e) under the Federal Climate Protection Act, by sector.   142

7.1 Total greenhouse gas (GHG) emissions (in MMT CO_2e) and percentage change in emissions in the Netherlands between 1990 and 2021, inclusive.   *163*

7.2 Petroleum (gas and oil) deposits in the Netherlands as of 2023. Although new gas fields are being discovered in the North Sea, the easternmost Slochteren field represents one of the largest land-based gas fields in Europe.   *165*

8.1 Total greenhouse gas (GHG) emissions (in MMT CO_2e) and percentage change in emissions in Poland between 1990 and 2021, inclusive.   *188*

9.1 Total greenhouse gas (GHG) emissions (in MMT CO_2e) and percentage change in emissions in Russia between 1990 and 2021, inclusive.   *215*

10.1 Total greenhouse gas (GHG) emissions (in MMT CO_2e) and percentage change in emissions in Czechia between 1990 and 2021, inclusive.   *245*

10.2 Czech climate obstruction activists' relationships/networks.   *257*

11.1 Total greenhouse gas (GHG) emissions (in MMT CO_2e) and percentage change in emissions in Italy between 1990 and 2021, inclusive. *269*

12.1 Total greenhouse gas (GHG) emissions (in MMT CO_2e) and percentage change in emissions in Spain between 1990 and 2021, inclusive.   *295*

13.1 Total greenhouse gas (GHG) emissions (in MMT CO_2e) and percentage change in emissions in the European Union between 1990 and 2021, inclusive.   *324*

13.2 Comparison of total lobby expenditures in 2019 for the 'green' and 'grey' alliances.   *330*

TABLES

1.1 The objectives, activities, and players in climate obstruction over three time frames, with examples of organized efforts opposed to climate action   8

4.1 Top three lobbying activities on 'climate' by organizations, September 2015–December 2022   92

4.2 Discourses of delay from agricultural sector and rural politicians in response to proposed emissions-reductions targets   99

6.1 Major German grey (nuclear/fossil) firms   146

6.2 Major Germany grey (nuclear/fossil) business associations The ranking rates lobbying transparency and positions taken on the Paris Agreement goals on a scale from A to F.   147

8.1 The Polish ministries responsible for climate, Energy, and the Environment between 1999 and 2024, inclusive   195

8.2 Ownership of energy and mining companies in Poland   196

9.1 Examples of how various actors contribute to the identified discursive framings   229

10.1 Czech Think tanks' 'hard' strategies and tactics   255

10.2 Types of discursive obstruction strategies and the actors involved   260

10.3 Types of discursive obstruction strategies and their frames   263

13.1 EU climate targets by year and status   326

CONTRIBUTORS

Núria Almiron is an associate professor with a full professorship accreditation at Universitat Pompeu Fabra. Her main research topics include the political economy of communication, the ethics of communication, and persuasion, among others.

Serena Beqja is an MSc candidate in the Energy, Environment and Society program at the University of Stavanger. Her research centres on the pervasive role of the fossil fuel industry in society.

Robert J. Brulle is a visiting professor of Environment and Society at Brown University and the Research Director of the Climate Social Science Network. His research focuses on US environmental politics, critical theory, and the political and cultural dynamics of climate change.

Achim Brunnengräber is a political scientist and an associate professor in the Department of Political and Social Sciences at Free University of Berlin. His research areas include energy, climate, and environmental governance; civil society participation; and socioecological transformation.

Ondřej Císař is a professor of Sociology at Charles University in Prague and the editor-in-chief of the Czech edition of *Czech Sociological Review* at the Institute of Sociology of the Academy of Sciences of the Czech Republic.

Freddie Daley is a research associate at the Centre for Global Political Economy at the University of Sussex. His research focuses on sustainable behaviour change, supply-side energy policies, and energy transitions.

William Dinan is a senior lecturer in the Division of Communication, Media and Culture in the Faculty of Arts and Humanities at the University of Stirling. His research interests include political communication and public affairs, environmental communication, and civil society advocacy.

Guus Dix is an assistant professor at the Center for Higher Education Policy Studies at the University of Twente. As an economic sociologist and sociologist of science, he studies the relationship between scientific knowledge and political power.

Martijn Duineveld is an associate professor in the Cultural Geography Group at Wageningen University. He is engaged in research that exposes how large industry obstructs transitions to a low-emission society and climate justice.

Kristoffer Ekberg is an assistant professor in Human Ecology at Lund University and part of the Centre for Studies of Climate Change Denialism at Chalmers University of Technology. His research focuses include the political history of the environment and climate change.

Victoria Esteves is a lecturer in Creative Industries in the Communications, Media and Culture division at the University of Stirling. Her research interests span online culture, internet memes, remix, DIY, and participatory culture.

Marco Grasso is a professor of Political Geography in the Department of Sociology and Social Research at the University of Milano-Bicocca. His research focuses on sustainability transitions, the role and responsibility of the fossil fuel industry in climate change, and climate denial and delay.

Tobias Haas is a research associate at the Research Institute for Sustainability (RIFS) in Potsdam. His research interests lie at the intersection of heterodox political economy and political ecology, with a regional focus on Germany and the EU.

Steven Harkins is a lecturer in Journalism Studies in the Division of Communication, Media and Culture at the University of Stirling. His research focuses on journalism and society and examines the role of news in shaping narratives on social, political, and environmental issues.

Milan Hrubeš is an assistant professor in the Philosophical Faculty at the University of Hradec Králové. His research interests include social movements, policy analysis, and political language analysis.

Martin Hultman is an associate professor in Science, Technology and Environmental Studies, and he directs the Centre for Studies of Climate Change Denialism at Chalmers University, where he leads three research groups.

Vatan Hüzeir is a sociologist and PhD candidate at the Erasmus School of Social and Behavioural Sciences, Erasmus University Rotterdam. He studies urban climate activists' critical discourse on the relationships between climate change and social order in northwestern Europe.

Orla Kelly is an assistant professor of Social Policy at the University College Dublin School of Social Policy, Social Work and Social Justice. She specializes in the social dimensions of climate change, with expertise in quantitative research and mixed methodologies.

Stella Levantesi is a freelance climate journalist, photographer, and author trained at New York University's Journalism Institute. She researches the history of climate change denial and climate action obstruction tactics.

Tomas Maltby is a reader in International Politics in the Department of Political Economy at King's College London. Maltby's research focuses primarily on the development of climate and energy policy.

Ellie Martus is a senior lecturer in Public Policy in the School of Government and International Relations at Griffith University. Her research explores the role of industry in shaping environmental politics, extractive industry responses to climate change, and climate politics in Russia.

Ruth McKie is a senior lecturer in Quantitative Criminology at De Montfort University and a member of its Institute for Research in Criminology, Community, Education and Social Justice. Her core research interests include climate change countermovements and anti-environmentalism.

Brenda McNally is a postdoctoral research fellow at the Centre for Climate and Society, Dublin City University. She is an interdisciplinary social scientist specializing in climate politics and communication.

Jose A. Moreno is a predoctoral researcher in the Department of Communication at the Universitat Pompeu Fabra. His research focuses on climate change communication, climate action obstructionism, journalism, and interest groups.

Moritz Neujeffski is a social scientist and Ph.D. student at the University of Tübingen. He has researched the transformation of the European cohesion policy and how a greater focus on regional competitiveness has changed development projects within the regions of the European Union.

Peter Newell is a professor of International Relations at the University of Sussex and co-founder of the Rapid Transition Alliance. He has undertaken research, advocacy, and consultancy work on climate change for more than twenty-five years.

James Painter is a research associate at Oxford University's Reuters Institute for the Study of Journalism, Department of Politics and International Relations and a senior teaching associate at Oxford's Environmental Change Institute, School of Geography.

Dieter Plehwe is a senior fellow at the Center for Civil Society Research at the WZB Research Group and a lecturer at the University of Kassel's Department of Political Science. His research interests include transformation and varieties of capitalism and neoliberalism, among others.

Gertjan Plets is an associate professor in Cultural History and Chair of the Cultural History section at Utrecht University. He coordinates and conducts research in the field of heritage politics, environmental history, and petro-nationalism.

Marianna Poberezhskaya is an associate professor in Politics and International Relations at Nottingham Trent University. Poberezhskaya's research interests include environmental communication, climate scepticism, geoengineering, and national climate governance.

J. Timmons Roberts is the Ittleson Professor of Environmental Studies and Sociology at Brown University and the executive director of the Climate Social Science Network (CSSN). His current research focuses on social drivers of action and inaction on climate change.

Miranda C. Spencer is a writer, editor, and project manager with forty years of experience in the publishing and media industries. She has worked primarily as a journalist and book editor focusing on environmental and health topics.

Jennie C. Stephens is the Dean's Professor of Sustainability Science & Policy at Northeastern University's School of Public Policy and Urban Affairs and a visiting professor at Trinity College Dublin. She is an internationally recognized expert on energy transformation, among other topics.

Julia Szulecka is a researcher in the Water and Society section at the Norwegian Institute for Water Research. Her research focuses on sustainability governance in the food, water, and forestry sectors as well as environmental politics.

Kacper Szulecki is a research professor of International Climate Governance at the Norwegian Institute of International Affairs and a fellow at Include, the University of Oslo's research centre for socially just energy transitions.

Kjell Vowles is a PhD candidate at Chalmers University of Technology. With a background as a climate journalist, he focuses his research on media discourses, climate obstruction as part of far-right ideology, and the use of nationalism to hinder climate action.

1

Introduction

The First Portrait of Climate Obstruction across Europe

ROBERT J. BRULLE AND J. TIMMONS ROBERTS

EUROPE'S GLOBAL ROLE IN CLIMATE ACTION

Decades of effort to address anthropogenic climate change have failed to decrease the greenhouse gas (GHG) emissions that are destabilizing Earth's life-support systems. Many theories of why we have failed have been advanced, but one reason has barely been studied: the well-organized efforts to obstruct climate action. Since the 2010s, an expanding body of investigative reporting and academic research has documented an extensive, well-organized enterprise, led by corporations and their affiliated trade associations, to interfere with progress on reducing carbon emissions.[1] Yet, for the most part, these impediments remain marginal to the public discussion on how best to address climate change.

Europe, as a cultural region and a political bloc, has taken the need to act on climate change more seriously than most other parts of the world. For decades, Europe has seen itself as a leader on climate action, and, in the more than thirty years of United Nations (UN) negotiations on the issue, the European Union (EU) has brought leading pledges and policy ideas to the table.[2] However even its efforts have been inadequate, uneven, and halting. Some climate policies in Europe have been rolled back, and others are threatened by economic crises, war, global competition, and authoritarianism. We must ask: Who are the actors and organizations obstructing climate action in Europe? What are their strategies, and how are those evolving? This volume seeks to advance our understanding of climate obstruction in the region as a whole and to learn from the significant variations across the continent.

To date, research into systematic efforts to obstruct climate action has focused primarily on the United States and been concentrated on the activities of a few major oil companies and a handful of publicly visible conservative think tanks. As a result, an inaccurate picture has emerged, centring on a few American industrial giants and organizations, particularly Exxon Mobil, the Koch brothers, the Competitive Enterprise Institute, and the Heartland Institute. However, the nature and extent of these organized efforts to obstruct climate action are far broader, more complex, and geographically dispersed than often portrayed. Recent scholarship has shown that they span multiple sectors, including agriculture, transportation, coal, and utilities, among others. As the essays in this book show, climate obstruction efforts take place across all of the European countries, each with its own particular characteristics. National industries and their trade organizations seek to slow climate action even in the 'greenest' countries. Research on the role of conservative think tanks, for example, reveals an increasingly coordinated and multinational effort to promulgate scientific misinformation and advocate against rapid and robust climate action by undermining confidence in renewable energy and other legitimate climate solutions.

The popular but inaccurate image of climate obstruction efforts as extremely limited in sectoral and geographic scope is both an academic and a practical concern. Addressing the lack of effective political action on climate change requires pulling back the curtain on the constellation of organized interests engaged in the contentious politics of climate change, the nature of their activities, and their impact on both public perception of the climate crisis and the policymaking process. It also requires an understanding of the actions climate advocates have taken to effectively overcome these efforts. As this volume shows for the first time, these constellations differ in important ways depending on national context, even within subregions of the continent.

Recently a growing number of scholars have moved beyond studying American obstructionism and have turned to researching various aspects of climate obstruction across Europe. This work opens up new perspectives on how various institutional actors in these nations influence climate policy based on their particular cultural and political structures. Capitalizing on this trend, the Climate Social Science Network's (CSSN) fall 2022 call for chapter proposals on the theme yielded eleven national case studies and one focused on the European Union. The cases are not exhaustive but do include countries in Europe's four major subregions: Northern Europe, the United Kingdom (UK), Eastern Europe, and Southern Europe. These studies show that entrenched interests vary significantly by country and region and that political structures create widely different opportunities for these interests to block, dilute, delay, or even reverse required action on climate change. And they show that, after exerting influence in their national arenas, these industrial interests frequently exploit a second opportunity to slow action by working to diffuse wider efforts in Brussels, the de facto capital of the European Union and the home of the European Commission (EC). Their collected findings form the basis of this book.

This introductory chapter lays out the basics of what we already know about how climate action is being obstructed. It begins with a review of the more than thirty years of scientific assessments by the Intergovernmental Panel on Climate Change (IPCC), and how the IPCC reports have handled obstructive action against climate solutions. We then outline the types of organizations involved and the main types of short-, medium-, and long-term strategies they have developed to avoid regulation and influence public opinion. These sections describe the 'structure of climate obstruction' in very general terms, and the insights from this issue area can be useful to those seeking to understand resistance to policy on other topics. After a brief review of Europe's emissions history, we introduce the forthcoming chapters and provide a brief overview of the core arguments presented. A fuller synthesis of the twelve case studies and the lessons they offer is covered in the concluding chapter.

IMPROVING ASSESSMENTS

One explanation for the minimization of obstruction efforts in humanity's inadequate response to climate change has been the failure of the IPCC to focus on this important factor. Formed in 1988 to bring together scientists from around the world to summarize scientific knowledge on climate change and possible solutions, the IPCC has produced six massive 'assessment reports', achievements that have vastly improved our understanding of the issue. Capping five years of intensive research by hundreds of authors and thousands of reviewers, each IPCC report is organized around the work of three Working Groups: one documenting changes in Earth's climate and their causes, the second looking at impacts around the world, and the third focused on mitigation, or efforts to reduce the emissions causing human-caused warming.[3] Therefore, Working Group III (WG III) would be expected to be the place where obstruction of emissions-reduction efforts is systematically reviewed and discussed.

Unfortunately, for the most part, the IPCC reports have minimized their attention to intentional efforts to hinder meaningful policy action to rapidly reduce GHG emissions. The latest IPCC Synthesis Report (AR6, released in 2023) contains no mention of organizational barriers to mitigation efforts in its Summary for Policymakers (SPM).[4] The Synthesis Report does mention unnamed 'institutional barriers' to mitigation efforts and a statement that 'developmental pathways create unintentional . . . barriers to accelerated mitigation'.[5] Both these statements lack mention of any actors and express an inevitability in the situation we face. In this way, thirty-five years on, IPCC reports still fail to clearly address intentional climate obstruction efforts in their leading summaries. Because the press and policymakers seldom examine the IPCC reports beyond the SPM, this limits the public discussion of climate obstruction.

It is hard to blame the IPCC authors for this omission in the report's summary, which is vetted by nearly every government on Earth and is frequently watered down, with key text struck from the final document before publication.[6] As de Pryck has documented, 'both authors and governments seek to have their perspectives reflected', and their interests and strategies are often in tension. As a result, we see 'the entanglement between the scientific and diplomatic rhetoric in the fabric of the SPM, which tends to construct climate change as a decontextualized and nonpolitical problem'. Still, the IPCC is under attack from right-wing organizations and media outlets that have advanced climate change denial.[7] Governments and other major economic actors would prefer to avoid attention to their failures and the ways they are being influenced. Sensitive to this, the scientists rewriting sections or wording of the report in response to government comments seek to avoid bringing up political issues and endangering the already fragile legitimacy of the organization.

Unlike the SPM, however, the full IPCC reports are not subject to government review, and important progress can be seen in their presentation of obstructive actors and practices in the most recent assessments. Though buried deep within the latest WG III report on mitigation (2022), a number of important conclusions regarding intentional efforts to oppose climate mitigation can be found. In the introductory chapter, the report concludes that 'Political and institutional dynamics shape climate change responses in important ways, not the least because incumbent actors have frequently blocked climate policy'.[8] Citing peer-reviewed studies of campaigns by oil and coal companies in the United States, Australia,

Brazil, South Africa, Canada, Norway, and Germany, the WG III authors concluded that 'One factor limiting the ambition of climate policy has been the ability of incumbent industries to shape government action on climate change'.[9] The authors also report that 'Countermovement coalitions work to oppose climate mitigation', and that 'A good number of corporate agents have attempted to derail climate change mitigation by targeted lobbying and doubt-inducing media strategies'.[10] Finally, the report notes that 'Accurate transference of the climate science has been undermined significantly by climate change counter-movements, particularly in the USA in both legacy and new/social media environments through misinformation, including about the causes and consequences of climate change'.[11]

The historic failure of the IPCC to accurately convey the extent and importance of organized efforts to obstruct climate action in its major public-facing documents hinders the global discussion of actions that can be taken to increase the pace and extent of mitigation efforts. As US Senator Sheldon Whitehouse put it, doing so is like telling the story of *Star Wars* without mentioning Darth Vader. This situation is improving, but not quickly enough. While it is clear from the full Working Group report that the IPCC is aware of this literature, the analyses of climate obstruction efforts remain buried in the text of the reports themselves and are not widely circulated in either media or policy discussions due to their absence from the key document, the SPM. Despite the aforementioned growing body of peer-reviewed research, awareness of these activities outside of the United States is limited, media coverage is still rare and mostly limited to a few news outlets, and scholarship remains scattershot throughout the social science literature. This knowledge gap limits the building of a coordinated research effort and inhibits the type of social movements and government policies that could remove major barriers to adequate and effective climate action. In turn, this lack of information allows climate obstruction efforts across the globe to continue uncontested.

This volume is the first effort in the much-needed task of collecting and disseminating existing knowledge on the scope and nature of obstruction efforts across the nations of the world. Because research on the Continent is accelerating and the analyses contained in this volume are likely to offer fruitful lessons for policymaking, Europe was the logical choice for a first region to examine in what we hope will become a series of books on climate obstruction across the globe. Before turning to the collected essays in this volume, we first review what is already known about the major sets of actors and strategies for obstructing action on climate change.

THE PRACTICE AND STRUCTURE OF CLIMATE OBSTRUCTION

The term 'climate obstruction' covers a wide variety of social, economic, and political practices. In this volume, we define climate obstruction as *intentional actions and efforts to slow or block policies on climate change that are commensurate with the current scientific consensus of what is necessary to avoid dangerous anthropogenic interference with the climate system.*[12]

Starting in the late 1980s, a broad range of actors with divergent interests entered into the public arena and engaged in a struggle to control public discussion and understanding of climate change and thus define appropriate policy responses. Extensive research has shown that, despite their knowledge of climate science and its implications, many corporations and trade associations, acting in coordination with conservative think tanks, foundations, and public relations firms, mounted a long-term effort to oppose action to mitigate the carbon emissions known to be responsible for climate change.[13] However, climate obstruction manifests differently in different parts of the world and by nation and can be compared with what we know about patterns of obstruction in the United States, where the most research to date has been conducted.

Moving from left to right in Figure 1.1, the network of organized opposition to climate change action begins with and is funded by wealthy individuals (and their philanthropic foundations), corporations, and foundations. These players fund and direct advocacy groups, advertising agencies, trade associations, think tanks, and university centres. These institutions then promulgate the positions of the funders through a network of blogs, social media, book publishing, sympathetic media outlets, lobbying firms, funding campaigns, and political action committees.[14] Climate change obstruction is often part of a broader political agenda, especially the effort to stop and roll back the power of the administrative state to address social issues. This libertarian and neoliberal movement has, since before the administrations of US President Ronald Reagan and British Prime Minister Margaret Thatcher, successfully shifted society's orientation away from governments and toward the rights of corporations. These various organizations act in different political and cultural arenas and employ different time horizons to achieve a range of objectives (Table 1.1). For these reasons, we cannot refer to the organized efforts to block or delay climate action in monolithic terms. Rather, these efforts stem from an amalgam of loosely coordinated groups that can be understood collectively as the climate change countermovement[15] (CCCM). Initially launched in the United States, the CCCM has taken root in other nations with histories of powerful fossil fuel industries and has been diffused

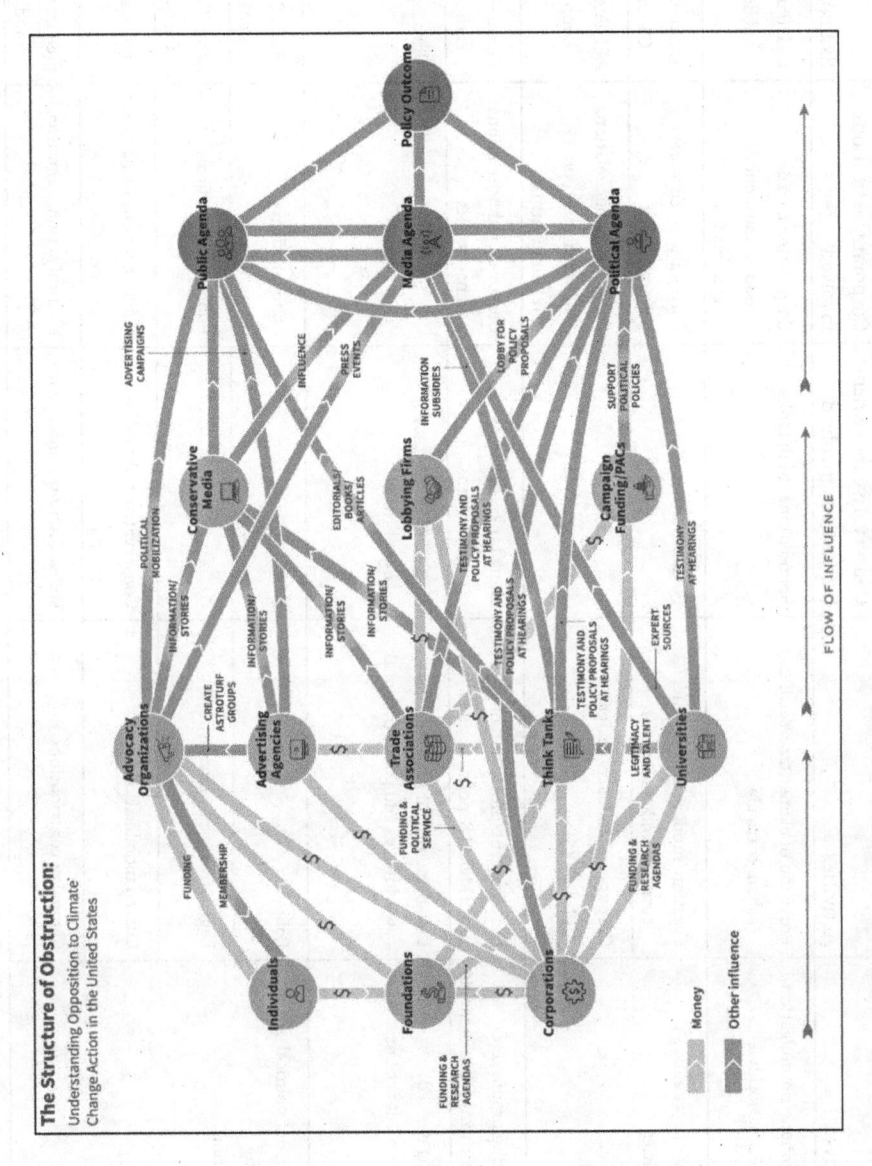

Figure 1.1 A visualization of the networks of relationships between agents and organizations seeking to obstruct climate action and the methods they use to do so (money and/or influence).

Table 1.1 THE OBJECTIVES, ACTIVITIES, AND PLAYERS IN CLIMATE OBSTRUCTION OVER THREE TIME FRAMES, WITH EXAMPLES OF ORGANIZED EFFORTS OPPOSED TO CLIMATE ACTION

Time frame	Objectives	Activities	Conservative movement institutions involved	Corporate institutions involved	Examples
Long-term 20 years to 5 years	Development/promulgation of specific worldview	Provision of elementary/secondary school curricula	Foundations think tanks	Corporations, trade associations, public relations firms	Heartland Institute publication, circulation
	Steering of academic activities	Creation/funding of academic curricula and research programs	Foundations	Corporations, corporate foundations	Stanford University 'Global Climate & Energy Project'
		Corporate & industry image promotion		Corporate public relations departments, public relations firms	API 'Fueling It Forward' campaign
Intermediate-term 5 years to 1 year	Circulating proposals & specific worldview in media	Development/promotion of specific worldview & policy actions	Think tanks, advocacy organizations	Corporate public relations departments	ExxonMobil proposed carbon tax
	Fostering desired government actions	Delegitimation of opposing worldviews and policy proposals	Think tanks, advocacy organizations		Climategate effort
Short-term 1 year to 6 months	Carrying out political action Elections Legislation	Lobbying		Lobbying firms	$8.6 million ExxonMobil spent on lobbying 2018
		Legislative issue advertising		Public relations firms	Accce 'Cold in the Dark Campaign'
		Citizen mobilization	Conservative political groups	Public relations firms, front groups	Americans for Prosperity
		Campaign contributions	Political action committees	Political action committees	Freedom Partners Action Fund

internationally primarily via networks of conservative think tanks. This countermovement, grounded in corporate interests seeking to maintain a fossil fuel-based energy system and its economic benefits, is augmented by a range of neoliberal ideological interests that are opposed to government regulations. Together, they are waging a concerted war against restrictions on carbon emissions.

A common tactic employed by these obstructive actors has been to deny the seriousness of anthropogenic climate change by manufacturing uncertainty about the scientific evidence, attacking climate scientists, and portraying climate science writ large as a controversial field—all of which are designed to undercut the perceived need for policies to address this crisis.[16] Starting in 1989, several conservative think tanks opposed to government regulatory action, often assisted by a small number of contrarian scientists, joined fossil fuel corporations in generating scientific misinformation about climate change. This information was then spread, and continues to be spread, by conservative media, sympathetic politicians, and other actors.[17] As climate impacts have accelerated, these efforts have placed more focus on delaying action and attacking proposed climate solutions such as renewable energy as expensive, unreliable, or even dangerous.

More recent scholarship aimed at understanding the forces that have thus far blocked effective efforts to reduce carbon emissions has broadened, focusing on funding for think tanks espousing denial and delay[18] and the larger network of actors involved in promoting climate change misinformation in which the think tanks and their funding sources are embedded.[19] Further research has shown that the promotion of scientific misinformation is only part of a much larger, integrated effort to develop and promulgate a consistent ideological message praising and defending fossil fuel use, which is then used to pressure decision-makers to limit efforts to reduce carbon emissions.[20]

From the beginning of organized opposition to climate action, coordinated information and influence campaigns, typically designed by advertising and lobbying firms, have been widely used by CCCM members (corporations, trade associations, and advocacy organizations) to achieve their political objectives—through either direct persuasion or generation of political pressure to influence the decision-making process.[21] This organizational strategy employs sophisticated public relations campaigns to simulate the appearance of a unified front that comprises diverse voices advocating for a uniform position. This perception is reinforced through the use of various communication strategies to reach different audiences, from members of parliament and prime ministers to influential media figures and key segments of the public at large.

In the next section, we provide an overview of climate obstruction by briefly reviewing the current literature on the practice and structure of climate obstruction efforts to establish a baseline from which to view the nature of climate opposition in Europe. The early research on the Global South indicates that different types of societies produce different forms of climate obstruction, including different discursive practices, organizational structures, and interactions among governing institutions.[22] This pattern is likely to emerge in the examination of European climate obstruction. Again, the material in this volume is based largely on the research conducted on climate obstruction efforts in the United States and offers only a preliminary perspective on the nature and extent of efforts to delay attempts to meaningfully address climate change.

The practice of climate obstruction

As noted earlier, key research has uncovered an integrated network of organizational relationships (sometimes termed the 'denial machine') that exists to influence the public, media, and political arenas to slow, stop, or reverse effective climate action. This countermovement is highly sophisticated, operates in multiple institutional arenas, and pursues a wide variety of coordinated strategies. These activities also operate in three distinct time frames: long term, intermediate term, and short term.[23] Table 1.1 provides an overview of these activities. As there are some variations in the activities of corporations and the conservative movement, that division of labour is noted in the figure.

Long-term activities

The first set of activities comprises long-term efforts ranging from five years to decades in duration. Their goal is to build and maintain a cultural and intellectual infrastructure of organizations that supports the development of ideas and policies favourable to conservative or industry viewpoints. One aspect of this effort is creating and maintaining academic programs at institutions of higher education, endowing academic chairs, and providing educational support for students in these programs.[24] In the United States, both corporations and the conservative movement engage in such efforts, which are only beginning to be documented in Europe. We can see their outcome in the proliferation of programs in economics and law that advocate Chicago School theories of neoliberal economics, which promote the value of a 'free market'.[25] Additionally, both fossil fuel corporations and conservative think tanks attempt to promulgate conservative ideas and support for fossil fuels in public schools, as exemplified by the Heartland Institute's circulation of misleading materials to secondary school science teachers.[26]

Another set of long-term activities in which corporations and affiliated trade associations engage is the development and implementation of corporate or industry-sector promotional campaigns to enhance their cultural legitimacy and thus defuse potential regulations. Such campaigns include sponsorship of cultural events and forums, one of the best-known examples of which is Mobil Oil's decades-long sponsorship of Masterpiece Theatre, the dramatic television series distributed by the Public Broadcasting Service.[27] This approach is known as 'affinity of purpose' advertising and seeks to improve the corporation's public image by associating it with scientific and cultural achievements.[28] Mobil also developed an aggressive public relations campaign. In 1970, the company began buying advertising space on *The New York Times'* editorial pages.[29] The campaign's overarching viewpoint was the purported need for growth in energy (oil) use and the economy.[30] Additionally, corporations engage in extended promotional advertising campaigns. To establish and enhance their legitimacy, these companies attempt to promote themselves as representing norms of rationality, progress, and appropriate conduct. Excellent examples of these sorts of campaigns are the American Petroleum Institute's 'Fueling It Forward' television and magazine ad campaign and BP's early 2000s' 'Beyond Petroleum' campaign. As of this writing, all of the major oil companies have ongoing major corporate promotion campaigns of this type, and, as this volume shows, other industries do as well.

Medium-term activities

The second set of climate opposition activities focuses on the intermediate time horizon of one to five years. This stage involves the translation and promulgation of scholarly ideas into concrete policies. One key example is Exxon's 2017 proposal for a carbon tax, which would have placed a small tax on carbon emissions while rolling back other regulations and indemnifying fossil fuel companies from civil suits related to their culpability for climate change. Such campaigns employ a wide range of channels to distribute their messages, from mass media to published books, and provide testimony at government hearings to influence legislation. The major institutional actors utilizing this time frame are think tanks, advocacy organizations, and public relations firms, which recruit credible third-party spokespersons to boost the legitimacy of their policy arguments.[31] Public relations firms play a further role in securing medium-term objectives by developing and disseminating materials that support policy objectives and by securing media coverage. Additionally, these same organizations seek to undermine the science of climate change by attacking the veracity of climate science and high-profile climate scientists. An example of this tactic was the 2009 'Climategate' affair, which involved denigrating several important climate scientists based on misinterpretations of their stolen emails.[32]

Short-term activities

The third set of climate obstruction activities focuses on short-term (six months to one year) political outcomes such as elections or pending legislation. Actors put considerable effort into influencing public opinion around climate change. One style of public opinion management is to promote positive perceptions of fossil fuel corporations through the extensive use of advertising campaigns. A second tactic involves citizen mobilization and/or the creation of front groups to demonstrate popular support for a political position. A third approach involves lobbying activities, either directly (by corporations or trade associations) or indirectly (through employing public affairs firms to influence legislative outcomes). In the United States, one notable example was the high levels of fossil fuel company spending in 2009 and 2010 to defeat the American Clean Energy and Security Act of 2009 (known as Waxman-Markey).[33] A fourth activity is targeted giving of political contributions via political action committees.

Information and influence campaigns are also used, which straddle the medium- and short-term time frames. Information and influence campaigns take the form of 'systemic, sequential and multifaceted effort[s] to promote information that orients the political decision-making process toward a desired outcome', either through direct persuasion or persuading other parties to bring pressure on decision-makers.[34] And as media outlets have proliferated, the bases of a public consensus have fragmented, and it can no longer be assumed that there is a commonly accepted position defining the basis of public discourse. 'Public discourse is fragmented structurally and culturally as different, incommensurable forms of interest come into competitive play'.[35] In this situation, organizations have powerful incentives to engage in activities to set the terms of the debate to favour their preferred policy outcomes.[36] Information and influence campaigns are highly sophisticated and coordinated actions that have now become a routine component of the political process. They are comprehensive, well-designed efforts that start with an analysis of the factors that impinge on the decision-making process and then bring pressure to bear to shift that decision in the desired direction. These campaigns involve communication, action, and relationship objectives all designed to manage the outcome to the advantage of the protagonist (client), in this case the opponents of climate action.

The structure of climate obstruction

The rapidly growing body of social science research reveals much about the major actors in the CCCM: not only who is responsible for obstructing efforts to mitigate climate change, but also their interrelationships and the strategies and tactics they employ. As discussed earlier, Figure 1.1 illustrates the organizations to which these actors belong and their relationships within the CCCM 'ecosystem'. These organizations seek to drive the overall policy agenda on climate change by influencing three arenas: (1) public opinion and what is seen to comprise the public agenda, (2) the media agenda; what and how the media cover climate change, and (3) the focus of political action and which actions politicians propose as their own political agendas. The following list describes the key actors in the US CCCM, the most-studied case against which the European CCCM players and national structures of obstruction can be compared.

1. *Corporations.* Since the early 1990s, individual corporations, especially fossil fuel companies like ExxonMobil, have engaged in efforts to obstruct climate action. These efforts include a wide range of activities, such as funding major misinformation campaigns[37] and large-scale corporate promotional advertising efforts,[38] along with traditional lobbying and political campaign contributions.
2. *Conservative foundations.* Several foundations have provided major funding to neoliberal think tanks that produce and disseminate climate change misinformation, challenging the need for government action on the crisis.[39] Research has shown that think tanks receiving foundation funding receive more attention in media and policymaking circles than do think tanks not receiving such funding.[40]
3. *Individuals.* When staging events in support of fossil fuels, the CCCM often uses corporate employees (and sometimes paid actors) as their spokespeople. However, some individuals exert enormous influence on their own, such as Charles Koch and the late David Koch. While

Koch family-affiliated foundations have played central roles in funding other actors, the brothers' personal and corporate networks provide numerous additional avenues of influence on policy issues such as energy and climate change.
4. *Advocacy coalitions.* Numerous corporations and trade associations from industry sectors facing threats of government regulation have banded together to form advocacy coalitions. These coalitions consolidate resources and engage in collective lobbying and public persuasion efforts to stop or slow regulatory action on climate change.[41]
5. *Advertising firms.* With the rise in concern over global climate change, fossil fuel interests have hired advertising firms to develop comprehensive public relations campaigns to both promote a positive image of their clients and discredit climate change mitigation efforts, including by designing campaigns against proposed legislation.[42]
6. *Trade associations.* Trade associations serve as mechanisms for corporations in similar industrial sectors to pursue collective political strategies by acting as command centres that help individual corporations pool resources, share information, and act as a collective political force.[43]
7. *Conservative think tanks.* As previously noted, by the early 1990s, many conservative think tanks had begun producing and disseminating climate change misinformation intended to sow doubt and confusion about global warming and the need to reduce carbon emissions. Global networks of think tanks—especially the Atlas Network—have also played a key role in diffusing denial internationally. Besides issuing press releases, policy reports, and books, think tanks' spokespersons have written op-eds, testified at congressional hearings, and given radio and television interviews to advance their goals.[44]
8. *Universities.* Major oil companies such as ExxonMobil, Shell, and Chevron Corporation fund large energy research programs at major universities over which they have considerable influence, leading these programs to take more industry-friendly approaches to addressing climate change.[45] Conservative foundations and individuals make major contributions that seek to support ideologies aligned with slowing climate action.
9. *Campaign funding/PACs.* Increasingly, corporations have been funding political action committees (PACs) as a way of influencing climate change legislation. Research has shown that targeted PAC funding significantly decreases the odds that candidates will take pro-climate stances. These committees have emerged as significant actors in shaping political discourse and potential legislation on climate change.[46]

10. *Lobbying firms.* There is an extensive and well-funded lobbying effort to prevent legislative action on climate change. Research in the United States has shown that fossil fuel interests outspend renewable energy corporations and environmental groups by a ratio of 10 to 1, providing these interests an overwhelming advantage in the crucial strategy of lobbying members of Congress.[47]
11. *Conservative media and denial bloggers.* Conservative media, including talk radio, TV and online sources, conservative newspapers, and widely circulated columnists, have become major amplifiers of climate change misinformation.[48] Users of these media show significantly lower levels of concern about the issue than individuals who didn't use those media outlets to learn about climate change.[49] A variety of social media and online outlets are also tools in the diffusion of climate change misinformation.

CLIMATE OBSTRUCTION IN EUROPE

The nations of Europe have their own distinct vested interests, coalitions, discourses, and strategies for blocking stronger climate action, but they have been scantily documented and never systematically compared with one another or with those of the United States. Europe is a critical area for the success or failure of global climate policy for several reasons. The countries that comprise it account for 8% of all production-based GHG emissions; more, if emissions are measured by the products consumed within a nation but produced elsewhere (a process known as *consumption-based emissions accounting*).[50] The European Union emits the fourth-largest quantity of GHGs in the world, followed by Russia, with the United Kingdom the eleventh-largest global emitter.[51] The distribution of GHG emissions among the countries examined in this book are illustrated in Figure 1.2.

As the figure shows, the largest quantity of emissions (3,460 metric tonnes of carbon dioxide equivalent, or MMT CO_2e) emanates from the twenty-seven EU countries collectively. Of this total, 2,254 MMT CO_2e, or 66% of the total emissions of the European Union, are covered by the EU members discussed in this volume. The remaining 1,206 MMT CO_2e in the EU countries not discussed in this book amounts to 33% of total emissions. So, although the book includes only nine of the twenty-seven countries in the European Union, it does cover most of the major emitting countries. On a country-by-country basis, Russia, with a total of 2,160 MMT CO_2e, is by far the single largest contributor. Germany is the second-largest GHG emitter, with a total of 763 MMT CO_2e; followed by the United Kingdom,

Total Greenhouse Gas Emissions 2021 (MMT CO2e)

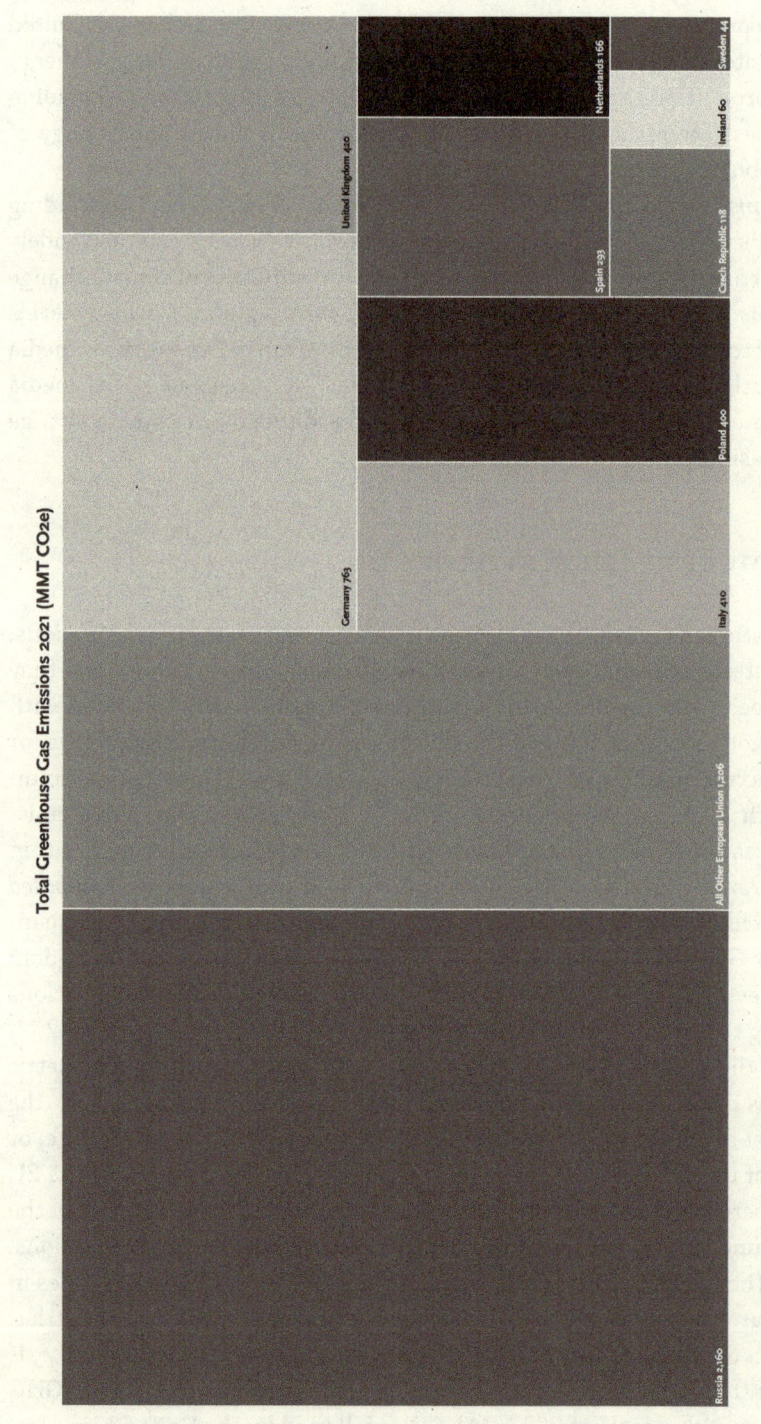

Russia 2,160 | All Other European Union 1,206 | Germany 763 | United Kingdom 410 | Italy 410 | Poland 400 | Spain 293 | Netherlands 166 | Czech Republic 118 | Ireland 60 | Sweden 44

Figure 1.2 The rectangles represent the total quantity of greenhouse gases (GHGs) for each of the eleven European countries studied and for the remaining twenty-seven European Union nations.

Italy, and Poland, with about 400 MMT CO_2e each. Spain, the Netherlands, the Czech Republic, Ireland, and Sweden all emit smaller quantities. Thus the overall emissions picture varies widely throughout Europe. In each chapter of this book, we therefore provide a discussion of each country's emissions trends, discuss their overall climate mitigation goals, and assess their success in reaching them.

Given their economic influence and political power, the countries of Europe also exert a major influence in global climate policy negotiations and decision-making. With the rise of the Green Party in Germany and then in other countries, Europe, as both a cultural region and a political bloc, has taken the need to act on climate change more seriously than most other parts of the world. For decades, Europe has seen itself as a leader on climate action, and in the over thirty years of UN negotiations on the issue, the European Union has brought leading pledges and policy ideas to the table.[52] Europe's support of the Kyoto Protocol, the development of the EU-Emissions Trading System, and the 2030 Climate Target Plan[53] are all significant (and flawed) achievements for this diverse region. Coordination and alignment of EU policies has not been easy, however, and early opposition by industry to region-wide climate policies led to targets not being met. Substantial effort has therefore been needed to address a 'credibility gap' between domestic climate policies and international proposals from the region.[54]

Indeed, although world-leading, even Europe's efforts on climate have been inadequate, uneven, and halting. In a 2014 article on EU climate policy, Jakob Skovgaard described a 'recurring pattern' by which proposals to increase the ambition of Europe-wide climate goals are 'quickly rejected, mainly by a coalition led by Poland and including Italy and some new Member States (Hungary and Romania among others) . . . a large group of actors either did not have a clear position for or against the step-up or oscillated between them'.[55] In a 2020 article, media scholar Juho Vesa and colleagues discussed how industrial lobbies work behind the scenes, outside of the media spotlight, to influence European climate policy by emphasizing the need for economic competitiveness.[56] Thus, some steps toward stronger EU climate policy have been rolled back and others threatened by economic crises, war, global competition, and authoritarianism. Pushback is growing as the region shifts from setting bold targets to implementing them.[57]

It is therefore an urgent task to explore the larger questions raised by the continuing difficulty of advancing ambitious climate policy and action in Europe. Three decades of halting progress suggest that a new understanding of the obstacles to climate action is needed. Who are the actors and organizations obstructing climate action in Europe? What are their strategies, and how are they evolving? This volume seeks to advance our understanding of climate obstruction in the region as a whole and to learn from the significant variations in such efforts across the continent.

ELEVEN NATIONAL CASES AND THE EUROPEAN UNION

Because the literature on climate obstruction efforts in Europe is scattered and sporadic, it was not apparent that there was a substantial enough body of research on which this book could be based. Therefore its development followed a unique approach. Working with the All Europe Academies of Science (ALLEA), the Climate Social Science Network (CSSN) solicited proposals for chapters on multiple geographic regions in Europe, hoping that the scholars in each country would be aware of sufficient material from which they could develop a review. From this solicitation, we received eighteen responses; twelve were selected for inclusion. While many of the major national actors are included here, regrettably, the editorial team was unable to develop analyses of climate obstruction in France, Greece, Norway, Portugal, or the other European nations. This gap indicates a need for further support to develop sufficient academic research to enable analyses commensurate with the twelve that appear in this book.

Additionally, by design, the analyses emphasize national-level politics. As such, they do not focus on the larger dynamics at play, such as the roles of multinational corporations, international agreements, or think tanks across international boundaries. They also do not focus on cities or other subnational regions. Some of these broader topical issues are taken up in the forthcoming *First Global Assessment of Climate Obstruction*, now under development. Each chapter of this European volume is intended as a stand-alone case study, as well as part of a larger unit. However, where appropriate, the authors discuss the broader dynamics in their individual chapters.

This volume thus represents the current state of social scientific knowledge on climate obstruction efforts. Given the above limitations, the chapters offer a relatively comprehensive and in-depth presentation of climate obstruction efforts across a wide range of countries in the four European regions and introduce the key actors in climate change mitigation in Europe. No overriding theoretical framework was promulgated to guide development of the manuscript: each team of scholars was left to its own creativity on how to approach their topic. The only guidance provided was to ask each team to provide four specific analyses related to climate obstruction in their specific geography: (1) a historical narrative on climate obstruction in the area, (2) a description and analysis of the major actors and type of institutions involved, (3) a discussion of the strategies and tactics they utilize, and (4) a description of the discursive framings they employ. While this approach yielded a variety of theoretical approaches to their topics, we hope it will stimulate further research and collaborative efforts that might ultimately refine a framework that can encompass the wide range of climate obstruction efforts described in this book. By assembling them in one volume, we hope to demonstrate the relevance of such analyses for expanding our understanding of climate change obstruction, especially its inherent links to social structure and societal dynamics.

The first three chapters focus on the British Isles. Chapter 2 focuses on climate obstruction in the United Kingdom. Through a historical account of the development of climate policy in the UK, the essay shows how incumbent interests utilize their structural, institutional, and discursive power to shape climate policy and obstruct ambitious climate action. This use of incumbent power has locked in future carbon emissions and will further restrain climate action. Chapter 3 provides a focused examination of climate obstruction related to gas and oil in Scotland. Rich with these fossil deposits in the North Sea, Scotland has been the site of an intense struggle over the development of these resources. This chapter provides a detailed analysis of this political struggle and how this effort has been centred on the protection of oil and gas jobs over mitigation of future climate change. Chapter 4 focuses on Ireland and the transformation of a primarily agricultural economy. Despite having a small fossil fuel-based economy, Ireland has a strong cultural tradition based on farming and the burning of high-carbon-emitting peat for home heating in rural areas. This analysis shows how the major agricultural interests act to obstruct climate regulations that might affect them.

The second set of chapters focuses on Northern Europe. It starts with Chapter 5, an analysis of climate obstruction in Sweden. This analysis centres on the Swedish notion of the 'middle way' when developing policy approaches to climate change. This way reflects a centrist approach to moderate and incremental policy adjustments to reduce Sweden's GHG emissions. To oppose these policies, the opposition to climate action utilizes indirect tactics involving delaying climate solutions and displacing impacts to other locations, such as by utilizing carbon offsets. The chapter concludes with a discussion of how climate obstruction is changing as politics have shifted to the right in Sweden. Chapter 6, on Germany, focuses on the strong neoliberal opposition to climate action there and how this opposition is realized through the use of think tanks and campaign organizations to shape public opinion against renewable energy. German climate opposition does not frequently engage in outright climate denial. Rather, the campaigns focus on delaying mechanisms such as advocating for carbon offsets, less ambitious vehicle emissions standards, and the use of natural gas as a 'transition fuel' on the path to a hydrogen economy. Chapter 7, on the Netherlands documents a history there of early ambition on climate change and the subsequent mobilization of strategic sceptics on the science, lobbying by the nation's largest corporations and trade groups, and a cultural offensive to keep fossil fuels as inevitable and positive contributors to solving the problem. Central to the story are the close ties between the Ministry of Economic Affairs and the fossil fuel industry, through lobbying and the revolving door. That ministry, in turn, undermined every effort to put in place ambitious climate policy in the Netherlands.

The third section of the book focuses on three former Soviet republics: Poland, Russia, and the Czech Republic. With the fall of the Soviet Union in 1989, all three of these nations experienced a dramatic drop in their GHG emissions after the collapse of their economies. Accordingly, there has been very little external pressure to reduce GHG emissions in these countries because the UNFCCC baseline year by which national reductions would be measured was set at 1990. In all three countries, state ownership of the energy industries and state investment led to a continuation of fossil fuels as the basis for these nations' economies. Chapter 8, on Poland, shows the central role that the coal industry plays in the Polish economy and how a coalition of government institutions, agencies, state-owned energy companies, and utilities works to perpetuate fossil fuel use. Similarly, Chapter 9 shows the centrality of fossil fuel use in the economy of Russia, where national defence and government stability ensure a taken-for-granted economic structure in which there are no significant ongoing efforts to meaningfully mitigate GHG emissions. Finally, Chapter 10's analysis of the Czech Republic shows how low 'issue saliency' and lack of pressure on politicians to reduce carbon emissions leads to a lack of meaningful climate policy in this country. Thus, all three former Soviet republic countries are burdened by a political system firmly linked to an economy based in fossil fuels, and little action on climate change is taking place because national priorities are focused elsewhere.

The final section of the book focuses on two southern European countries (Italy and Spain) and the European Union. Italy is the focus of Chapter 11, which shows the enduring opposition to climate change in that country is based in a strong climate countermovement. This effort, which is linked to conservative think tanks in the United States, appears to engage in very similar tactics to oppose climate action. This effort is backed by Italian oil and gas companies, their lobby groups, and corporate and institutional allies. Chapter 12, on Spain, shows that the legacy of Spanish authoritarian government has obstructed action on climate change and allowed vested interests to maintain the status quo in energy policy, which favours large corporations and fossil fuel use. Additionally, its strong agricultural industry limits action to address carbon emissions from cattle production. Chapter 13 concludes with an analysis of obstruction at the level of the European Union. The chapter provides a historical perspective on the development of climate policy in the European Union and the conflict between its core mandate to develop an integrated European economy and a secondary effort to reduce its overall carbon emissions. The opposition to ambitious climate action at the EU level is not based on climate denial. Rather, it consists of lobbying efforts to reduce the ambition of climate policy initiatives and make them more market friendly. The chapter presents an empirical analysis to show how fossil fuel interests exercise a significant advantage in lobbying capacity and how this resource advantage leads to a systematic weakening of EU climate policy.

LOOKING AHEAD

However belatedly, the social sciences are finally turning more intently to examine human-caused climate change, a welcome trend critical for both the development of a series of other subspecialties in our fields—and the very survival of our species. In this first-of-its-kind volume, multidisciplinary social science teams seek to understand the ways in which the primary drivers of global climate change are social-structural and sociocultural phenomena. These eleven national case studies and the review of climate obstruction at the level of the European Union therefore represent a major leap forward in our understanding of climate obstruction efforts in the region, provide a good sense of what social science can contribute to this enterprise, and underscore the urgency of incorporating social-science perspectives into future research, action, and policy on climate change. Finally, a concluding chapter distills the book's main findings into a series of ten lessons to suggest new avenues for policy and action. We look forward to a new era of useful research on climate obstruction across Europe.

NOTES

1. See Robert Brulle and Riley Dunlap (2021), 'A Sociological View of Efforts to Obstruct Climate Action', *Footnotes* 49: 3, https://www.asanet.org/footnotes-article/sociological-view-effort-obstruct-action-climate-change/.
2. C. F. Parker, C. Karlsson, and M. Hjerpe (2015), 'Climate Change Leaders and Followers: Leadership Recognition and Selection in the UNFCCC Negotiations, *International Relations*, 29, 4: 434–454, https://doi.org/10.1177/0047117814552143.
3. See https://www.ipcc.ch/.
4. IPCC (2023), 'Summary for Policymakers'. In: IPCC (ed.), *Climate Change 2023. Synthesis Report: A Report of the Intergovernmental Panel on Climate Change. Contribution of Working Groups I, II and III to the Sixth Assessment Report of the Intergovernmental Panel on Climate Change* (Core Writing Team, H. Lee and J. Romero [eds.]). Geneva: IPCC.
5. IPCC (2023), 'Current Status and Trends'. In: IPCC (ed.), *Climate Change 2023. Synthesis Report: A Report of the Intergovernmental Panel on Climate Change. Contribution of Working Groups I, II and III to the Sixth Assessment Report of the Intergovernmental Panel on Climate Change* (Core Writing Team, H. Lee and J. Romero [eds.]). Geneva: IPCC, p. 61.
6. Kari De Pryck (2021), 'Intergovernmental Expert Consensus in the Making: The Case of the Summary for Policy Makers of the IPCC 2014 Synthesis Report', *Global Environmental Politics*, 21, 1: 108–129, doi:https://doi.org/10.1162/glep_a_00574.
7. Riley E. Dunlap and Aaron M. McCright (2011), 'Organized Climate Change Denial'. In: Dryzek, J. R. Norgaard and D. Schlosberg (Eds.), *The Oxford Handbook of Climate Change and Society*, pp. 144–160. J. Painter, J. Ettinger, D. Holmes, et al. (2023), 'Climate Delay Discourses Present in Global Mainstream Television Coverage of the IPCC's 2021 report', *Communications Earth & Environment*, 4, 118, https://doi.org/10.1038/s43247-023-00760-2.
8. IPCC, 2022: Climate Change 2022: Mitigation of Climate Change. Contribution of Working Group III to the Sixth Assessment Report of the Intergovernmental Panel on Climate Change [P.R. Shukla, J. Skea, R. Slade, A. Al Khourdajie, R. van Diemen, D. McCollum, M. Pathak, S. Some, P. Vyas, R. Fradera, M. Belkacemi, A. Hasija, G. Lisboa, S. Luz, J. Malley, (eds.)]. Cambridge University Press, Cambridge, UK and New York, NY, USA. doi: 10.1017/9781009157926 p. 186.
9. Ibid., p. 170.
10. Ibid., p. 557.
11. Ibid., p. 1377.
12. See cssn.org for a further discussion of climate obstruction.
13. Aaron McCright and Riley Dunlap (2015), 'Challenging Climate Change: The Denial Countermovement', In Dunlap, Riley E., and Robert J. Brulle, (Eds.) *Climate change and society: Sociological perspectives*. Oxford University Press, 300–332.
14. Ibid.
15. Riley Dunlap and Robert Brulle (2020), 'Sources and Amplifiers of Climate Change Denial', *In* Holmes, D. C., & Richardson, L. M. (Eds.). *Research handbook on communicating climate change*. Edward Elgar Publishing., 49–61.
16. McCright and Dunlap, 'Challenging Climate Change'.

17. James Bohr (2022), 'The Structure and Culture of Climate Change Denial', *Footnotes* 49: 3
18. Robert Brulle (2014), 'Institutionalizing Delay: Foundation Funding and the Creation of US Climate Change Counter-Movement Organizations', *Climatic Change*, 122: 681–694.
19. Justin Farrell (2016), 'Corporate Funding and Ideological Polarization About Climate Change', *Proceedings of the National Academy of Sciences*, 113, 1: 92–97.
20. Robert Brulle (2020), 'Denialism: Organized Opposition to Climate Change Action in the United States', In: Konisky, D. M. (Ed.), *Handbook of US Environmental policy*. Edward Elgar Publishing., pp. 328–341. .
21. James Manheim (2011), *Strategy in Information and Influence Campaigns: How Policy Advocates, Social Movements, Insurgent Groups, Corporations, Governments and Others Get What They Want*. Milton Park, UK: Routledge.
22. Carlos Milani, Guy Edwards, James Walz, et al. (2021), *Is Climate Obstruction Different in the Global South? Observations and a Preliminary Research Agenda*. CSSN Position Paper 2021:4. Climate Social Science Network. Providence, RI: Brown University.
23. Sally Covington (1997, July), *Moving a Public Policy Agenda: The Strategic Philanthropy of Conservative Foundations*. Washington, DC: National Committee for Responsive Philanthropy.
24. Jane Mayer (2016), *Dark Money: The Hidden History of the Billionaires Behind the Rise of the Radical Right*. New York: Anchor Books.
25. Steven Teles (2008), *The Rise of the Conservative Legal Movement: The Battle for Control of the Law*. Princeton, NJ: Princeton University Press.
26. Jennifer Washburn (2010), *Big Oil Goes to College: An Analysis of Ten Research Collaboration Contracts between Leading Energy Companies and Major U.S. Universities*. Washington, DC: Center for American Progress. Merrill Goozner and Eryn Gable (2008), *Big Oil U*. Washington, DC: Center for Science in the Public Interest.
27. Robert Kerr (2005), *The Rights of Corporate Speech: Mobil Oil and the Legal Development of the Voice of Big Business*. New York: LFB Scholarly Publishing.
28. Herb Schmertz (1977, 12 January), 'Problems in Communicating With and Through the Media', Public lecture to the Business International Chief Executives Round Table. Public Relations Society of America, University of Wisconsin, Madison WI.
29. Kerr, *The Rights of Corporate Speech*; Schmertz, 'Problems in Communicating'; Charles Brown and Herbert Waltzer (2005), 'Every Thursday: Advertorials by Mobil Oil on the op-ed page of *The New York Times*', *Public Relations Review*, 31: 197–208. See also Bill St. John III (2014), 'Conveying the Sense-Making Corporate Persona: The Mobil Oil "Observations" Columns, 1975–1980', *Public Relations Review*, 40, 4: 692–699.
30. Mobil Oil (1982), *Evolution of Mobil's Public Affairs Programs 1970–81*. Fairfax VA: Mobil Oil Company.
31. Robert Brulle, Melissa Aronczyk, and Jason Carmichael (2020), 'Corporate Promotion and Climate Change: An Analysis of Key Variables Affecting Advertising Spending by Major Oil Corporations, 1986–2015', *Climatic Change*, 159: 87–101.
32. Fred Pearce (2010, 4 July), ''Climategate' Was "A Game-Changer" in Science Reporting, Say Climatologists', *The Guardian*, 2010.

33. Robert Brulle (2018), 'The Climate Lobby: A Sectoral Analysis of Lobbying Spending on Climate Change in the USA, 2000 to 2016', *Climatic Change*, 1493, 4: 289–303.
34. Manheim, *Strategy in Information and Influence Campaigns*.
35. James Greenberg, William Knight, and Edward Westersund (2011), 'Spinning Climate Change: Corporate and NGO Public Relations Strategies in Canada and the United States', *International Communication Gazette*, 73, 1–2: 65–82.
36. Charles Cooper and Andrew Nownes (2004), 'Money Well Spent? An Experimental Investigation of the Effects of Advertorials on Citizen Opinion', *American Politics Research*, 32, 5: 546–569.
37. Union of Concerned Scientists (2007), *Smoke, Mirrors & Hot Air: How ExxonMobil Uses Big Tobacco's Tactics to Manufacture Uncertainty on Climate Science*. Washington, DC: Union of Concerned Scientists.
38. Brulle, Aronczyk, and Carmichael, 'Corporate Promotion and Climate Change'.
39. Robert Brulle (2021), 'Networks of Opposition: A Structural Analysis of US Climate Change Countermovement Coalitions 1989–2015', *Sociological Inquiry*, 913: 603–624.
40. Justin Farrell (2016), 'Network Structure and Influence of the Climate Change Counter-Movement', *Nature Climate Change*, 64: 370–374.
41. Robert Brulle, Galen Hall, Loredana Loy and Ken Schell-Smith (2021), 'Obstructing Action: Foundation Funding and US Climate Change Counter-Movement Organizations', *Climatic Change*, 166: 1–7.
42. Robert Brulle and Cartie Werthman (2021), 'The Role of Public Relations Firms in Climate Change Politics', *Climatic Change*, 169, 1–2: 8.
43. Robert Brulle and Christian Downie (2022), 'Following the Money: Trade Associations, Political Activity and Climate Change', *Climatic Change*, 175: 3–4, 11.
44. Aaron McCright and Riley Dunlap (2000), 'Challenging Global Warming As a Social Problem: An Analysis of the Conservative Movement's Counter-Claims', *Social Problems*, 47, 4: 499–522.
45. Ben Franta and Geoffrey Supran (2017, 13 March), 'The Fossil Fuel Industry's Invisible Colonization of Academia', *The Guardian*. https://www.theguardian.com/environment/climate-consensus-97-per-cent/2017/mar/13/the-fossil-fuel-industrys-invisible-colonization-of-academia.
46. Kerry Ard, Nancy Garcia, and Peter Kelly (2017), 'Another Avenue of Action: An Examination of Climate Change Countermovement Industries' Use of PAC Donations and Their Relationship to Congressional Voting over Time', *Environmental Politics*, 266: 1107–1131.
47. Brulle, 'The Climate Lobby'.
48. James Bolin and Lawrence Hamilton (2018), 'The News You Choose: News Media Preferences Amplify Views on Climate Change', *Environmental Politics*, 27, 3, 455–476.
49. Jason Carmichael and Robert Brulle (2018), 'Media Use and Climate Change Concern', *International Journal of Media & Cultural Politics*, 14, 2: 243–253.
50. Julia K. Steinberger, J. Timmons Roberts, Glen P. Peters, and Giovanni Baiocchi (2012), 'Pathways of Human Development and Carbon Emissions Embodied in Trade', *Nature Climate Change* 2, 2: 81–85.
51. IPCC, 'Climate Change 2021'.

52. Sebastian Oberthür and Claire Roche Kelly (2008), 'EU Leadership in International Climate Policy: Achievements and Challenges', *The International Spectator*, 43, 3: 35–50.
53. European Commission (2020), *2030 Climate Target Plan*, https://climate.ec.europa.eu/eu-action/european-green-deal/2030-climate-target-plan_en.
54. Oberthür and Kelly, 'EU Leadership'.
55. Jakob Skovgaard (2014), 'EU Climate Policy after the Crisis', *Environmental Politics*, 23, 1: 1–17.
56. Juho Vesa, Antti Gronow, and Tuomas Ylä-Anttila (2020), 'The Quiet Opposition: How the Pro-Economy Lobby Influences Climate Policy', *Global Environmental Change*, 63: 102–117, https://doi.org/10.1016/j.gloenvcha.2020.102117.
57. William Booth and Anthony Faiola (2023, 6 August), 'Europe Blinks in Its Commitment to a Great Green Transition', *The Washington Post*. https://www.washingtonpost.com/world/2023/08/06/europe-britain-carbon-cost/.

2

Climate Obstruction in the United Kingdom

Charting the Resistance to Climate Action

FREDDIE DALEY, PETER NEWELL, RUTH MCKIE, AND JAMES PAINTER

INTRODUCTION: MAKING SENSE OF CLIMATE OBSTRUCTION IN THE UNITED KINGDOM

The United Kingdom punches above its weight in climate politics for several historical reasons. Although currently ranked seventeenth globally regarding greenhouse gas (GHG) emissions, the United Kingdom is the world's sixth-largest economy. As the birthplace of the Industrial Revolution, which entrenched the power of 'fossil capital'[1] and marked the beginning of anthropogenic climate change, as well as the heart of the British Empire and home to one of the world's major financial centres (London, a centre of both carbon trading and fossil fuel financing), what happens in the United Kingdom continues to have global consequences for climate action. The United Kingdom's experience of a shift away from coal as part of the 'dash for gas' also holds important lessons for supporting just transitions, given the industrial conflict and regional decline that followed in the wake of this transition[2]—experiences that serve as a reference point for obstructionist actors seeking to delay climate ambition today.

Freddie Daley, Peter Newell, Ruth McKie, and James Painter, *Climate Obstruction in the United Kingdom* In: *Climate Obstruction across Europe*. Edited by: Robert J. Brulle, J. Timmons Roberts and Miranda C. Spencer, Oxford University Press. © Oxford University Press 2024. DOI: 10.1093/oso/9780197762042.003.0002

Through the size and reach of UK-based multinational companies, its role in global institutions such as the United Nations (as a Security Council member) and the World Bank (as a major contributor), as well as the size of its aid budget, the United Kingdom is also a significant player in global climate politics. Precisely because of this peculiar global profile, the United Kingdom has been subject to intense pressure to lead on climate change and fierce resistance to reject this responsibility from a powerful and dynamic network of incumbent actors highly influential within the United Kingdom's climate policy regime.

Celebrated for its climate leadership, the UK had slowly cut its domestic emissions to 41% of 1990 levels by the end of 2021 (Figure 2.1). In 2019, the United Kingdom became the first Group of Seven (G7) economy to enshrine a 'net zero' target into law, committing the nation to reduce its GHGs by 100% by 2050, a proportion equal to or less than the amount of GHG emissions the country removes from the environment. The United Kingdom has also made significant progress in decarbonizing its energy system, with 2020's emissions 51% lower than 1990 levels.[3] While fossil gas remains as a dominant sources of electricity generation, by March 2023, renewable energy had slowly increased to a record share of 44.1%. Moreover, the United Kingdom's nuclear capacity has progressively increased since the 1990s, with commitments made to expand nuclear power capacity to 25% of the United Kingdom's electricity supply by 2050.[4] However, some see the prioritization of nuclear energy in the United Kingdom as a costly, time-consuming distraction from investments into renewable energies that could make a more immediate contribution to reaching net zero.[5]
Despite the government's triumphant declaration that Britain is 'halfway to net zero' and to achieving its nationally determined contributions (NDC) under the 2015 Paris Agreement,[6] UK emissions rebounded by 6% in 2021, and its policy responses to the global energy crisis look set to lock in further fossil fuel production and consumption.[7] In 2022, a UK High Court ruling found the government's 'Net Zero Strategy' breached the nation's 2008 Climate Change Act and needed reassessment.[8] In January 2023, an independent review into the Net Zero Strategy, conducted by former Conservative energy minister Chris Skidmore MP, concluded that the government is 'not matching world-leading ambition with world-leading delivery' and, on the current trajectory, Britain would miss out on 'the growth opportunity of the 21st century'.[9]

While the United Kingdom has curtailed territorial emissions by phasing out coal and embracing cleaner forms of manufacturing, the energy mix, transportation system, and built environment are still deeply intertwined with fossil fuel usage. The United Kingdom's currently 'insufficient'

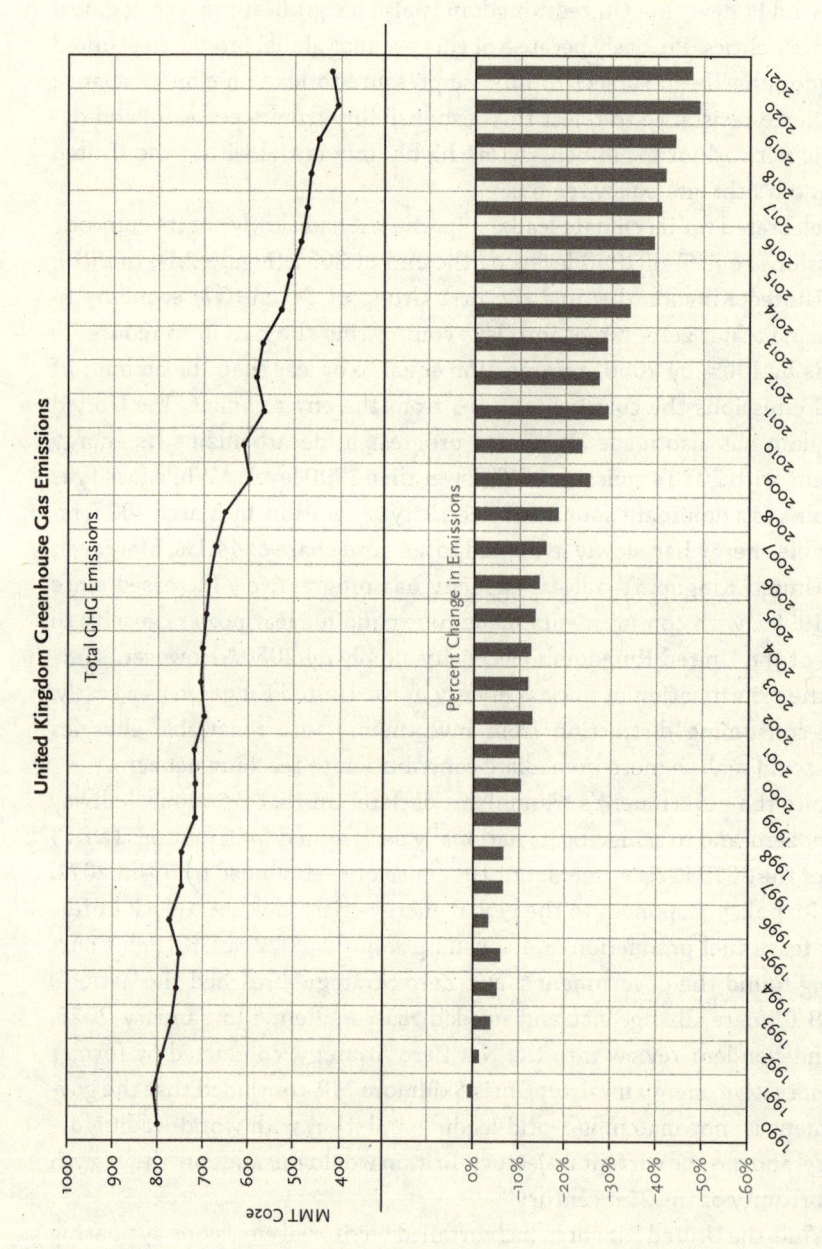

Figure 2.1 Total greenhouse gas (GHG)¹ emissions (in MMT CO₂e) and percentage change in emissions in the United Kingdom between 1990 and 2021, inclusive.

Source: Total GHG emissions based on data provided by Gütschow and Pflüger (2023) for Kyoto Six Greenhouse Gas Totals.

progress toward meeting its NDCs highlights the need for further government action and the need to tackle the forces of climate obstruction in the United Kingdom.[10] This reality, paired with the current lack of progress on delivering net zero, suggests that the United Kingdom's image as an international climate leader may be fading. Indeed, in June 2023, the United Kingdom's independent Climate Change Committee (CCC) expressed public concern that government efforts to scale up climate action were 'worryingly slow', leaving it 'markedly' less confident that the United Kingdom would achieve its legally binding emissions-reduction targets.[11]

Multiple strands of academic literature speak directly and indirectly to the question of climate obstruction.[12] Alongside long-standing literature on business lobbying, some of which focuses on the United Kingdom and actors such as the Confederation of British Industry (CBI),[13] there is growing interest in exploring the role of incumbent actors in resisting decarbonization and industrial conversion.[14] Besides documenting specific strategies adopted by businesses, from contesting the science to exaggerating the costs of climate action, these accounts draw on different theoretical perspectives, from neo-pluralism (emphasizing how business actors are one actor among many)[15] to more structural accounts that explain the power of incumbents in relation to their ability to serve broader state aims of growth and competitiveness.[16]

Here we adopt a broadly neo-Gramscian approach to account for the ways in which material, institutional, and discursive forms of power reinforce one another to maintain structural power throughout society and resist pressures for more transformative climate action.[17] For Antonio Gramsci, the Italian revolutionary upon whose work the approach draws, hegemony was the key focus of analysis. For us, exploring these dimensions of power helps to shed light on the ways incumbent power is upheld and how challenges to it are dissipated and contained through strategies of obstruction.[18] We apply this multidimensional understanding of power to identify the different sites and functions of climate obstructionism in the United Kingdom and how a network of actors converges to sustain climate obstructionism and thwart more ambitious climate action.

We recognize that definitively attributing influence to incumbent actors is an almost impossible endeavour. Yet careful analysis, process tracing, interviews with actors involved in key decision-making moments, and triangulation with multiple sources can help to build a well-rounded, more comprehensive, and, we think, convincing account of the political economy of climate obstructionism in the United Kingdom, as manifested in and through government policy, governance arrangements, and the media.

A HISTORICAL DESCRIPTION OF CLIMATE OBSTRUCTION IN THE UNITED KINGDOM

In 1989, Margaret Thatcher, then prime minister of the United Kingdom, delivered a famous speech to the United Nations General Assembly, stressing the dangers of climate change and the need for international action. Thatcher also outlined the role the United Kingdom would play in advancing climate science, coordinating what would become the Intergovernmental Panel on Climate Change (IPCC) in 1988. Thatcher argued that multinational corporations were part of the solution to mitigating climate change and environmental degradation and that Britain's liberalized economic model would generate wealth to pay for these aims.

As part of this transition, several previously government-controlled sectors went through a process of privatization over a fifteen-year period, including the oil, gas, and energy sectors. British Petroleum (BP) had been privatized in 1979, when the government became the minority shareholder. Similarly, the British Gas Cooperation began its transition to privatization under the Gas Act of 1986. A close relationship between government and corporations cemented by interlocking relationships between government officials and corporations, as well as lobbying activities, ensured that businesses would be central to the development of the United Kingdom's climate-related policy, creating the framework through which the private sector could significantly affect and even shape climate policy developments.

Following the New Labour Party's electoral victory in 1997, then-Prime Minister Tony Blair laid out plans to align ecological concerns with social and economic considerations.[19] This approach engendered a form of 'ecological modernization' whereby economic growth was not deemed antithetical to protecting the environment, but rather something that could enhance it.[20] Over time, New Labour would go on to introduce a variety of measures advancing reduction targets for climate-related emissions. Nevertheless, it was not until 2006 that the party introduced a more ambitious strategy that would come to include explicit emissions-reduction targets following the implementation of the 2008 Climate Change Act.[21] This act institutionalized the United Kingdom's plans to significantly reduce its GHG emissions by 2050 and set up the independent CCC mentioned above.

In 2010, a coalition government was formed between the Conservative Party and the Liberal Democrats, led by Prime Minister David Cameron, who promised to create the 'greenest ever' cabinet[22] and cultivated a positive image of an environmentally conscious Conservative Party.[23] Yet, in 2011, his chancellor of the exchequer, George Osborne, expressed concerns around green policies for the United Kingdom and European Union, citing the burden on British businesses.[24] In the same year, the government outlined its Carbon Plan, maintaining its commitments under the revised Climate Change Act. Demonstrating further commitment to reducing emissions, in 2016, the United Kingdom signed the Paris Agreement before committing to net zero and a green industrial plan before the 2020 COVID-19 pandemic.

The 2016 'Brexit' referendum, when the United Kingdom voted to sever ties with the European Union, provided a rupture in British politics that obstructionist forces would capitalize upon. Indeed, some of the major actors in furthering climate obstructionism (expanded upon in this chapter) began pushing for the dilution of environmental regulations and standards in prospective Brexit trade deals,[25] building on strong transatlantic links with climate obstructionist forces in the United States, such as the Charles Koch Foundation, the Atlas Network, and the Heritage Foundation.[26] There is a clear and continuing alignment of interests between climate obstructionist forces and Brexit-supporting actors. For instance, multimillionaire Jeremy Hosking continues to donate to anti-European political parties such as the Reclaim Party and Reform UK, which publicly denounce the UK government's net zero push. Hosking's company, Hosking Partners, held over £100 million in fossil fuel investments in 2022.[27]

While each subsequent UK government from the 1980s onward has ostensibly committed to meeting national and international emissions-reduction targets, the forces of climate obstruction have been ever-present. Some of the earliest evidence of climate obstruction can be seen in the campaigns of fossil fuel corporations. Early episodes centred on efforts to disrupt EU proposals for a carbon tax, in which lobbying by the coal industry successfully mobilized the House of Lords' EU Committee (a committee chaired by Lord Ezra, a former chairman of British Coal) to produce a report echoing their concerns about the effect of the tax on the competitiveness of the UK coal industry. Targeting more powerful ministries, such as trade, was a key strategy for the World Coal Institute, which acknowledged 'Our influence probably goes further because the Trade Ministries are more powerful than the Environment ministries'.[28] From the 1990s, these actors would closely track the UK government's approach to emissions reductions and seek to obstruct them, first through lobbying and later through greenwashing activities.[29]

UK CLIMATE OBSTRUCTIONISM: MAJOR ACTORS AND INSTITUTIONS INVOLVED

The network of actors that sustains climate obstructionism in the United Kingdom is diverse and highly developed. It is also inherently fluid and shaped by exogenous forces, such as electoral cycles and geopolitical shifts, while also shaping public debate on climate around domestic and international phenomena such as the Russian invasion of Ukraine in 2022. This section explores a selection of key actors and institutions within five categories to sketch the fault lines and dynamics of British climate obstructionism, highlight how these actors interact, and explore how networks converge around specific policy moments and political opportunities to further entrench obstructionism (see Figure 2.2).

The key types of actors and institutions include (1) organized sceptic groups and think tanks, such as the Global Warming Policy Foundation (GWPF) and Net Zero Watch (NZW); (2) news outlets within the British media, such as right-leaning newspapers *The Daily Telegraph* and *Daily Mail*; (3) business lobby groups and trade associations, such as the CBI and UK Petroleum Industry Association (UKPIA); (4) government actors and institutions, like the Department of Business, Energy and Industrial Strategy (BEIS) (now the Department of Energy Security and Net Zero); and (5) 'floating' organizations, such as the GMB trade union, a significant number of whose members are based in the energy sector, and local chapters of the Campaign to Protect Rural England (CPRE), a 'countryside charity' that has local arms across the United Kingdom that often campaign against renewable energy projects.

Organized sceptic groups and think tanks

Organized sceptic groups and think tanks have long been essential players in incumbent actors' information and influence campaigns on climate change and beyond.[30] In Britain, the GWPF has been an instrumental actor in climate obstructionism since it was founded in 2009 by former Conservative Chancellor, Nigel Lawson. The GWPF has published reports, made multiple media interventions challenging the climate policy regime in the United Kingdom, forged strong links with key ministers and policymakers within government, and fostered a transnational network of actors seeking to circumvent climate action, working closely with climate sceptic groups and think tanks in the United States.[31] The GWPF says it does not receive donations from anyone with an interest in an energy company, but its

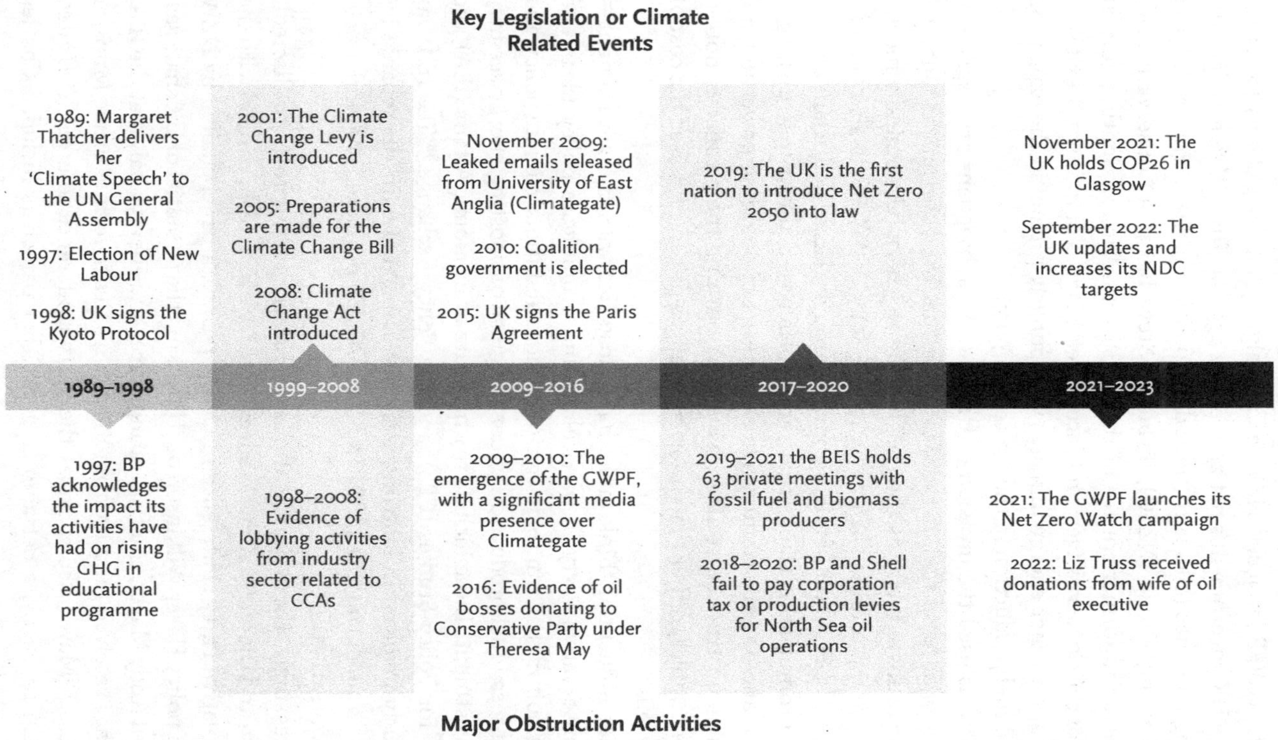

Key Legislation or Climate Related Events

1989–1998
- 1989: Margaret Thatcher delivers her 'Climate Speech' to the UN General Assembly
- 1997: Election of New Labour
- 1998: UK signs the Kyoto Protocol

1999–2008
- 2001: The Climate Change Levy is introduced
- 2005: Preparations are made for the Climate Change Bill
- 2008: Climate Change Act introduced

2009–2016
- November 2009: Leaked emails released from University of East Anglia (Climategate)
- 2010: Coalition government is elected
- 2015: UK signs the Paris Agreement

2017–2020
- 2019: The UK is the first nation to introduce Net Zero 2050 into law

2021–2023
- November 2021: The UK holds COP26 in Glasgow
- September 2022: The UK updates and increases its NDC targets

Major Obstruction Activities

1989–1998
- 1997: BP acknowledges its activities have had on rising GHG in educational programme

1999–2008
- 1998–2008: Evidence of lobbying activities from industry sector related to CCAs

2009–2016
- 2009–2010: The emergence of the GWPF, with a significant media presence over Climategate
- 2016: Evidence of oil bosses donating to Conservative Party under Theresa May

2017–2020
- 2019–2021 the BEIS holds 63 private meetings with fossil fuel and biomass producers
- 2018–2020: BP and Shell fail to pay corporation tax or production levies for North Sea oil operations

2021–2023
- 2021: The GWPF launches its Net Zero Watch campaign
- 2022: Liz Truss received donations from wife of oil executive

Figure 2.2 A timeline of significant obstruction activities in the United Kingdom, 1988–2022. The upper part of the figure shows key legislative changes or events; the lower section provides examples of obstruction activities and the actors and institutions involved.

funding remains opaque. GWPF's initial public interventions challenged the veracity of climate science, yet its more recent interventions have questioned the necessity of the net zero agenda and the associated costs.[32] In 2021, GWPF launched NZW, a campaign group that claims to scrutinize the UK government's net zero policy agenda. The GWPF manages the campaign and has links with elected UK politicians that comprise the Net Zero Scrutiny Group (NZSG), including Steve Baker, a Conservative MP, Brexiteer, and former trustee of the GWPF; Iain Duncan Smith, a former leader of the Conservatives; and Jacob Rees-Mogg, a prominent Brexit supporter and former energy minister who publicly promoted the expansion of fossil fuel production in the United Kingdom.[33]

NZW has used the invasion of Ukraine and subsequent price spikes in international fossil fuel markets to argue that the government must 'recommit to fossil fuels', as the crisis is due to 'thermodynamically incompetent energy sources such as wind and solar'.[34] NZW has also been an effective disseminator of talking points through elected politicians and the media, promulgating claims that renewables pose a threat to British agricultural land and the countryside as a means of engaging powerful rural groups as obstructionist allies, as well as pushing the erroneous notion that renewables, rather than international gas prices, increased the cost of energy bills.[35]

Other influential think tanks have close ties with the government—the Conservative Party in particular—and promulgate policy ideas and talking points that aim to delegitimize more ambitious climate policy and expand fossil fuel production under the guise of boosting energy security and investment. These include the Institute of Economic Affairs (IEA), the Centre for Policy Studies (CPS), and the Policy Exchange.[36] The IEA was previously involved in lobbying activities associated with the tobacco industry[37] and, during the 1990s and early 2000s, promoted the work of climate contrarians.[38]

Increasingly, these organizations coordinate their efforts to discredit net zero and the wider climate movement. For example, in parallel with the Conservative Party conference in October 2022, a panel chaired by NZW head of policy Harry Wilkinson on 'Unlocking the potential of UK shale gas' featured Andy Mayer, chief operating officer and energy analyst at the IEA, which has received funding from BP and has ties to former Prime Minister Liz Truss.[39] Mayer also spoke at the party conference to urge the United Kingdom to scrap its net zero target. Other panellists included Charles McAllister, director of policy, government, and public affairs at the trade body United Kingdom Onshore Oil and Gas (UKOOG).[40] In 2023, Just Stop Oil protestors targeted the headquarters of the think tank Policy Exchange after Prime Minister Rishi Sunak praised the organization for helping to inspire draconian new laws to suppress climate protests.[41] Shortly before this confrontation, Policy Exchange received a sizeable donation from US fossil fuel major ExxonMobil.[42]

The British media

The UK media enjoys global reach and influence among policymakers[43] and diverse audiences. Many of these media outlets are vital for the maintenance of climate obstructionism both within Britain and internationally. The websites of right-wing outlets such as *The Daily Mail*, *The Daily Telegraph*, and *The Sun* are among the most visited news websites in the world.[44] At times, *The Daily Mail* online has topped the list[45] and is particularly influential within the British climate debate and beyond by regularly giving space to columnists offering various obstructionist arguments to justify inaction on climate change. Many of the right-leaning media actors in the United Kingdom are at the centre of a dynamic relationship between a group of right-wing politicians, organized sceptic groups and think tanks, and the public. Indeed, some argue that the ideological or conviction contrarianism promoted by the GWPF, rather than contrarianism driven by fossil fuel interests, has aimed to influence the legacy media's framing of the climate challenge[46] via its ideological bedfellows in the right-wing media. Indeed, GWPF representatives have enjoyed a prominent presence there.[47]

As both networked actors and enablers for other obstructionist actors, UK media outlets are essential to legitimizing and amplifying a variety of obstructionist discourses and provide platforms for prominent sceptical figures. UK media actors are highly responsive to current public debates around climate change and often pivotal in shaping public and political discourse. For example, in response to a proposal by the leader of the Labour party, Sir Keir Starmer, to end new oil and gas infrastructure in the United Kingdom's North Sea, right-leaning media outlets published a series of editorials condemning the policy as 'dangerous'[48] and 'economic suicide'.[49] This concerted and coordinated media push played a major role in forcing Starmer to soften the policy and insist that fossil fuels will continue to be used in Britain 'for many, many years'.[50] Media actors are also effective at linking social and economic issues, such as the cost of living, to climate policy despite the fact that these links are often tenuous, overstated, and sometimes misleading. One erroneous line pushed consistently by actors such as *The Daily Telegraph* and *The Daily Mail* is that the United Kingdom's net zero policy is pushing up household energy bills,[51] despite documented evidence to the contrary.

There are numerous examples of links between media actors and other recognized obstructionist actors. The *Telegraph*'s chief interviewer and columnist, Allison Pearson, joined the GWPF as a trustee in 2023.[52] Pearson has written a number of articles questioning the necessity of climate justice,[53] criticizing the United Kingdom's net zero policy programme,[54] reprimanding British climate activist groups like Just Stop Oil,[55] and condemning policy proposals to phase out fossil fuel production in the United Kingdom.[56] Other notable media personalities associated with obstructionism include James Delingpole,[57] formerly of *The Daily Telegraph*, who currently writes for the right-wing outlet *The Spectator* and the far-right outlet Breitbart, as well as Julia Hartley-Brewer,[58] a TalkRadio host and columnist for *The Daily Telegraph*.

While the strategies and tactics of UK media actors have evolved over time (discussed later), their prominent role in furthering climate obstruction endures. These actors play a critical part in bolstering and mainstreaming discourses against climate action, demonizing climate activists, shaping public discourse around specific climate policies, and providing platforms for climate sceptic voices.

Business lobby groups and trade associations

Business lobby groups and trade associations have played prominent roles in sustaining climate obstruction by influencing government policy and granting incumbent actors access to policymakers. Their role in furthering climate obstructionism has been noted in the United States,[59] Finland,[60] and the European Union,[61] and at the international level.[62] While several prominent trade associations within the United Kingdom have played such roles, two organizations of particular interest are the CBI and UKPIA.

The CBI, founded in 1965, is a British membership organization acting on behalf of trade associations and UK businesses, both domestically and overseas, with international offices in Brussels, Washington, DC, and Beijing. The CBI's stance on climate change has transformed over two decades, moving from open scepticism about unilateral climate action[63] to an increasingly positive public position on climate action and net zero. The CBI is particularly supportive of developing 'strategies and incentives' to unlock investment into 'green growth',[64] which aligns economic growth with environmental protection. While endorsing some climate policies, the CBI continues to take a more cautious approach to the introduction of a carbon tax or carbon border adjustments, which create tariffs on carbon-intensive products such as cement and fossil fuel-generated electricity, over fears of 'taxing "bad" too much'.[65]

Another leading business organization, the UKPIA, represents eight of the largest internationally operating fossil fuel companies, including ExxonMobil, Total, and the UK-based majors BP and Shell. Like many trade associations, UKPIA describes its role as supporting its members through engagement to achieve 'a regulatory and legislative environment that secures results both for our sector and for wider society'.[66] UKPIA has been highly effective at utilizing the British government's All-Party Parliamentary Groups (APPG) system, comprising 'informal cross-party groups that have no official status within Parliament'[67] that are run 'by and for Members of the Commons and Lords, though many choose to involve individuals and organizations from outside Parliament in their administration and activities.'[68] While involving individuals and organizations from outside the government can provide opportunities for civil society engagement and foster transparency, it can also make APPGs vulnerable to regulatory and elite capture. The institutional rules governing APPGs inhibit public scrutiny, which can further entrench the power of incumbents. For example, their rules allow trade associations, and the companies they represent, to be omitted from official parliamentary transparency logs as only benefits in kind above £1,500 a year must be declared—a threshold many industry bodies claim not to meet.[69]

UKPIA plays a leading role in facilitating the APPG on Downstream Energy and Fuels, despite the trade association and its members not appearing on the Official Register.[70] Through this group, fossil fuel executives have gained access to policymakers, and, in one meeting chaired by UKPIA, politicians received presentations from BP and Phillips Petroleum, both of which are members of the trade association.[71] Similar dynamics have been observed in other APPGs, such as the one on hydrogen, where the lobbying firm Connect PA—which represents fossil fuel majors Shell and Equinor, as well as gas boiler manufacturers Baxi and Bosch—operates as its secretariat.[72] The seventeen politicians who are members of the APPG on hydrogen have repeatedly lobbied the government to support the installation of 'hydrogen-ready boilers' that burn fossil gas, and the generation of 'blue hydrogen', which is made from fossil gas.[73] While these APPGs have been set up according to parliamentary procedure and there is no suggestion of impropriety, the access and discretion trade associations and business groups afford to incumbent actors through these fora make them an important element of climate obstruction in the United Kingdom.

Government actors and institutions

Government departments and individual ministers play pivotal roles in sustaining climate obstruction in the United Kingdom. Due to the siloed policy responsibilities of government departments, efforts from incumbent interests to further climate obstruction typically target multiple actors simultaneously. This can be achieved through a variety of means.

The now-reformed BEIS has long been a target for actors seeking to further climate obstructionism. Between 2019 and 2021, ministers from BEIS held sixty-three private meetings with fossil fuel and biomass producers according to public records.[74] Public records over the same period show that BEIS ministers also attended 309 larger group meetings with fossil fuel companies and their representatives, while ministers attended only sixty such meetings with renewable energy companies.[75] While collaboration and coordination between government and industry is commonplace, the frequency and number of such meetings are noteworthy.

Also, government actors and institutions in the United Kingdom often fill key roles within departments by hiring former employees of major energy firms, with some individuals moving between industry and government multiple times. The 'revolving door' between fossil fuel companies and government agencies is a notable feature of climate obstructionism, with similar patterns documented in Australia,[76] Poland,[77] and Spain,[78] among other countries. BEIS has been particularly prolific at hiring from the fossil fuel industry, with ex-employees of Shell being given influential roles within the department overseeing energy market policy, hydrogen strategy, and heat pump deployment, and ex-senior executives from BP given non-executive director roles.[79] While mapping a direct path from revolving-door hires to specific government policies is difficult, the poor regulation of departmental hiring practices and the often-overlooked conflicts of interest point to how incumbent actors can embed themselves within government to further climate obstructionism.

'Floating' organizations

The term 'floating organizations' refers to actors mobilized and operationalized by incumbent interests to further climate obstructionism. Often, these actors are not motivated primarily by support for or opposition to climate policy but become engaged in such public and political discourses when they affect their interests or the interests of those they represent. For example, in the UK context, some trade unions have supported the expansion of high-carbon infrastructure such as airports, posed questions about the effectiveness of low-carbon technologies such as heat pumps, and called for further investment into domestic production of fossil fuels on the grounds that it would ensure job security and industry longevity. While these concerns are valid, and policymakers must integrate principles of a just transition into climate and industrial policy, the public stances of these groups provide opportunities for incumbent actors to use them to further climate obstruction.

The third-largest union in Britain, GMB Union, has made multiple public interventions in support of the UK's domestic fossil fuel industry and against elements of the government's net zero policy programme. Representing more than 600,000 workers, about 8% of whom work directly or indirectly in the energy sector, GMB Union aligned with business interests to support hydraulic fracturing (fracking) for shale gas in the United Kingdom, calling on the British government to 'take a firm line with anti-fracking protests against shale gas suppliers'.[80] GMB has made multiple public interventions in support of the United Kingdom's domestic fossil fuel industry on the grounds of improving energy security[81] and concerns over losing out to international competition.[82] In turn, GMB's position on domestic gas production has been exploited by obstructionist actors, including NZW. During a public media debate in 2016, on UK shale gas production, GWPF director Dr Benny Peiser commented that 'trade unions have a choice between a policy based on the eco-dogmatism of green campaigners and the GMB Union's energy policy that focuses first and foremost on safeguarding UK manufacturing and tackling fuel poverty'.[83] More recently, in 2023, the leader of the GMB Union, Gary Smith, publicly attacked the Labour Party's proposed policy of ending oil and gas licences in the North Sea, stating that the United Kingdom needs 'plans not bans'.[84] Smith's comments were subsequently highlighted throughout the media.[85] These examples illustrate how obstructionist actors can use labour concerns and recruit their spokespeople to bolster resistance to ambitious climate policy.

Another example of a 'floating organization' is the CPRE. At a national level, this countryside charity has made a series of statements on how the 'climate emergency is the biggest threat facing our countryside and planet'[86] and highlighted the need for rapidly scaling up renewable energy sources.[87] However, at the local level, the CPRE operates differently. For example, its Devon chapter has been highly effective at mobilizing communities against renewable energy projects.[88] The Devon CPRE trustee and energy spokesperson, Dr Philip Bratby, has espoused climate denialists viewpoints on several occasions, stating that 'man-made global warming does not exist'.[89] Similar patterns have been repeated elsewhere. Cornwall CPRE has publicly opposed solar power projects on the basis that Cornish farmland should be used for food production and carbon capture,[90] calling on local politicians to 'modify or refuse planning applications in light of the true environmental impact of every proposed development'.[91] While these groups purport to be acting in the interests of the local population and countryside, many of the talking points and issues cited mirror those repeatedly pushed by other obstructionist actors.

THE STRATEGIES AND TACTICS UTILIZED BY OBSTRUCTIONIST ACTORS

This section focuses on obstructionist actors' specific strategies and tactics and how they relate to the three dimensions of power described in the Introduction.

Material and structural

This first pillar supports the range of ways in which material and structural power is translated into specific strategies of obstruction, particularly by energy firms whose centrality to the economy gives them disproportionate power in the United Kingdom.[92] These strategies include threats of capital flight and fears of 'carbon leakage' in the face of regulatory changes, as when businesses fought a sustained and successful battle against a 1992 proposal for an EU- (then EC-) wide carbon tax; claims about lost tax revenue (from reduced oil and gas production); raising concerns over the risks of overreliance on fossil fuel imports (and hence the need for expanded domestic production); and warning about the social implications of unemployment allegedly caused by more ambitious climate measures. The granting of systematic advantages, privileges, and exemptions underscores the nature of this structural power. For example, between 2020 and 2023, Shell and BP paid almost no corporation taxes or production levies on oil and gas production in Britain's North Sea, yet benefitted from tax breaks and other forms of government support.

Critical accounts of corporate obstructionist strategies often point to the transnational way in which they are organized and funded.[93] Of the £1.45 million that the GWPF has received in charitable donations since 2017, around 45% has come from the United States.[94]

Paying for access, influence, and disproportionate representation in the policy process is another means to generate obstruction. The UK climate minister Graham Stuart received £12,000 in donations from one of the United Kingdom's largest fuel distributors and an aviation consultancy.[95] Former Prime Minister Liz Truss raised more than £420,000 in donations for her successful Conservative leadership bid, with the biggest single contribution made by the wife of a former BP executive.[96] Oil executives also gave more than £390,000 to the Conservative Party after Theresa May became prime minister in July 2016, hoping for 'further government support if the Conservatives are returned to power'.[97] During her brief spell as prime minister, Liz Truss, a former employee of the oil giant Shell,[98] adopted the demands of the NZSG to remove so-called green levies from energy bills, lifted the United Kingdom's ban on fracking for shale gas where there is 'local support', and laid out plans to increase extraction of oil and gas in the North Sea.[99] At the time, her ally at the head of BEIS, Jacob Rees-Mogg, a NZSG member, boasted he would extract 'every cubic inch of gas from the North Sea'.[100] The Brexit donor Jeremy Hosking, who bankrolls anti-net zero campaigns, was revealed to have more than £101million invested in fossil fuels.[101]

Institutional

Institutional power is a second pillar of power that undergirds obstructionist strategies. This power is often manifested in funding for parties and candidates resisting climate action, using access to key committees that exercise power over policymaking; secondments, where employees from private industry temporarily work within government or think tanks; and revolving doors, where politicians take directorships in major energy companies or energy executives join the government. For example, in 2010, Lord Browne, former CEO of BP, was appointed by then-Prime Minister David Cameron as the 'lead non-executive director' at the Cabinet Office, the department responsible for supporting the Prime Minister and the Cabinet, which comprises the most senior ministers in government. At the same time, Browne was also chair of fracking company Cuadrilla and pledged to do 'whatever it takes' to promote shale gas.[102] As minister of state for energy, Charles Hendry secured £3,333 a day as a consultant for Vitol, the world's biggest oil trader, handling 270 million tonnes of oil in 2016.[103] Former head of BP Peter Mather was appointed to the government's BEIS board as a non-executive board member in March 2022, despite triggering 'allegations of a "revolving door" between the government and the energy sector it is supposed to oversee' after being paid to chair a BP shareholders meeting.[104]

Engagement in shaping the details of policy design provides a key means of slowing ambition on and delaying the implementation of climate policy. Looking at the voluntary Climate Change Agreements (CCAs) made between British industry and the government, research has found that trade associations exerted 'considerable influence on the design of the Climate Change Levy [CCL] and CCAs'.[105] Here, authors Bailey and Rupp suggest, 'Many associations have maintained a defensive approach, focusing on immediate threats to their industry' and 'have conceived their role as that of braking, even derailing, environmental protection and sustainable development initiatives'.[106] Their negotiation was a product of compromise whereby the government responded 'to industry's protestations about the competitive effects of the CCL' by announcing a series of alleviating measures, such as an 80% reduction in the CCL 'in exchange for' legally binding CCAs for energy-intensive sectors.[107] Other research indicates that while the Treasury originally offered only a 50% CCL reduction to a few energy-intensive sectors, trade association lobbyists persuaded it to increase this percentage to 80% and make CCAs available to more sectors.[108] Through these policy instruments and the United Kingdom's flagship Emissions Trading Scheme (ETS), a cap and trade system to create a carbon price and subsequent market, we see how the reliance on market mechanisms noted earlier enables corporate obstruction. Scholars have observed in the story of the ETS in the United Kingdom how 'an attempt to secure industry support and cooperation became far too reliant on industry guidance, subsequently leading to regulatory capture, and the extraction of concessions for industry cooperation'.[109]

Delay, dilution of policy content, and resistance to implementation provide a further way to obstruct more ambitious climate action. Britain's largest housebuilders, including Barratt Developments, Berkeley Group, and Taylor Wimpey, 'privately lobbied for the government to ditch rules requiring the installation of electric car chargers in every new home' announced in November 2021.[110]

Discursive

Discursive power forms the third pillar of power that can be mobilized to obstruct climate action. It includes popular media strategies to discredit protests, exaggerate the costs of action, underplay the threat of climate change, overestimate the viability of technologies such as carbon capture and storage (CCS), blame other countries (especially China) to downplay the United Kingdom's role and responsibility, and use geopolitical events (such as the war in Ukraine) to question net zero ambitions.

Many of these obstructionist discourses have been identified in work on organized climate contrarian groups.[111] This research identifies what the authors call 'super-claims', such as 'climate solutions won't work' and 'the climate movement/science is unreliable', and then 'sub-claims', such as 'policies are ineffective', and 'sub-sub-claims' like 'green jobs don't work'. While this analysis was based on the outputs of contrarian groups in the United States, similar discourses are used by certain groups and right-wing media opposed to climate action in the United Kingdom. However, discursive power operates through the cultural spheres of arts and sports sponsorship, consistent with neo-Gramscian accounts which view media, education, and the arts as key sites of social struggle and in policing the terrain of debate.[112]

Co-option, dilution, and misappropriation of the language of advocates of climate action form part of a commonly used strategy. For example, despite foes of climate action sometimes adopting the language of transition and even transformation, there is a gulf between bold proclamations, such as BP CEO Bernard Looney's statement that the energy transition 'will require nothing short of reimagining energy as we know it',[113] and the company's ongoing support for CCS to prolong the life of fossil fuels. Such ameliorative and 'plug and play' measures do not require any systemic or infrastructural changes to implement.[114] This example supports other researchers' findings that even the most ambitious firms are engaging in hedging by mitigating risk through diversifying business operations, rather than rapid decarbonization.[115]

One key discursive framing that obstructionists use is describing gas as a 'transition fuel'. For example, ExxonMobil, BP, and Shell lobbied a UK government minister to keep burning natural gas for years on the basis that fossil fuel use is 'a necessary compromise', even though the United Kingdom is committed to reaching net zero by 2050.[116] A 2015 investigation by *The Guardian* revealed BP 'helped spur a concerted industry push to curb EU policy support for renewable energies such as wind and solar in favour of gas'.[117] The head of BP's Brussel's office, Howard Chase, stated, 'Large-scale use of natural gas could secure immediate emissions reductions on an economic basis',[118] also reiterating the claim that the gas will help to displace coal. The EU Climate Commissioner Connie Hedegaard then repeated BP's phrase that natural gas was 'an indispensable component' of EU climate strategy.[119]

The enduring prevalence of the discursive power of climate obstructionism in the United Kingdom's public sphere can be explained by the presence of right-wing politicians espousing some variation of climate scepticism, the existence of organized interests that feed sceptical coverage (particularly the GWPF), and partisan, right-wing media receptive to this message. Much of the right-leaning media in the United Kingdom stand at the centre of a dynamic relationship between a group of right-wing politicians and organized obstructionism, which has shifted its discourse from questioning climate science to questioning policy. While some historical strategies such as attacks on unreliable science or climate alarmism persist, by 2022, the key terrain had moved to discrediting net zero ambitions by exaggerating the costs and taking advantage of the cost-of-living crisis and geopolitical events, such as the war in Ukraine, to push for a re-evaluation of the United Kingdom's emission pledges. The discourse may change, but the effect is the same: climate action and ambitious policy aims are deprioritized or delegitimized, and fossil fuel production and incumbent actors remain entrenched.

The UK media has followed a similar path to that of other countries (particularly the United States) where the terrain of climate debates has shifted from evidence scepticism (which doubts the trend of warming temperatures or the real and potential impacts) toward 'discourses of delay' that focus on undermining support for policies meant to address climate change.[120] Analysis of more than 1,000 editorials in right-wing papers from 2011 to 2021 showed a substantial increase toward the end of that period in those calling for climate action (from just one in 2011 to sixty-five in 2021), and 2022 was the fourth year in a row that evidence scepticism had been completely absent, partly because it had become increasingly difficult to doubt the existence of climate change.[121] However, what remained was 'response' or 'policy' scepticism, particularly arguments around the 'cost' of taking action, 'whataboutism' (why should the United Kingdom take action when other countries were not?), and strongly worded attacks on climate activist organizations such as Extinction Rebellion and Just Stop Oil.

On television, which in most countries is by far the most consulted source for climate information,[122] the BBC seems to have followed a similar journey, particularly after it changed its editorial policy regarding climate sceptics in 2018.[123] Detailed content analysis of the television coverage of the IPCC Sixth Assessment Report (AR6) of August 2021 from Working Group One (WG I) shows that the mainstream news programmes included in the analysis (BBC1, ITV, and Channel 4 evening news) did not include any sceptic voices.[124] This was in sharp contrast to the reporting on the

IPCC Fifth Assessment Report (AR5) from WGs I and II, where criticism of some of the science was prominently presented.[125]

However, in 2022, the number of editorials in right-wing titles that supported reversing the United Kingdom's ban on fracking shot up, driven partly by the argument that domestic shale gas development would reduce UK dependence on expensive gas from Russia and other foreign countries. *The Sun* alone published thirty-two pro-fracking editorials during the year—a larger number than all UK newspapers collectively had ever published before.[126] Similarly, an unprecedented number of right-wing editorials attacked Just Stop Oil, arguing that the climate movement was unreliable and alarmist.

Perhaps the most important shift was the drop in the number of editorials supporting climate action and a corresponding rise in the number of attacks on the United Kingdom's commitment to net zero by 2050, apparently prompted by the high costs of the energy crisis. For example, on 6 March 2022, the *Mail* described the goal as 'utopian and impracticable',[127] and on the same day published a comment piece by broadcaster and former politician Nigel Farage[128] criticizing the high costs of the commitment and arguing for fracking along with opening a coal mine in Cumbria. He used the piece to launch a campaign for a referendum on what he called 'the Net Zero delusion' under the banner of the organization 'Britain Means Business'.

Indeed, the growing opposition to the Net Zero strategy rekindled the powerful nexus between right-wing politicians, organized obstructionism, and right-leaning media in 2021–2022. First, twenty members of the parliamentary NZSG signed an open letter to the *Daily Telegraph* in January 2022 calling for removing environmental levies on domestic energy. The Conservative MP Craig Mackinlay announced that the NZSG would rely on the GWPF for its campaign information. His main argument was that the potential cost of capital investment needed for Net Zero would be £1.4 trillion, a figure dismissed as an exaggeration due to his failure to present the cost savings of £991 billion during the 30 years to 2050, particularly by avoiding the purchase of fossil fuels.[129]

Second, Reform UK (previously the Brexit Party) changed its political focus to campaigning against Net Zero.[130] Third, several right-wing Tory politicians were given space in some right-leaning titles to promote their anti-Net Zero views throughout the year.[131] A Google site search shows that the *Mail*, *Telegraph*, and *Sun* were particularly strident in their sustained opposition to the Net Zero plans, giving voice to arguments or studies mainly emphasizing the high cost of meeting the 2050 goal. The *Telegraph* was egregiously partisan: of the first twenty results from a Google site search for 2022, sixteen articles were predominately against the policy, three neutral, and only one in favour.[132] The cost was variously headlined as 'huge',[133] 'too high',[134] 'vast',[135] or even 'dirty';[136] in short, the policy was 'mad' and should be 'torn up to save individuals' wealth'.[137] The constant publication of such one-sided articles is a particularly significant example of media-mediated climate obstructionism given the *Telegraph*'s status as the right-wing broadsheet with the most readers[138] and probably the most influential among Conservative Party members.[139]

CONCLUSION

The forces of climate obstructionism in the United Kingdom are alive and well, though the intensity and vociferousness of their interventions may reflect growing demands for more ambitious climate action. This chapter has argued that the United Kingdom offers numerous insights into the dynamics of climate obstructionism, the actors involved, and the strategies and tactics deployed to thwart climate action. What happens in the United Kingdom often has wider consequences for global climate action because of the financial power concentrated in the city of London, the fact that the United Kingdom is home to some of the carbon majors,[140] and because of its historical role as the home of fossil capital and the birthplace of the Industrial Revolution. It is also a self-declared leader in climate diplomacy, despite recent evidence of the fragility of this claim.[141] We have also noted how the United Kingdom is embroiled in global efforts aimed at climate obstructionism through transnational funding and advocacy networks.

Our discussion has surveyed a range of key actors and explored the strategies employed and the dimensions of power expressed to obstruct more ambitious climate action. Climate obstructionism within Britain has created several fault lines across the climate governance landscape and wider public discourse that could prove crucial for determining the speed and scale of decarbonization in the United Kingdom. Net zero policies and growing pressure to adopt supply-side policies to limit new gas and oil production have served as flashpoints around which obstructionist actors have mobilized.

While this chapter has sought to sketch out the current alignments of obstructionist actors and where the current fault lines exist, additional research and advocacy is needed from academia and civil society to explore the specific dynamics of each site of climate obstructionism and the mechanisms available to reduce obstructionist actors' ability to shape policy and public sentiment to further entrench their power. To guide this future work, we have identified several instances of obstructionist activity that speak to the different pillars of power analysed in this chapter. Countermoves and advocacy aimed at checking each of these sources of power could include the following:

First, the material power of incumbents needs to be addressed by clearer rules and limits to and transparency around party donations, directorships, and the financial ties between elected politicians and those they are meant to be governing.

Second, disproportionate access to key committees and departments through internships, secondments, and revolving doors between government and business must be subject to greater scrutiny and control. The current rules governing revolving doors are insufficient, with regulators overlooking conflicts of interest to the extent that it has become a systemic issue. Policies that could close revolving doors include a 'cooling-off period' whereby placements, secondments, and sabbaticals between government and the fossil fuel industry are stopped for a certain number of years to prevent conflicts of interest during a crucial phase of UK's efforts to decarbonize its economy. Additionally, the 'rules of engagement' for APPGs should be transformed. Narrowing the opportunities provided to vested interests to donate, coordinate, and run APPGs could help curb climate obstructionism, especially around key parliamentary votes or at crucial political junctures.

Finally, sources of mis- and disinformation about climate change and a lack of media transparency about on whose behalf people are speaking need to be addressed. This task requires greater oversight of traditional and social media. Taken together, these strategies amount to a conscious rebalancing of politics favouring those who are underrepresented and have the most to lose from runaway global heating.

NOTES

1. Andreas Malm (2016), *Fossil Capital: The Rise of Steam Power and the Roots of Global Warming*. London: Verso.
2. Guri Bang, Knut Einar Rosendahl, and Christoph Böhringer (2022), 'Balancing Cost and Justice Concerns in the Energy Transition: Comparing Coal Phase-Out Policies in Germany and the UK', *Climate Policy*, 22, 8: 1000–1015.
3. Simon Evans (2021, 18 March), 'Analysis: UK Is Now Halfway to Meeting Its "Net-Zero Emissions" Target, Carbon Brief', https://www.carbonbrief.org/analysis-uk-is-now-halfway-to-meeting-its-net-zero-emissions-target/.
4. UK Government, 'Nuclear Energy: What You Need To Know', GOV.UK, https://www.gov.uk/government/news/nuclear-energy-what-you-need-to-know.

5. Paul Dorfman (2022, 13 December), 'Nuclear Power Is Just a Slow and Expensive Distraction', *The New Statesman*, https://www.newstatesman.com/spotlight/climate-energy-nature/2022/12/nuclear-power-slow-expensive-distraction.
6. Gov.UK (2021, 22 April), 'PM Statement at the Leaders' Summit on Climate', https://www.gov.uk/government/speeches/pm-statement-at-the-leaders-summit-on-climate-22-april-2021/.
7. Department for Business, Energy & Industrial Strategy (BEIS). '2021 UK Greenhouse Gas Emissions, Provisional Figures', *National Statistics*, https://assets.publishing.service.gov.uk/government/uploads/system/uploads/attachment_data/file/1064923/2021-provisional-emissions-statistics-report.pdf/.
8. ClientEarth (2022, 18 July), 'Historic High Court ruling finds UK government's climate strategy "unlawful"', *ClientEarth*, https://www.clientearth.org/latest/press-office/press/historic-high-court-ruling-finds-uk-government-s-climate-strategy-unlawful/.
9. Chris Skidmore, 'MISSION ZERO: Independent Review of Net Zero 2023', GOV UK, https://assets.publishing.service.gov.uk/government/uploads/system/uploads/attachment_data/file/1128689/mission-zero-independent-review.pdf.
10. Climate Action Tracker. 'United Kingdom', CAT, https://climateactiontracker.org/countries/uk/.
11. Justin Rowlatt and Greg Brosnon (2023, 28 June), 'Climate Change Committee Says UK no Longer a World Leader', *BBC News*, https://www.bbc.co.uk/news/science-environment-66032607.
12. William K. Carroll (2020), *Regime of Obstruction How Corporate Power Blocks Energy Democracy*. Edmonton, AB: AU Press; J. Jacquet (2022), *The Playbook: How to Deny Science, Sell Lies, and Make a Killing in the Corporate World*. London: Penguin Books.
13. Wyn Grant (1984), 'Large Firms and Public Policy in Britain', *Journal of Public Policy*, 4: 1–17.
14. Adrian Ford and Peter Newell (2021), 'Regime Resistance and Accommodation: Toward a Neo-Gramscian Perspective on Energy Transitions', *Energy Research & Social Science*, 79: 102163; F. Geels (2014), 'Regime Resistance against Low-Carbon Transitions: Introducing Politics and Power into the Multi-level Perspectives,' *Theory, Culture & Society*, 31, 5: 21–40.
15. Jonas Meckling (2011), *Carbon Coalitions: Business, Climate Politics, and the Rise of Emissions Trading*. Cambridge, MA: MIT Press.
16. David Levy and Peter Newell (2005), *The Business of Global Environmental Governance*. Cambridge, MA: MIT Press.
17. Peter Newell (2019), 'Trasformismo or Transformation? The Global Political Economy of Energy Transitions', *Review of International Political Economy*, 26/1: 25–48.
18. Adrian Ford and Peter Newell (2021), 'Regime Resistance and Accommodation: Toward a Neo-Gramscian Perspective on Energy Transitions', *Energy Research & Social Science*, 79: 102163.
19. Irene Lorenzoni, Sophie Nicholson-Cole, and Lorraine Whitmarsh (2007), 'Barriers Perceived to Engaging with Climate Change among the UK Public and Their Policy Implications', *Global Environmental Change*, 17: 445–459.
20. Arthur Mol (2003), *Globalization and Environmental Reform: The Ecological Modernization of the Global Economy*. Cambridge: MIT Press.

21. Neil Carter and Michael Jacobs (2014), 'Explaining Radical Policy Change: The Case of Climate Change and Energy Policy under the British Labour Government 2006–10', *Public Administration*, 92: 125–141.
22. Damian Carrington (2012, 26 April), 'David Cameron: This Is the Greenest Government Ever', *The Guardian*, https://www.theguardian.com/environment/2012/apr/26/david-cameron-greenest-government-ever.
23. Neil Carter and Ben Clements (2015), 'From "Greenest Government Ever" to "Get Rid of All the Green Crap": David Cameron, the Conservatives and the Environment', *British Politics*, 10: 204–225.
24. Katherine Dommett (2015), 'The Theory and Practice of Party Modernisation: The Conservative Party under David Cameron, 2005–2015', *British Politics*, 10: 249–266.
25. Chloe Farand and Mat Hope (2018, 18 November), 'Matthew and Sarah Elliott: How a UK Power Couple Links US Libertarians and Fossil Fuel Lobbyists to Brexit', *Desmog*, https://www.desmog.com/2018/11/18/matthew-sarah-elliott-uk-power-couple-linking-us-libertarians-and-fossil-fuel-lobbyists-brexit/.
26. Ibid.
27. Seth Thevoz (2022, 22 March), 'Revealed: Brexit Donor Behind Net-Zero Backlash Has $130m in Fossil Fuels', *Open Democracy*, https://www.opendemocracy.net/en/dark-money-investigations/jeremy-hosking-brexit-donor-net-zero-invest-fossil-fuels/.
28. Peter Newell (2000), *Climate for Change: Non-State Actors and the Global Politics of the Greenhouse*. Cambridge: Cambridge University Press.
29. Mei Li, Gregory Trencher, and Jusan Asuka (2022), 'The Clean Energy Claims of BP, Chevron, ExxonMobil and Shell: A Mismatch Between Discourse, Actions and Investments', *PLoS ONE*, 17, 2: e0263596.
30. Timo Busch and Lena Judick (2021), 'Climate Change—That Is Not Real! A Comparative Analysis of Climate-Sceptic Think Tanks in the USA and Germany', *Climatic Change*, 164: doi:10.1007/s10584-021-02962-z; T. K. Burki (2019), 'Links Between Think Tanks and the Tobacco Industry', *The Lancet Oncology*, 20: doi:10.1016/s1470-2045(19)30063-4.
31. Mat Hope (2023, 23 November), 'US Donors Gave $177k to UK Climate Science Denying Global Warming Policy Foundation', *Desmog*, https://www.desmog.com/2018/11/23/us-donors-give-177k-uk-climate-science-denying-global-warming-policy-foundation/.
32. Adam Barnett and Michaela Herrmann (2022, 18 March), 'Mapped: How the Net Zero Backlash Is Tied to Climate Denial – and Brexit', *Desmog*, https://www.desmog.com/2022/03/18/mapped-how-the-net-zero-backlash-is-tied-to-climate-denial-and-brexit/.
33. Desmog. 'Net Zero Watch' (2022, 4 October), *Desmog*, https://www.desmog.com/net-zero-watch/.
34. Net Zero Watch (2022, 4 October), 'In a Bad Situation There Are No Good Moves: What the UK Should (But Will Not) Do To Address the Energy Crisis', *Net Zero Watch*, https://www.netzerowatch.com/in-a-bad-situation-there-are-no-good-moves-what-the-uk-should-but-will-not-do-to-address-the-energy-crisis/.
35. Barnett and Herrmann, 'Mapped'.
36. R. McKie (2023), *The Climate Change Counter Movement: How the Fossil Fuel Industry Sort to Delay Climate Action*. London: Palgrave Macmillan; R. McKie (2021), 'Obstruction, Delay, and Transnationalism: Examining the Online

Climate Change Counter-Movement', *Energy Research & Social Science*, 80: 102217.
37. Tobacco Tactics (2021, 13 August), 'IEA: Working with RJ Reynolds, BAT and Philip Morris on Environmental Risk', *Tobacco Tactics*, https://tobaccotactics.org/wiki/iea-working-with-rj-reynolds-bat-and-philip-morris-on-environmental-risk/.
38. McKie, *The Climate Change Counter Movement*.
39. Lawrence Carter and Alice Ross (2018, 30 July), 'Revealed: BP and Gambling Interests Fund Secretive Free Market Think Tank', *UnEarthed*, https://unearthed.greenpeace.org/2018/07/30/bp-funding-institute-of-economic-affairs-gambling/A. Bychawski (2023, 11 January), 'Revealed: Truss-Allied Think Tank Met Dozens of MPs Prior to Leadership Win', *Open Democracy*, https://www.opendemocracy.net/en/institute-of-economic-affairs-liz-truss-2022-accounts/.
40. Adam Barnett (2022, 4 October), 'Old School Climate Science Denial Lingers on Outskirts of Tory Conference', *Desmog*, https://www.desmog.com/2022/10/04/old-school-climate-science-denial-lingers-on-outskirts-of-tory-conference/.
41. Adam Bychawski (2023, 20 July), 'Just Stop Oil Targets Think Tank over Role in Protest Crackdown', *Open Democracy*, https://www.opendemocracy.net/en/policy-exchange-just-stop-oil-exxon-mobil-rishi-sunak/.
42. Peter Geoghegan (2023, 20 July), 'The Tory Donor, Soviet-born Billionaire and Fossil Fuel Interests Bankrolling British Politics', *Byline Times*, https://bylinetimes.com/2023/07/20/the-tory-donor-soviet-born-billionaire-and-fossil-fuel-interests-bankrolling-british-politics/.
43. Saffron O'Neill, Hywel T. P. Williams, Tim Kurz, et al. (2015), 'Dominant Frames in Legacy and Social Media Coverage of the IPCC Fifth Assessment Report', *Nature Climate Change*, 5: 380–385.
44. Aisha Majid (2023, 15 March), 'Top 50 News Websites in the World in February 2023: New York Times Sees Decline for First Time in a Year', *Press Gazette*, https://pressgazette.co.uk/media-audience-and-business-data/media_metrics/most-popular-websites-news-world-monthly-2/.
45. Roy Greenslade (2012, 25 January), 'Mail Online Goes Top of the World', *The Guardian*, https://www.theguardian.com/media/greenslade/2012/jan/25/dailymail-internet.
46. Richard Black (2018), *Denied: The Rise and Fall of Climate Contrarianism*. London: Real Press, pp. 16–23.
47. James Painter (2011), *Poles Apart: The International Reporting on Climate Scepticism*. Oxford: Reuters Institute for the Study of Journalism, p. 105.
48. Daily Mail (2023, 5 June), 'Daily Mail Comment: Starmer's Dangerous Dash to Net Zero', *The Daily Mail*, https://www.dailymail.co.uk/debate/article-12158887/DAILY-MAIL-COMMENT-Starmers-dangerous-dash-net-zero.html.
49. Allison Pearson (2023, 31 May), 'The Six Major Flaws of Starmer's Ridiculous Plan to Stop Drilling in the North Sea', *The Daily Telegraph*, https://www.telegraph.co.uk/columnists/2023/05/31/labour-north-sea-drilling/.
50. Andrew Quinn (2023, 5 June), 'Keir Starmer Insists North Sea Oil and Gas Will Be Used 'For Many Years' Following Union Backlash', *The Daily Record*, https://www.dailyrecord.co.uk/news/politics/keir-starmer-insists-north-sea-30156434.
51. Mark Duell and James Tapsfield (2023, 30 March), 'How £100-a-Year Gas 'Penalty' Will Affect YOU: Experts Warn Government 'Green Drive' onto Electricity Will Cost Households up to £13,000 to Fit Heat Pumps They Need to Avoid Hiked Gas Bills', *The Daily Mail*, https://www.dailymail.co.uk/news/

article-11920815/How-100-year-gas-penalty-affect-YOU.html.; Telegraph View (2023, 26 June), 'The Levy on Energy Bills Is a Net-Zero Poll Tax', *The Daily Telegraph*, https://www.telegraph.co.uk/opinion/2023/06/26/the-levy-on-energy-bills-is-a-net-zero-poll-tax/.

52. Sam Bright (2023, 2 May), 'Telegraph Columnist Allison Pearson Becomes Director of Climate Science Denial Group', *Desmog*, https://www.desmog.com/2023/05/02/telegraph-columnist-allison-pearson-becomes-director-of-climate-science-denial-group/.
53. Allison Pearson (2022, 8 November), 'We Don't Owe Developing Countries "Climate Reparations"—They Owe Us', *The Daily Telegraph*, https://www.telegraph.co.uk/columnists/2022/11/08/cop27-dont-owe-developing-countries-climate-reparations-owe/.
54. Allison Pearson (2022, 20 August), 'Boris Can Never Be Forgiven for Sacrificing Britain to His Net Zero Fantasy', *The Daily Telegraph*, https://www.telegraph.co.uk/columnists/2022/08/30/boris-johnson-can-never-forgiven-sacrificing-britain-net-zero/.
55. Allison Pearson (2023, 18 April), 'Being a Just Stop Oil Activist Only Makes Sense If You've Been Pampered by Mummy and Daddy', *The Daily Telegraph*, https://www.telegraph.co.uk/columnists/2023/04/18/just-stop-oil-snooker-protest-activists-know-nothing/
56. Pearson, 'The Six Major Flaws'.
57. Desmog, 'James Delingpole', https://www.desmog.com/james-delingpole/.
58. Desmog, 'Julia Hartley-Brewer', https://www.desmog.com/julia-hartley-brewer/.
59. Robert Brulle and Christian Downie (2022), 'Following the Money: Trade Associations, Political Activity and Climate Change', *Climatic Change*, 175: doi:10.1007/s10584-022-03466-0.
60. Juho Vesa, Antti Gronow, and Tuomas Ylä-Anttila (2020), 'The Quiet Opposition: How the Pro-Economy Lobby Influences Climate Policy', *Global Environmental Change* 63: 102117.
61. Wyn Grant, Duncan Matthews, and Peter Newell (2000), *The Effectiveness of European Union Environmental Policy*. New York: Springer.
62. Peter Newell and Matthew Paterson (1998), 'A Climate for Business: Global Warming, the State and Capital', *Review of International Political Economy*, 5: 679–703.
63. Mark Milner and Larry Elliot (2006, 23 November), 'Unilateral Action on Climate Change Could Ruin Economy, Says CBI Chief', *The Guardian*, https://www.theguardian.com/business/2006/nov/23/ukeconomy.climatechange.
64. CBI (2022, 1 March), 'Chancellor's Spring Statement: Must Be the Time to Act or Economy Will Drift Back to Low Growth—CBI Chief', *CBI*, https://www.cbi.org.uk/media-centre/articles/chancellor-s-spring-statement-must-be-the-time-to-act-or-economy-will-drift-back-to-low-growth-cbi-chief/.
65. CBI (2021, March), 'Greening the Tax System: How Tax Policy Could Support Net-Zero', *CBI*, https://www.cbi.org.uk/media/6332/2021-03-greening-the-tax-system.pdf.
66. UKPIA (2024, March). 'Who We Work With', *UKPIA*, https://www.ukpia.com/about/who-we-work-with/.
67. UK Parliament (2024, March), 'All-Party Parliamentary Groups', *UK Parliament*, https://www.parliament.uk/about/mps-and-lords/members/apg/.
68. Ibid.

69. Shanti Das (2022, 20 February), 'Oil and Gas Firms Have Unlisted Links to Westminster', *The Guardian*, https://www.theguardian.com/politics/2022/feb/20/lobbyists-for-oil-and-gas-companies-shell-bp-exxonmobil.
70. UK Parliament (2022), 'Register of All-Party Parliamentary Groups [as at 19 October 2022]', *UK Parliament*, https://publications.parliament.uk/pa/cm/cmallparty/221019/downstream-energy-and-fuels.htm.
71. Das, 'Oil and Gas Firms Have Unlisted Links'.
72. Ben Webster (2023, 18 January), 'Gas Industry Paid Lobbyists £200,000 to Get MPs' Support for "Blue Hydrogen"'. *Open Democracy*, https://www.opendemocracy.net/en/blue-hydrogen-appg-alexander-stafford-lobbying-shell/.
73. Ibid.
74. Fiona Harvey (2021, 10 September), 'UK Ministers Met Fossil Fuel Firms Nine Times As Often As Clean Energy Ones'. *The Guardian*, https://www.theguardian.com/environment/2021/sep/10/uk-ministers-met-fossil-fuel-firms-nine-times-more-often-than-clean-energy-companies.
75. BEIS (2022), 'BEIS: Ministerial Gifts, Hospitality, Travel and Meetings', *Department for Business, Energy and Industrial Strategy*, https://www.gov.uk/government/collections/beis-ministerial-gifts-hospitality-travel-and-meetings.
76. Adam Lucas (2021), 'Investigating Networks of Corporate Influence on Government Decision-Making: The Case of Australia's Climate Change and Energy Policies', *Energy Research & Social Science*, 81: 102271.
77. Kacper Szulecki (2018, May), 'The Revolving Door Between Politics and Dirty Energy in Poland: A Governmental-Industrial Complex', *Greens/EFA Group in the European Parliament*, https://www.greens-efa.eu/files/assets/docs/report_of_revolving_doors_digital_-min.pdf.
78. Óscar Reyes (2018, May), 'Revolving Doors in Spanish Climate and Energy Policy', *Greens/EFA Group in the European Parliament*, https://www.greens-efa.eu/files/assets/docs/report_of_revolving_doors_digital_-min.pdf.
79. Adam Bychawski (2022, 26 August), 'Fears Ex-BP and British Gas Bosses Could "Sway" Government Energy Policy', *Open Democracy*, https://www.opendemocracy.net/en/ofgem-energy-price-cap-revolving-door-bills-government-oil-bp/.
80. Ruth Hayhurst (2017, 20 March). '"Police and Judges Should Take a Firm Line with Anti-Fracking Protests Against Shale Gas Suppliers" – Union', *Drill or Drop?*, https://drillordrop.com/2017/03/20/police-and-judges-should-take-a-firm-line-with-anti-fracking-protests-against-shale-gas-suppliers-union/.
81. GMB (2019, 11 June), 'Two Thirds Worried About Foreign Countries Keeping Britain's Lights On', *GMB Union*, https://www.gmb.org.uk/news/two-thirds-worried-about-foreign-countries-keeping-britains-lights.
82. GMB (2015), 'GMB Congress 2015—CEC Statement on "Fracking" (Hydraulic Fracturing for Shale Gas)', *GMB Union*, https://gmb.org.uk/sites/default/files/GMB15-Fracking.pdf.
83. Net Zero Watch (2016, 8 August), 'GWPF Calls on Government To Deliver on UK Shale Development', *Net Zero Watch*, https://www.netzerowatch.com/gwpf-calls-on-government-to-deliver-on-uk-shale-development/.
84. GMB (2013, 6 June), 'UK Energy Needs "Plans Not Bans"', *GMB Union*, https://www.gmb.org.uk/news/uk-energy-needs-plans-not-bans.
85. Greg Heffer (2023, 29 May), 'One of Labour's Biggest Union Backers Urges Keir Starmer to Rethink Party's "Self-Defeating" Plan to Ban All New North Sea Oil and Gas Drilling', *The Daily Mail*, https://www.dailymail.co.uk/news/arti

cle-12135709/One-Labours-union-backers-urges-rethink-plan-ban-new-North-Sea-oil-gas-licences.html.
86. CPRE (2024, March), 'Climate Emergency', *CPRE*, https://www.cpre.org.uk/what-we-care-about/climate-change-and-energy/climate-emergency/. 8 February 2011.
87. CPRE (2022, 8 November), 'Why Do We Need Renewable Energy?', *CPRE*, https://www.cpre.org.uk/explainer/why-do-we-need-renewable-energy/.
88. Sharon Goble (2021, 23 September), 'Significant Win for Devon CPRE Campaign as Mid Devon Refuses Langford Solar Farm Proposal', *The Devon Daily*, https://www.thedevondaily.co.uk/news/local-news/significant-win-devon-cpre-campaign-mid-devon-refuses-langford-solar-farm-proposal.
89. Mike Hulme (2009, 16 March), 'What Message, and Whose, from Copenhagen?', *BBC News*, http://news.bbc.co.uk/1/hi/sci/tech/7946476.stm.
90. Lisa Young (2022, 7 March), 'Solar Farms in Cornwall Threaten Farming Claims CPRE', *The Falmouth Packet*, https://www.falmouthpacket.co.uk/news/19972474.solar-farms-cornwall-threaten-farming-claims-cpre/
91. Paul Armstrong (2021, 10 May), 'CPRE, Cornwall Urges New Council to Act over Countryside', *The Falmouth Packet*, https://www.falmouthpacket.co.uk/news/19290219.cpre-cornwall-urges-new-council-act-countryside/.
92. Peter Newell and Matthew Paterson (2011, 8 February), 'A Climate for Business: Global Warming, the State and Capital', *Review of International Political Economy*, 5: 679–703.
93. Ibid.
94. Helena Horton and Adam Bychawski (2022, 4 May), 'Climate Sceptic Thinktank Received Funding from Fossil Fuel Interests', *The Guardian*, https://www.theguardian.com/environment/2022/may/04/climate-sceptic-thinktank-received-funding-from-fossil-fuel-interests.
95. Jamie Grierson (2023, 26 January), 'UK Climate Minister Received Donations from Fuel and Aviation Companies', *The Guardian*, https://www.theguardian.com/politics/2023/jan/26/uk-climate-minister-received-donations-fuel-aviation-companies.
96. Aubrey Allegretti and Henry Dyer (2022, 8 September), 'Liz Truss Reveals Campaign Donation of £100,000 from Wife of Ex-BP Executive', *The Guardian*, https://www.theguardian.com/politics/2022/sep/08/liz-truss-reveals-campaign-donations-including-100k-from-wife-of-ex-bp-exec..
97. Rajeev Syal (2017, 23 May), 'Oil Bosses Have Given £390,000 to Tories under Theresa May', *The Guardian*, https://www.theguardian.com/business/2017/may/23/oil-bosses-have-given-390000-to-tories-conservatives-under-theresa-may..
98. The Ferret (2022, 13 September), 'Fact Check: Zarah Sultana Claims on Oil and Gas Links of Liz Truss', *The Ferret*, https://theferret.scot/zarah-sultana-energy-industry-tories-truss/.
99. Jessica Elgot, Peter Walker, and Alex Lawson (2022, 8 September), 'Liz Truss to Freeze Energy Bills at £2,500 a Year Average, Funded by Borrowing', *The Guardian*, https://www.theguardian.com/politics/2022/sep/08/liz-truss-to-freeze-energy-bills-price-at-2500-a-year-funded-by-borrowing.
100. Adam Forrest (2022, 4 April), 'UK Should Drill "Every Last Drop" of North Sea Oil and Gas, Says Jacob Rees-Mogg', *The Independent*, https://www.independent.co.uk/climate-change/news/oil-gas-north-sea-rees-mogg-b2050519.html.
101. Seth Thévoz (2022, 22 March), 'Revealed: Brexit Donor behind Net-Zero Backlash Has $130m in Fossil Fuels', *Open Democracy*, https://www.opendemocr

acy.net/en/dark-money-investigations/jeremy-hosking-brexit-donor-net-zero-invest-fossil-fuels/.
102. Molly Scott Cato (2018, 8 May), 'Revealed: The Revolving Door Between Westminster and the Fossil Fuel Industry', *Left Foot Forward*, https://leftfootforward.org/2018/05/the-revolving-door-between-westminster-and-the-fossil-fuel-industry/.
103. Ibid.
104. Adam Bychawski (2023, 4 January), 'Government Energy Adviser Quits BP after "Conflict of Interest" Fears Revealed', *Open Democracy*, https://www.opendemocracy.net/en/peter-mather-bp-europe-beis-board-member.
105. Ian Bailey and Susanne Rupp (2006), 'The Evolving Role of Trade Associations in Negotiated Environmental Agreements: The Case of United Kingdom Climate Change Agreements' *Business Strategy and the Environment* 15: 47.
106. Ibid., p. 44.
107. Ibid., p. 45.
108. Adrian Smith (2002), 'Policy Transfer in the Development of UK Climate Policy for Business', *SPRU Science and Technology Policy Research Electronic Working Paper Series*, 75: 1–24.
109. Fredrik von Malmborg and Peter Strachan (2005), 'Climate Policy, Ecological Modernization and the UK Emission Trading Scheme', *European Environment* 15: 157.
110. Jasper Jolly (2022, 12 September), 'Housebuilders Lobbied against Plan for Electric Car Chargers in New Homes in England', *The Guardian*, https://www.theguardian.com/business/2022/sep/12/housebuilders-lobbied-against-plan-for-electric-car-chargers-in-new-homes-in-england.
111. William F. Lamb, Giulio Mattioli, Sebastian Levi, et al. (2020), 'Discourses of Climate Delay', *Global Sustainability*, 3, 17: https://doi.org/10.1017/sus.2020.13; Travis G. Coan, Constantine Boussalis, John Cook, et al. (2021), 'Computer-Assisted Classification of Contrarian Claims about Climate Change', *Communications & Media Studies*, 1, 11: doi:10.1038/s41598-021-01714-4.
112. David Levy and Peter Newell (2002), 'Business Strategy and International Environmental Governance: Toward a Neo-Gramscian Synthesis', *Global Environmental Politics*, 2, 4: 84–101.
113. Hugo Britt (2020, 13 November), 'BP "Fundamentally Changing the Organization" to Become Net-Zero', *Thomas Insights*, https://www.thomasnet.com/insights/bp-fundamentally-changing-the-organization-to-become-net-zero/.
114. Peter Newell and Abigail Martin (2020, November), 'The Role of the State in the Politics of Disruption & Acceleration', *Climate KIC Working Paper*.
115. Jessica Green, Jennifer Hadden, Thomas Hale et al. (2021), 'Transition, Hedge, or Resist? Understanding Political and Economic Behavior Toward Decarbonization in the Oil and Gas Industry', *Review of International Political Economy*, 29, 6: 2036–2063.
116. Jane Dalton (2021, 11 July), 'Oil Giants Lobbied Minister to Keep UK Burning Fossil Fuels', *The Independent*, https://www.independent.co.uk/climate-change/news/oil-gas-climate-fossil-exxon-shell-bp-lobby-b1879948.html.
117. Arthur Neslen (2015, 20 August), 'BP Lobbied against EU Support for Clean Energy to Favour Gas, Documents Reveal', *The Guardian*, https://www.theguardian.com/environment/2015/aug/20/bp-lobbied-against-eu-support-clean-energy-favour-gas-documents-reveal?CMP=share_btn_link.

118. Ibid.
119. Ibid.
120. William F Lamb, Giulio Mattioli, Sebastian Levi et al., 'Discourses of Climate Delay'; Hannah Schmid-Petri, Silke Adam, Ivo Schmucki, et al. (2015), 'A Changing Climate of Skepticism: The Factors Shaping Climate Change Coverage in the US Press', *Public Understanding of Science*, 26, 4: doi:10.1177/0963662515612.
121. Josh Gabbatiss, Sylvia Hayes, Joe Goodman et al. (2022), 'Analysis: How UK Newspapers Changed Their Minds About Climate Change', *Carbon Brief*, https://interactive.carbonbrief.org/how-uk-newspapers-changed-minds-climate-change/; Carbon Brief (2023, 11 January), 'Analysis: How UK Newspapers Commented on Energy and Climate Change in 2022', *Carbon Brief*, https://www.carbonbrief.org/analysis-how-uk-newspapers-commented-on-energy-and-climate-change-in-2022/.
122. Simge Andi (2020), 'How People Access News about Climate Change', *Digital News Report*, https://www.digitalnewsreport.org/survey/2020/how-people-access-news-about-climate-change/.
123. Mike Schäfer and James Painter (2021), 'Climate Journalism in a Changing Media Ecosystem: Assessing the Production of Climate Change-Related News around the World', *WIREs Climate Change*, 12: doi:10.1002/wcc.675.
124. James Painter, Josh Ettinger, David Holmes, et al. (2023), 'Climate Delay Discourses Present in Global Mainstream Television Coverage of the IPCC's 2021 report', *Communications Earth and Environment*, 4, 118: doi.org/10.1038/s43247-023-00760-2.
125. Saffron O'Neill, Hywell T.P. Williams, Tim Kurz, et al. (2015), 'Dominant Frames in Legacy and Social Media Coverage of the IPCC Fifth Assessment Report', *Nature Climate Change*, 5: 380–385; James Painter (2014), 'Disaster Averted? Television Coverage of the 2013/14 IPCC's Climate Change Reports', *Reuters Institute for the Study of Journalism*, https://reutersinstitute.politics.ox.ac.uk/our-research/disaster-averted-television-coverage-201314-ipccs-climate-change-reports.
126. Carbon Brief, 'Analysis'.
127. The Daily Mail (2022, 6 March). Mail on Sunday Comment, 'The Astronomical Costs of Pursuing a Net Zero Utopia', *The Daily Mail*, https://www.dailymail.co.uk/debate/article-10582167/MAIL-SUNDAY-COMMENT-astronomical-costs-pursuing-Net-Zero-utopia.html.
128. Nigel Farage (2022, March), 'The Net Zero Zealots Are the Same Elitists Who Sneered at Brexit and Don't Have to Worry about Paying Their Gas Bills', *The Daily Mail*, https://www.dailymail.co.uk/debate/article-10581281/NIGEL-FARAGE-Net-Zero-zealots-elitists-sneered-Brexit.html.
129. Bob Ward (2021, 10 September), 'Misinformation and Propaganda Campaign on Net Zero', *Grantham Research Institute on Climate Change and the Environment*, https://www.lse.ac.uk/granthaminstitute/news/misinformation-and-propaganda-campaign-on-net-zero/.
130. Neil Carter and Mitya Pearson (2022), 'From Green Crap to Net Zero: Conservative Climate Policy 2015–2022', *British Politics*, doi:10.1057/s41293-022-00222-x.
131. Peter Lilley (2022, 12 October), 'Net Zero Puritans Don't Want You to Hear the Truth', *The Telegraph*, https://www.telegraph.co.uk/news/2022/10/12/net-zero-puritans-don't-want-hear-truth/.

132. Zac Goldsmith and Chris Skidmore (2022, 10 July), 'Ditching Net Zero Would Be Electoral Suicide for Conservatives', *The Telegraph*, https://www.telegraph.co.uk/news/2022/07/10/ditching-net-zero-would-electoral-suicide-conservatives/.
133. Louis Ashworth (2022, 25 January), 'Counting the Huge Cost of Net Zero—and Who's Going To Pay It', *The Telegraph*, https://www.telegraph.co.uk/business/2022/01/25/green-future-counting-cost-net-zero.
134. Matthew Lynn (2022, 6 February), 'Why the Cost of Net Zero Is Too High for the Country—and for Boris Too', *The Telegraph*, https://www.telegraph.co.uk/environment/0/cost-net-zero-high-country-boris/.
135. Telegraph View (2022, 1 November), 'The Cost of Net Zero Is Now Becoming Clear', *The Telegraph*, https://www.telegraph.co.uk/opinion/2022/11/01/cost-net-zero-now-becoming-clear/.
136. Hayley Dixon (2022, 17 June), 'Exclusive: Dirty Cost of Keeping the Government's Net Zero Strategy Alive Revealed', *The Telegraph*, https://www.telegraph.co.uk/politics/2022/06/16/exclusive-dirty-cost-keeping-governments-net-zero-strategy-alive.
137. Laura Almeida (2022, 18 September), 'The Net Zero Policies That Should Be Torn Up to Save Your Wealth', *The Telegraph*, https://www.telegraph.co.uk/money/consumer-affairs/mad-net-zero-policies-threatening-wealth.
138. Statista (2022, 27 July), 'Monthly Reach of Leading Newspapers in the United Kingdom from April 2019 to March 2020', *Statista*, https://www.statista.com/statistics/246077/reach-of-selected-national-newspapers-in-the-uk/.
139. Amol Rajan (2022, 20 July), 'Tory Leadership: Why Newspapers Matter in Race to Be Next PM', *BBC*, https://www.bbc.co.uk/news/uk-politics-62238069.
140. Richard Heede (2013). *Carbon Majors: Accounting for carbon and methane emissions 1854–2010: Methods & Results Report*. Washington, DC: Climate Justice Programme and Greenpeace International.
141. Fiona Harvey (2023, 28 June), 'UK Missing Climate Targets on Nearly Every Front, Say Government's Advisers', *The Guardian*, https://www.theguardian.com/technology/2023/jun/28/uk-has-made-no-progress-on-climate-plan-say-governments-own-advisers?CMP=Share_iOSApp_Other.

3
Climate Obstruction in Scotland

The Politics of Oil and Gas

WILLIAM DINAN, VICTORIA ESTEVES,
AND STEVEN HARKINS

INTRODUCTION: WESTMINSTER, HOLYROOD, AND THE BLACK, BLACK OIL

Obstruction of climate action is a pressing policy issue in territories where the oil and gas industries are established economic and political actors. Scotland remains part of the United Kingdom but, since the reopening of the Scottish Parliament in 1999, has devolved powers in relation to environment and planning. Given Scotland's abundance of natural energy resources, policy concerning their exploitation has been central to ongoing debates about the nation's economic prospects and constitutional future. This distinctive political and economic context provides a unique case study for examining the evolution of climate change obstructionism in Europe.

Since the discovery of significant fossil fuel reserves in the North Sea in the late 1960s, the UK Parliament (Westminster) has been keen to foster and support investment in oil and gas exploration. The Scottish Parliament (Holyrood) has charted a more ambiguous course since 1999. Successive Scottish governments have tried to balance economic and environmental concerns: seeking to protect jobs based on oil and gas exploration and refining while acknowledging the growing climate crisis and the need for a new industrial strategy based on a 'greener' economy. Outside of

William Dinan, Victoria Esteves, and Steven Harkins, *Climate Obstruction in Scotland* In: *Climate Obstruction across Europe*. Edited by: Robert J. Brulle, J. Timmons Roberts and Miranda C. Spencer, Oxford University Press.
© Oxford University Press 2024. DOI: 10.1093/oso/9780197762042.003.0003

government, a constellation of economic interests (oil and gas companies, trade associations, and trade unions) emphasize the economic costs of disinvestment from fossil fuels, such as the loss of well-paid jobs in the oil and gas sector. These actors are at the forefront of climate delay in Scotland. Their arguments are reflected in policies that often favour short-term protection of oil and gas jobs, including those in related supply chains. This is the main practical form of climate obstruction in Scotland and represents a compromise consensus that climate advocates have struggled to disrupt, despite growing awareness of the climate emergency in policy circles.

Scotland's contribution to the United Kingdom's historical greenhouse gas (GHG) emissions and projected future emissions (based on current nationally determined contributions [NDCs]) is significant (Figure 3.1). Agriculture, business, and manufacturing are key contributors to Scotland's overall GHG emissions profile. Energy supply in Scotland in 2020 accounted for 5.368 million metric tonnes of carbon dioxide equivalents (MMT CO_2e), with oil and gas extraction accounting for 1.134 MMT CO_2e (compared with 0.936 MMT CO_2e for England and 0.056 MMT CO_2e for Wales; Figure 3.1).[1]

A focus on oil and gas is instructive because extraction is a policy arena 'reserved' to the UK government. The respective powers relating to climate held by the UK and Scottish parliaments complicate policy analysis but also open political opportunities to those lobbying against climate mitigation. Energy policy and regulatory powers are reserved to Westminster, which means the Scottish government possesses few policy instruments to control licencing and extraction. Holyrood does have 'devolved' powers over planning (including new infrastructure) and environmental standards (air quality, pollution, etc.) The UK government has been criticized for approving new exploration in the North Sea. According to the independent Climate Action Tracker, 'Developing new oil and gas reserves is incompatible with the 1.5°C temperature limit and will not help address the current energy crisis'.[2] The policy trajectory in London supports continuing to exploit the reserves available from the UK continental shelf (UKCS). Moreover, the UK Treasury remains reluctant to impose windfall taxes on oil and gas companies despite soaring energy prices and attendant profits in the sector during 2022–2023. The UK Treasury recently offered 90% tax relief to companies investing in North Sea extraction. In 2022, the UK government also briefly lifted the moratorium on unconventional gas extraction (UGE) from shale, commonly known as hydraulic fracturing or fracking,[3] guided by a policy objective of boosting energy security in the United Kingdom. This trend illustrates the dynamics of energy politics within the United Kingdom, with different factions within the same

Scotland Greenhouse Gas Emissions

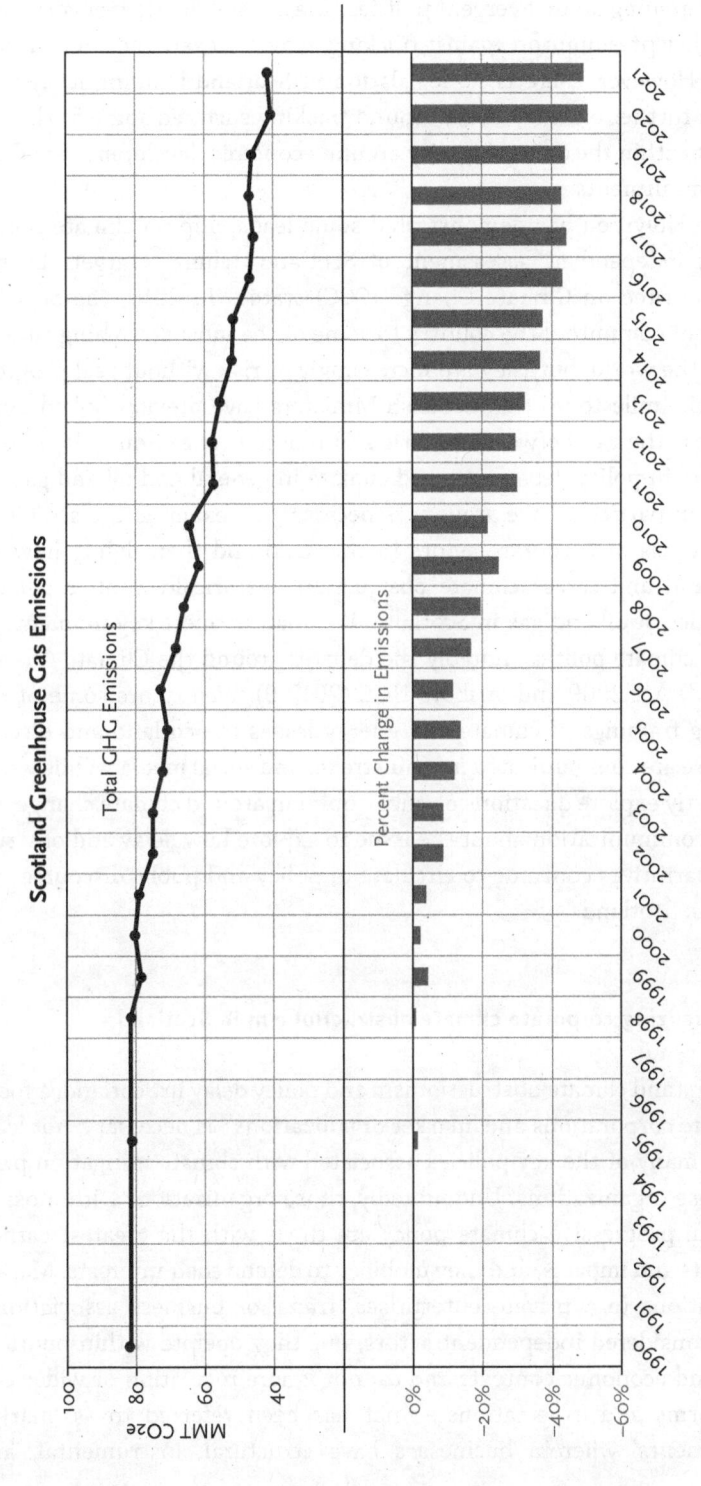

Figure 3.1 Total greenhouse gas (GHG) emissions (in MMT CO₂e) and percentage change in emissions in Scotland between 1990 and 2021, inclusive.
Source: Total GHG emissions based on data provided by Scotland's Nationally Determined Contribution submission for Kyoto Six Greenhouse Gas Totals.

parties pursuing quite divergent policies. The current Scottish government has a policy presumption against fracking, which is exercised via planning powers. However, there is no legislation in Scotland banning it, and, as outlined further on, the debate around fracking surfaced some of the key tensions within the Scottish polity around economic development and climate commitments.

While Holyrood has demonstrated some leadership on climate policy, a recent independent assessment of Scotland's climate targets by the UK Committee on Climate Change (CCC) states, 'In 2019, the Scottish Parliament committed the country to some of the most stretching climate goals in the world, but they are increasingly at risk without real progress toward the milestones that Scottish Ministers have previously laid out'.[4] To address the gap between rhetoric and practice we examine the role of key actors in policy debates around climate in general and oil and gas extraction in particular. We argue it is necessary to examine the strategies and activities of corporate actors in this field and then define how we understand and assess climate obstruction, historically contextualizing the politics of oil and gas in Scotland. We analyse select key moments in Scottish climate politics, notably the debates around the Climate Change (Scotland) Act 2009 and onshore UGE (2012-9). We explore some of the recurring framings of climate and energy issues in Scotland and discuss how these are communicated in mainstream and social media. While we do not directly explore questions of public opinion around climate change, we use the communication about the issue to explore how delay and obstructionist narratives continue to circulate in policy and public discourses on climate in Scotland.

Contextualizing corporate climate obstructionism in Scotland

To understand climate obstructionism and policy delay in Scotland, a focus on private corporations and 'market organizations'[5] is necessary, not least because many of the key policies associated with climate mitigation privilege these organizations. Undoubtedly, those organizations with most to lose from progressive climate policy are those with the greatest carbon footprints and impacts, and they mobilize to defend their interests. Market organizations (e.g. private enterprises, trade, or business associations) can be considered independent actors, but they operate within political, social, and economic contexts and cannot ignore regulation or wider cultural norms and expectations—what has been referred to as 'market environments' wherein businesses have structural, instrumental, and discursive power.[6] As such, the political and business strategies that market organizations pursue reflect a balance of their economic interests and their assessments of what is achievable in relation to policy and legislation. For those operating at different levels of governance, there is a need to consider the political opportunities and risks associated with their lobbying activities in different polities (referred to as 'forum shopping' in political science literature). Indeed, one of the functions of corporate communications strategies is to ensure there is some alignment and consistency across political boundaries and that corporate positioning on, for example, climate issues appear credible to different stakeholders and publics at different levels of governance.

Scotland could be considered a market environment where there are some possibilities to advance climate mitigation and reduce GHG emissions, given the stated ambitions of the Scottish Parliament to respond seriously to the climate emergency. Our analysis complicates this picture by drawing attention to the significant policy constraints that the current devolution settlement presents to policymakers in Edinburgh. But the analysis also demonstrates the enduring power (structural, instrumental, and discursive) of the oil and gas industry operating in Scotland. For climate obstructionism and delay to be a successful strategy pursued by oil and gas interests, they do not need to convince all policymakers, civil society, or wider public opinion. Fostering short-term conditions under which meaningful climate policies are seen as too politically difficult or economically costly has proved to be a remarkably resilient approach in Scotland.

A BRIEF HISTORY OF THE POLITICS OF OIL AND GAS IN SCOTLAND

As Marriott and Macalister observed, 'The discovery of substantial reserves in the British North Sea changed the fortunes of BP, Shell, northeast Scotland and the British state'.[7] Since the discovery of oil and gas reserves in the North Sea, UK government policy has rapidly developed and exploited these resources, with investment flowing into the northeast of Scotland throughout the 1970s and 1980s. These reserves were a boon for the UK Treasury. During the 1970s, political sentiment in Scotland was shifting, and the campaign for Scottish independence gained ground. The case for independence rested, in part, on the opportunity to build a Scottish state on the proceeds of oil and gas exploration. While the political project of Scottish independence stalled in the late 1970s, North Sea exploration and production grew significantly. The key political and policy developments relating to energy in Scotland since the late 1980s have included the decline and demise of coal as a source of employment and, later, power generation.

Privatization, the flagship policy of the neoliberal Thatcherite government at the time, has been perhaps the key to explain the power dynamics of current UK energy production. According to Marriott and Macalister 'the disposal of state-held North Sea oil and gas assets—specifically, the upstream interests of both the British National Oil Corporation (BNOC) and British Gas and, arguably, the British government's majority shareholding in British Petroleum (BP) proved to be the spearhead of a privatization wave that was to sweep Great Britain, first, and then much of the rest of the world, during the 1980s and early 1990s'.[8] Prominent among the major energy corporations based or operating in Scotland were those brought into being by privatization. BP (Scotland), Scottish Power, Scottish and Southern Energy, Scottish Gas (part of Centrica), and British Energy all were leading players in the Scottish business scene.

Historically, the Scottish policy arena has been viewed as more corporatist and consensual than that in the rest of the United Kingdom.[9] The style of governance in Scotland has been described as a 'negotiated order' between business and political elites. The tight networks that comprise this quasi-corporatist negotiated order have tended to be business friendly. Prior to devolution, large industrial interests in Scotland enjoyed political access to decision making via the Scottish Office, the Westminster department charged with managing and administering Scotland. During the 1980s and 1990s, the energy sector including oil and gas producers could use the Scottish Office to press their case with other UK government departments, particularly the Treasury. Little of this political activity attracted media attention. According to Professor Paul Stevens, oil and gas interests in Scotland 'steered away from publicity. They preferred to be covert rather than overt because oil companies had never been popular. And they didn't really have to lobby too hard because [of their big tax revenues] they were pushing an open door with government'.[10]

With the advent of Scottish devolution, initially little changed. The extractive industries located and operating in Scotland responded to the twin strategic threats of growing policy awareness of climate change (after the collapse of the Global Climate Coalition in 1998) and the devolution of political power to the newly created Scottish Parliament in 1999 by retaining a focus on maintaining relations and lines of communication with key political contacts. In parallel with significant political mobilization by business globally to respond to the wider policy challenges associated with sustainable development agendas, there was also increased political activism by business organizations to shape the policy agenda of the new Scottish Parliament.[11] The New Labour-Liberal Democrat coalition administration in Edinburgh quickly set about scoping the policy challenges associated with climate change confronting the new institutions.[12]

The energy sector continued to enjoy privileged access to decision-makers in Scotland. In 2000, a special Scottish Utilities Forum was created to address key issues for the sector. The impetus behind its creation was to bring high-level politicians and business leaders together to exchange views under Chatham House rules (a convention that protects the privacy of participants, in which media cannot identify the sources of statements). According to participants, it evolved into a forum for briefing and sectoral lobbying. While some members soon began to question the purpose of the forum, its very existence illustrates an enduring style of corporatist politics. The participating energy companies remained keen to keep policy dialogue going, particularly if it were to address trade issues. A compromise was reached wherein the forum would continue to run while the companies might more usefully address their other concerns through the Scottish Parliament's cross-party group on oil and gas. It was agreed that this cross-party group should therefore be encouraged to widen its scope to consider the kinds of 'downstream' (i.e. consumer) issues that would be of benefit to Scottish Utilities Forum members.

Cross-party groups are one thread of the fabric of business and politics in Scotland. Peak business organizations such as the Confederation of British Industry, the Institute of Directors and Chambers of Commerce are networks that provide venues and opportunities for lobbying. In addition, policy debates hosted by think tanks and interest groups also offer space for exchanges between public affairs professionals and the political class in Scotland. Attendance at party conferences and fringe events (sponsored side events not part of the official conference) is also a key feature of the lobbying scene in Scotland and is routine for the big energy companies. As one industry source recounted:

> We go to all the party conferences here ... and we will also do cross party fringe meetings ... it's very good because they are all well attended by the politicians and the activists and so on, ... I think *particularly in Scotland ... politics permeates everything, you know, it is very high profile, it's very close. So you can't stand aside from that*, so we're heavily involved in ... socially responsible subjects such as fuel poverty [that] are closely linked, obviously to politics large and small 'p', and are linked to our business' [emphasis added].[13]

BP: Corporate strategy in Scotland

The case of BP illustrates some of the strategies oil and gas companies have used to secure their social and political licence to operate in Scotland. These strategies are ultimately focused on delaying and obstructing climate mitigation policy: as well as having dedicated communications and public affairs personnel in Aberdeen (upstream), Grangemouth (downstream), Edinburgh (managing the wider swathe of political relations in Scotland), and, of course, London (its headquarters, but managing relations with Scottish politicians at Westminster), the company had a relatively early focus in the late 1990s on social investment and strategically understanding sentiment among political influentials, including critics.[14] BP's rebranding as 'Beyond Petroleum' in 2000 coincided with growing concern about its social licence and a drive to position the company as becoming increasingly serious about climate.

Locally, the company was often accused of lacking commitment to Scotland. To help tackle this perception, in 2001, the company commissioned private economic consultants to independently attest to BP's economic importance to Scotland.[15] An advisory board in Scotland was formed, seen as a means for the company's management to take the temperature of the Scottish political class and develop its corporate strategy in Scotland, drawing in advice and expertise.[16]

BP invested significantly in promotion and communications. Corporate social responsibility (CSR) featured heavily, with a sophisticated political and community strategy implemented in Scotland. For example, in April 2001, BP supported a think tank, the International Futures Forum (IFF), to bring strategic thinking and policy analysis to bear on real-world problems. The public affairs value of IFF's Scottish work was that it gave BP the opportunity to demonstrate its engagement with the broader policy agenda of the Scottish and UK governments. BP's stated long-term aspiration was for the IFF to eventually become a resource for Scottish policymaking on economic regeneration.

Indeed, the IFF morphed a few years later into the Scottish Parliament's cross-party Futures Forum. In 2006, the Forum invited perhaps the highest-profile climate sceptic at the time, Bjorn Lomborg, to the Scottish Parliament. Lomborg's visit was hosted by Fergus Ewing, a Scottish National Party (SNP) Member of the Scottish Parliament (MSP) who also chaired the cross-party group on oil and gas in Holyrood. Ewing praised Lomborg's work, describing it as 'rigorous', 'dispassionate', and 'non-polemical'. He rejected criticism of Parliament's hosting Lomberg from members of the Green Party as 'puerile'. Lomborg's message was that climate change would benefit the flora and fauna of Scotland and that 'cutting carbon emissions believing that will have make much of a difference is almost illusory'.[17]

BP had been keen to position themselves as realist policy actors, willing to innovate and invest to explore business-friendly climate solutions. Some of that corporate profiling was undertaken in Scotland. The company pioneered an early version of emissions trading (often referred to as 'cap and trade' schemes) as a means of driving down CO_2 emissions and increasing efficiency. For example, the BP refinery business at Grangemouth sold some of its internal corporate carbon credits to the extraction business in the North Sea Forties field. That transaction released investment in the facility at Grangemouth, all at very low cost to the company.[18] It had the advantage of reducing emission from the production process, driving business focus on this issue, whilst also giving BP tangible evidence that it was serious about its commitment to addressing climate. The company developed the scheme alongside other speculative climate solutions, including carbon capture and storage (CCS) in Scotland and 'a commitment to spend £4.5 billion over the next decade on wind power, solar energy, and hydrogen and gas-fired power stations'.[19]

To that end, in addition to its new 'Beyond Petroleum' tagline, BP purchased some renewables companies. At the same time as it was supporting the IFF in mapping a new socially responsible twenty-first century, it was also seeking to dispose of much of its refinery business, including the facility at Grangemouth, in central Scotland. The shorter-term PR and policy benefits of 'the second Scottish enlightenment' were made quite clear. The political impact of this strategy and its social investment for community regeneration helped build goodwill for the corporation, which maintained its wider social license to operate and eventually helped facilitate its corporate exit from the community. Based on BP's own statements, 'The objective was to keep talking with decision makers, don't let [media coverage] get in the way of the conversations you're having with people who can really make things happen'.[20] BP's refinery business was ultimately sold in 2005, to INEOS, a company that would be at the forefront of promoting fracking in Scotland a few years later.

Scotland acts on climate?

A key milestone in Scottish climate politics was the passage of the Climate Change Act (Scotland) in 2009. This act set ambitious targets for emissions reductions and put Scotland at the forefront of polities seeking to underpin their decarbonization transition policies with demanding targets dictated by primary legislation. The 2007 SNP manifesto committed to create a ministerial post for climate change and to introduce a climate change bill 'with mandatory carbon reduction targets of 3% per annum and also set a long-term target of cutting emissions by a minimum of 80% by 2050 above the UK target of 60%'.[21] The manifesto reflected the lobbying efforts of environmental nongovernmental organizations (NGOs) who pressed the party to adopt a radical climate agenda. The passage of the manifesto into legislation within two years involved considerable political manoeuvring, resulting in a significant feat of cross-party cooperation as all parties in Holyrood supported the legislation.

While it is tempting to see this high point of climate policymaking as ushering in a new era in Scotland, such a view is confounded by the political powers available to the Scottish Government, where the legislative levers to regulate North Sea production are reserved to London. Moreover, it underestimates the existing social movements and economic interests in Scotland keen to slow and obstruct the transition to a decarbonized economy. While these interests lacked a strong parliamentary voice in Holyrood and enjoyed little media prominence or support, they re-emerged when the issue of fracking became a matter of planning and political controversy in 2012.

Fracking the transition: Advocating for unconventional gas

In Spring 2011, Dart Energy announced plans to develop a coalbed methane (CBM) project in central Scotland, a location identified as a promising site for exploration in the 1990s.[22] In response to a planning application in August 2012 for permission to drill twenty-two wells at fourteen different sites to explore for CBM, the local authorities in Stirling and Falkirk (also home to the Grangemouth refinery) set in motion a protracted planning, regulatory, and political dispute that illustrated the challenges Scotland faced in meeting the targets set in the Climate Change (Scotland) Act of 2009. Tensions between short-term growth and long-term sustainability came into play, with those promoting fracking emphasizing the economic benefits that unconventional gas extraction could bring to individuals (jobs), communities (a share of shale gas revenues), and the nation (a sustainable future) if managed responsibly. The framing of fracked gas as a desirable transition resource aligned with climate objectives was a key plank of the business and trade association platform advocating for permission to drill, which was designed to assuage local planners and national politicians.

The planning application for CBM attracted many objections and widespread opposition in local communities near the proposed drilling sites, many of which had experienced the environmental costs of methane flaring, noise, and transport pollution associated with coal mines, along with the social cost of their closures during the 1980s. The application became the subject of a prolonged public inquiry. The actors involved in the planning inquiry (and subsequent public debate over the viability and desirability of fracking in Scotland) illustrate the complex interlinkages between scientific expertise, policy, planning, and economic interests.[23] While the key advocates for fracking have been careful not to be associated with denialist tropes, there is nevertheless unmistakable evidence of delay and the use of transition arguments to secure further fossil fuel development. For example, in 2014, an early independent scientific assessment commissioned by the Scottish government delivered a decidedly ambiguous verdict on what many observers and experts saw as a clearly problematic technology in terms of climate.

> The impact of unconventional oil and gas resources in Scotland on the Scottish Government's commitment to reduce greenhouse gases is not definitive. There could be minimal impact from unconventional hydrocarbons if they are used as a petrochemical feedstock, but lifecycle analysis of an unconventional hydrocarbon industry is required to inform the debate and provide a clearer view on the impact of their development.[24]

This statement was particularly useful to proponents of fracking, such as INEOS and the trade association UKOOG, who could now rely on scientific uncertainty around emissions and climate targets to press for unconventional gas development across Scotland. The argument that shale gas could be considered a green bridging technology recured frequently. While the nuance was lost in most public debate, shale gas was agreed to be greener than coal,[25] though coal is known to be the dirtiest of fossil fuels. This comparison does not imply that shale gas warrants the sustainability lustre that corporate spin attempted to bestow on fracking technology or the resource itself.

While the licencing of onshore gas extraction remained a Westminster power (until 2018, when these powers were devolved to Scotland), the Scottish government used its planning powers to implement a de facto moratorium on fracking in Scotland between 2015 and 2018. Nevertheless, the efforts of those holding Westminster-granted licences to exploit shale resources in Scotland were confronted with growing public opposition. INEOS had acquired exploration licences across central Scotland in 2014, and it set about an extensive public relations and public affairs campaign to convince both Scottish policymakers and the public of the necessity of shale exploration. Climate change was notably downplayed in this communications campaign, with the emphasis firmly on the economic arguments for development.

> INEOS has already committed to full and open consultation with local communities and has also promised to share 6% of the revenue from its wells with homeowners, landowners, and local authorities. INEOS . . . believes a combination of community consultation and a fair share of the profits could lead to much greater understanding and acceptance of this important technology.[26]

INEOS organized community meetings and roadshows across central Scotland to sell their vision of shale exploration as a joint venture between corporation and community, wherein CEO Jim Ratcliffe claimed 'we would see INEOS giving away £2.5 billion in the next 10 to 15 years'.[27] The public relations consultancy Mediazoo were hired to help manage media relations and community consultation programmes, the latter described by critics as an effort to 'love bomb Scottish communities to stop worrying and love fracking'.[28]

As part of its campaign, INEOS produced audio-visual content that was shared online and across social media channels to promote shale exploration: 'There is widespread concern about the environmental risks of fracking based on misinformation, which is at odds with the scientific consensus that extraction is safe and compatible with our climate change ambitions'.[29] Neither the scientific consensus nor the 'our' whose climate ambitions are invoked were explained. Instead, the narrative emphasized the transformational impact shale gas could have for local employment in extraction and high-skills manufacturing. The visual rhetoric in one promotional video downplayed the impacts on local amenities and traffic disruption. Panning shots of green fields and largely unspoilt farmland were used to convey an image of a relatively unobtrusive and unproblematic technology. Renewables technologies were compared unfavourably with the proposed shale gas wells. Using a hypothetical example of a four-well drilling site, INEOS claimed that 'In the first decade it would take about thirty-two wind turbines to generate the same energy created by these four gas wells'. None of the assumptions that underpin these rhetorical claims is substantiated (the periodization, the recoverable gas from shale reserves in central Scotland, the size and efficiency of wind relative to shale, etc.). The video superimposes a large wind farm over the shale gas well pad to emphasize the visual impact of turbines and then displays a panning shot of a wind farm in silhouette with more than fifty turbines, subtly exaggerating the comparison. Ted Crotty, a director at INEOS, claimed 'the local residents would be up in arms about it'.

In fact, residents were up in arms about the proposed fracking development rather than renewables. There had been no applications to site large windfarms in central Scotland, but the renewables comparison served as a red herring to distract from the wider climate questions. While issues associated with fracking, such as earth tremors and water contamination, are addressed in INEOS's promotional video, nothing is said about the climate impacts of the technology. The issue is framed exclusively in terms of a narrow range of poor choices: rising energy prices, decreasing North Sea reserves, and coal as an undesirable substitute are core messages. The necessity of importing fracked gas if shale development was not permitted in Scotland is presented as the only other policy alternative—and was indeed the business decision pursued by INEOS in 2016.

In another video INEOS commissioned, the case for gas is made by highlighting all the current consumer goods (e.g. plastics and synthetics) and creature comforts that are made possible by gas. The narrative is constructed around a young couple looking at their energy bill and debating whether to 'get rid of gas'. As household items disappear, the lights go out, and eventually the couple are left naked, the viewer is invited to consider that current lifestyles and civilization are not possible without gas.[30]

According to seasoned industry observers, INEOS 'signifies a new type of institution in the industry and gives a picture of the UK oil and gas world as it now is and is set to be in the future. It has scant need for journalists, unlike the corporations which used the media to build a positive profile even as they largely lobbied ministers behind the scenes. The likes of INEOS are straightforwardly hidden and largely closed to scrutiny, except via their own public presentations'.[31]

The fracking debate exposed fault lines not only in the geology of central Scotland, but within the ruling political party in Holyrood, the SNP. While the Scottish government publicly disavowed UGE development (with many SNP candidates elected on an anti-fracking platform in 2016), INEOS CEO Ratcliffe claimed that, privately, senior SNP ministers were supportive of fracking.[32] This account tallies with the political intelligence anti-fracking campaigners were picking up as they lobbied against UGE development. Key ministers like Fergus Ewing were known to be consistent supporters of the oil and gas industry and sceptical about climate change more generally. It is likely they used their ministerial clout to resist an early ban on UGE developments. Key elements of SNP policy around climate now directly undermined the arguments the party had advanced in the 2014 independence referendum, which relied on projected oil and gas revenues to help make the economic case for independence and underwrite a new Scottish exchequer in an independent Scotland.

While INEOS was at the forefront of the public relations effort to secure acceptance of shale development in Scotland (including a failed court case against the Scottish Government in 2017–2018, which sought damages if fracking in Scotland was banned), other companies and vested interests were also actively promoting fracking. Their preferred framing of UGE never addressed the wider climate impacts of fossil fuels and instead focused on the necessity of gas as the optimal transition fuel. A 'balanced sustainable approach' to energy sourcing became code for short-term exploitation of shale reserves. Both large landowners (such as the Duke of Buccleuch's estates) and smaller oil and gas companies (Dart Energy, IGas, Cluff Natural Resources, BCG Energy, and Aurora Energy Resources) were involved in lobbying for unconventional gas extraction. The proliferation of new exploration and supply-chain companies highlights another feature of the contemporary oil and gas industry in Scotland: it is increasingly differentiated, financed by private equity and interests not domiciled in Scotland, and no longer dominated by the 'majors'. Here, the roles of trade associations and peak business organizations become significant in understanding how climate policies can be delayed and obstructed.

North Sea for net zero: A new climate obstructionism

In 2000, BP and Shell accounted for nearly 40% of the UK's oil production. By 2019, this share of the market was halved, with new entrants buying up concessions and licences.[33] While an estimated 20 billion barrels of oil could yet be extracted (from the UKCS), the economic viability of such development will depend on global oil prices and taxation policy, neither of which can be controlled by Holyrood. The economic opportunity is clearly articulated by the UK's Oil and Gas Authority (OGA), while any 'economic/environment trade-off appears to be absent from the OGA's deliberations to date'.[34]

The UK Government commissioned the Wood Review (2014) to examine the future exploitation of North Sea reserves. The recommended maximizing economic recovery strategy that emerged was completely shaped by industry preferences. Indeed, the industry association Oil and Gas UK was explicitly thanked in the foreword to the final report.[35] The review barely acknowledges the climate crisis. In recommending a new regulator for the mature UKCS, the report justifies an industry-friendly regulatory model by arguing that such a body would not have to compete with the other priorities of the Department of Energy and Climate Change (DECC), not least a global deal on climate change.[36] The OGA was created very shortly after this review in 2015 and rebranded as the North Sea Transition Authority (NSTA) in March 2022. NSTA claims to be 'fully committed to enabling the achievement of the UK government's commitment to reach net zero emissions by 2050',[37] which is to be realized by 'licensing of exploration and development of the UK's offshore and onshore (England only) oil and gas resources, gas storage and unloading activities in accordance with the [Net Zero] Strategy'.[38] This UK government strategy to incentivize North Sea extraction stands at odds with scientific advice on addressing climate change and the policy trajectory of the Scottish government.

The origins of NSTA as an industry-friendly regulator created to promote fossil-fuel exploration is symptomatic of a form of policy denialism that pretends continued exploration is compatible with commitments the UK government has made during the United Nations Framework Convention on Climate Change (UNFCC) Conference of the Parties (COP) negotiations, let alone climate science. Simply rebranding a pro-exploration regulator as a transition body represents a deeply cynical communicative logic. A similar repositioning strategy can be seen in the private sector. Oil and Gas UK, the trade association for the fossil fuel industry in the North Sea, has been rebranded as Offshore Energies UK (OEUK).[39] The OEUK website is vague about its foundation and rebranding, simply asserting a (misleading) pedigree stretching back almost half a century.[40]

OEUK stresses the role its members can play in reducing emissions from their operations. It is silent about the emissions to be generated from the billions of barrels its members are busily exploiting from the UKCS. Much play is made of investments in carbon capture, usage and storage (CCUS), and hydrogen as a means of contributing to net zero. Echoes of BPs social investment strategy in Scotland are to be found in OEUK's commitment to maintaining the skills base of communities around Aberdeen and promising up to 40,000 new energy jobs (many of which appear to be linked with as-yet untried and unproven technologies).[41]

The promise of new jobs is one enthusiastically embraced by trade unions representing gas fitters and offshore workers. The GMB trade union had been vocal supporters of fracking in Scotland and have consistently argued for the protection of jobs in the oil and gas industry, despite recognizing that climate change is happening. Gary Smith, GMB's general secretary, argued, 'If we want to avoid the double disaster of a climate crisis and a jobs crisis, then we need a balanced energy policy across our regions and nations that supports workers and communities on the journey to "net zero".[42] Here the union adopts the language of climate mitigation, stating that 'GMB recognizes that we are in the grip of a climate crisis created by man-made global warming, and that global warming is the gravest long-term threat that faces the planet'.[43] However, the union (and others in the sector) also warn that an accelerated or arbitrary journey to net zero risks mass job losses and 'exporting' demand. The union supports the UK government's net zero target for 2050 but has focused on campaigning against transition policies that substitute away from gas and oil: 'Some economists have even argued that "renewable energy conveniently requires less labour for operation and maintenance" than traditional energy sources, and that the UK should speed the transition to renewables to save on long-term labour costs. GMB rejects this cynical attempt to undermine good quality employment'.[44]

Notably, the GMB represents members in industries with the highest gross carbon emissions in the United Kingdom and has repeatedly prioritized the protection of jobs ahead of climate goals. While this is an understandable policy position, it is well aligned with industry's repeated preference for indefinite and short-term prioritization of business as usual, which has been a key theme running through climate obstructionist discourse in Scotland. It also often means that trade unions are at the forefront of opposing climate policies in public debate.

CLIMATE OBSTRUCTIONISM IN SCOTLAND: DISCOURSES OF DELAY AND DENIAL

Outright denialism in the mainstream Scottish press, casting doubt over the causes of climate change, is increasingly rare: 'To assume that our behaviour is a primary cause of planetary climate change seems to me an expression of hubris, an overweening pride, that is matched only by our arrogance in forecasting its effects for a century ahead when we are unable to do much more than improve on chance in predicting next week's weather', as one opinion writer put it.[45] The legacy press in Scotland is no different from other media organizations in featuring voices from business and politics in their construction of events. While this practice largely excludes extreme climate deniers, there are spaces for such arguments to surface. In 2013, for example, the Alliance Party was launched to campaign against windfarms in Scotland. As an official party, its arguments received mainstream media coverage. The most prominent and consistent climate criticism in the press is sourced from industry groups, although their position is now couched in the language of net zero. Recently, outgoing OEUK Chief Executive Deirdre Michie argued against the 'environmental populism' of windfall taxes on oil and gas at the end of 2022 and for further fossil fuel development.

> Projects like the Cambo field [an off-shore oil field in the North Atlantic] are part of a low-carbon journey that will support energy security, jobs, the economy and the net-zero future that everyone wants to see. Like all future UK oil and gas projects, the Cambo field is designed with lower-operating emissions in mind. It has been built 'electrification-ready', with the potential to import renewable power when it becomes feasible in the future.[46]

This line of argument represents an important strand of mainstream opinion in Scotland. It appears realistic and reasonable, particularly in the context of more fringe voices who circulate denialist speaking points, misinformation, and disinformation.

Online discourses of delay and denial

Some frequent letter writers campaigning against net zero policies still get published occasionally in the mainstream press. Whilst outright climate denial seems to be losing ground in online spaces, the undermining of climate solutions—resulting in delay—is growing, and public opinion appears split. A third of British people surveyed currently believe it is impossible to forego fossil fuels.[47] While Scotland is not home to climate denial think tanks, their work is reported to the Scottish public via British media outlets. Although Scottish public debate is increasingly distinct from the 'national' British policy conversation, their respective public spheres do overlap. Further untangling our current understanding of the relationship between the online sphere and opinion on climate issues, research shows that British social media consumers are also potentially exposed to misinformation, like their TV-watching counterparts.[48] The online space remains a problematic realm of discussion in relation to climate denial.

A recent development in the United Kingdom has been the advent of right-wing news channels, like GB News. While the live audiences for these channels are very small compared with mainstream news of established broadcasters, content from GB News circulates widely on social media. A leading presenter on GB News, Neil Oliver, recently resigned from the Royal Society Edinburgh (RSE). 'In discussion with Mr Oliver, he understood that his current views on various matters, widely aired on television, put him at odds with scientific and broader academic learning within the Society. Following discussions, he offered to resign his association with the RSE with immediate effect'.[49] Oliver's broadcasts have featured vaccine scepticism and climate denial.

Whilst broad public discussion of climate change can be scattered across social media, particularly Twitter (now X) and YouTube, efforts to influence debate in the form of independent online publications and blogs are present. These visible spaces of climate sceptic opinion include *The Scottish Sceptic* blog as well as UK-based spaces with Scottish-specific subsections or articles, such as *Climate Scepticism*, *The Daily Sceptic*, and *The Conservative Woman*.

The Daily Sceptic[50] is a British online publication that often criticizes environmental initiatives. It features two Scottish writers, Andrew Montford and Richard Lyon. Montford is a Scottish writer who compiled *The Climategate Inquiries* for the Global Warming Policy Foundation (GWPF) in 2010, and is a well-known name in climate sceptic circles via his online publications. The UK's *Daily Sceptic* features essays on renewable energy and Scottish climate policy by Richard Lyon, a former senior oil and gas operations manager who also runs *The State of Britain* blog.[51] One of the most common critiques within these publications relates to renewable energy in Scotland. Its core arguments allege that renewables destroy wildlife, offer unreliable supply, are too expensive, and will harm the economy.

The overarching rhetoric that permeates broader discussions around climate denial or delay are reflected within these online publications, which twist facts using misinformation and exaggeration. Lyon criticizes 'weather-dependent energy scavenging devices',[52] further amplifying misinformation regarding renewable energy. While this might be tempting to dismiss, worryingly, 27% of Britons surveyed say they share this belief.[53]

The Conservative Woman[54] is a British publication that addresses right-leaning views and concerns. Two of its contributors are Scottish climate sceptics Clark Cross and William Loneskie. In line with *The Daily Sceptic*, Cross's pieces for *The Conservative Woman* are also critical of electric vehicles and wind energy. Criticism of Scotland's net zero goals are a favourite theme.[55] This outlet adopts a less scientific approach, emphasizing economic arguments to undermine climate solutions. Cross and Loneskie's positioning vis-à-vis identity is markedly different from Montford's and Lyon's: while the latter lean into their credentials (both in terms of expertise and work experience), the former do not, but potentially offer more relatable content. This divergence in terms of identity and communicative tactics allows delay and denialist messages to reach a much broader audience whilst also demonstrating the diversity of climate change sceptics in terms of their sociopolitical alignments.

CONCLUSION

As shown, discursive framings of climate obstruction and delay in Scotland include economic arguments against action, the use of transitioning tactics (including positioning gas as a critical bridging resource), and critiques around the reliability of renewable energy. Such economic arguments include the short-term importance of the oil industry and the lack of secure and well-paid jobs to replace those associated with extraction and refining. Critiques of renewable energy are framed in terms of energy instability and impacts on local fauna and flora. Energy security has been a theme of climate obstruction for many years, and it has been revived to promote fracking in Scotland, and more recently, to address wider geopolitical concerns about energy supply and the associated price increases and 'cost of living' crisis.

The political influencing strategies of the oil and gas industry in Scotland appear to largely avoid engaging in media and public debate and seek instead to build relationships and support with key political advisors and decision-makers—UK government departments in particular. In addition to developing CSR initiatives to maintain social 'license to operate', the oil and gas industry in Scotland has sophisticated public affairs programmes that track political sentiment toward individual companies and issues-management strategies that closely monitor political and regulatory agendas associated with climate policy. More research on these influencing strategies is urgently needed if the public and policymakers are to understand the scale of corporate led climate-delay efforts. The woefully weak lobbying disclosure regulations in Westminster and Holyrood are a significant barrier to public understanding of policymaking in general, and climate politics in particular.

The recent enthusiastic embrace of net zero rhetoric by key trade associations and trade unions representing oil and gas interests in Scotland illustrates their repositioning as realist actors in climate policy networks. Net zero is a remarkably business-friendly approach to addressing climate issues as it allows fundamental changes to business practices and strategies to be postponed almost indefinitely. While oil and gas interests have offered symbolic concessions to climate concerns, they have also been highly effective in securing their own short-term economic interests. In the Scottish context, this success is due largely to the pro-exploration policy position of the UK government.

While the Scottish government has championed climate mitigation, it would be a mistake to assume that there is either wide or deep consensus that such policy goals can easily be pursued in the short to medium term, even if the respective division of powers between Scotland and the United Kingdom were to change. Many within the SNP are supportive of the oil and gas industry, and, while onshore extraction was hugely unpopular with the electorate, factions within most of Scotland's political parties (save the Greens and Scottish Socialists) have been prepared to consider such development. With offshore extraction, the British political class has repeatedly sought to protect investment and employment in oil and gas. The hard decisions around fossil fuel disinvestment and transition have been continually postponed. This is the practical effect of the widespread political lobbying efforts to sustain the inherently unsustainable extractive industries. While there is a rhetorical recognition of the climate emergency in Scottish politics, the oil and gas industry in Scotland continues to operate and expand.

The connection between discourses of climate delay and denial and public opinion also requires further research. While establishing the oil and gas industries' preferred framings of climate issues is reasonably straightforward, the impacts of these framings requires greater exploration. That work needs to examine the attitudes of policy elites as well as those of the public. While online climate deniers can be dismissed as unserious and uninformed, it would be a mistake to assume that their ideas and arguments have no effects. Understanding the diffusion and circulation of climate denial on social media is a prerequisite to effectively informing publics about climate science, improving understanding of climate policy, and motivating climate action.

NOTES

1. Estimates drawn from BEIS, 'Greenhouse Gas Inventories for England, Scotland, Wales and Northern Ireland: 1990–2020', 20 September 2022, https://naei.beis.gov.uk/reports/reports?report_id=1080.
2. https://climateactiontracker.org/countries/uk/. Accessed 22 March 2023.
3. https://www.gov.uk/government/news/uk-government-takes-next-steps-to-boost-domestic-energy-production. This policy was quickly reversed when Rishi Sunak became UK Prime Minister in October 2022.
4. Climate Change Committee (2022, 7 December), 'Scotland's Climate Targets Are in Danger of Becoming Meaningless', https://www.theccc.org.uk/2022/12/07/

scotlands-climate-targets-are-in-danger-of-becoming-meaningless/. Accessed 22 March 2023.
5. C. Perrow and S. Pulver (2015), 'Organizations and Markets', In: Riley E. Dunlap, and Robert J. Brulle (eds.), *Climate Change and Society: Sociological Perspectives*, online edition. New York: Oxford Academic, https://doi-org.ezproxy-s2.stir.ac.uk/10.1093/acprof:oso/9780199356102.003.0003. Accessed 22 March 2023.
6. Ibid., and S. Pulver (2007), 'Making Sense of Corporate Environmentalism: An Environmental Contestation Approach to Analyzing the Causes and Consequences of the Climate Change Policy Split in the Oil Industry', *Organization and Environment*, 20, 1: 44–83, https://doi.org/10.1177/1086026607300246.
7. J. Marriott and T. Macalister (2021), *Crude Britannia: How Oil Shaped a Nation*. London: Pluto Press, p. 81.
8. Ibid.
9. A. Brown and D. McCrone (1999 March), *Business and the Scottish Parliament Project: Report*. Edinburgh: Governance of Scotland Forum, p. 71.
10. Professor Paul Stevens, Professor of Petroleum Policy and Economics, University of Dundee, cited in Marriott and Macalister, *Crude Britannia*, p. 5.
11. P. Schlesinger, D. Miller, and W. Dinan (2001), *Open Scotland? Journalists, Spin Doctors and Lobbyists*. Edinburgh: Polygon.
12. A. Kerr, S. Shackley, R. Milne, and S. Allen (1999), *Climate Change: Scottish Implications Scoping Study*. Edinburgh: Scottish Executive Central Research Unit.
13. Interview, author archives.
14. Interview, author archives.
15. DTZ Pieda Consulting (2001 March), 'The Economic Impact of the Activities of BP in Scotland'. DTZ. Edinburgh, 99/70325.
16. Interview, author archives.
17. https://www.scotsman.com/news/climate-change-good-scotland-professor-2456075.
18. Marriott and Macalister, *Crude Britannia*.
19. Ibid., p. 201.
20. Interview, author archives.
21. SNP (2007), 'It's Time: Manifesto for Scottish Parliament Elections', p. 29. https://blog.stevenkellow.com/scottish-parliament-election-manifesto-archive/
22. David K. Smythe (2020), 'Inadequate Regulation of the Geological Aspects of Shale Exploitation in the UK', *International Journal of Environmental Research and Public Health*, 17, 19: 6946, https://doi.org/10.3390/ijerph17196946.
23. A. Watterson and W. Dinan (2016), 'Health Impact Assessments, Regulation, and the Unconventional Gas Industry in the UK: Exploiting Resources, Ideology, and Expertise?', *New Solutions*, 25, 4: 480–512, https://doi.org/10.1177/1048291115615074; and A. Watterson, and W. Dinan (2020), 'Lagging and Flagging: Air Pollution, Shale Gas Exploration and the Interaction of Policy, Science, Ethics and Environmental Justice in England', *International Journal of Environmental Research and Public Health*, 17, 12: art. no. 4320, https://doi.org/10.3390/ijerph17124320.
24. C. Masters, Zoe Shipton, R. Gatliff, et al. (2014), *Independent Expert Scientific Panel: Report on Unconventional Oil and Gas. Project Report*. Edinburgh: Scottish Government, p. v.
25. Ibid., p. 63.

26. https://www.ineos.com/inch-magazine/articles/issue-7/shale-gas-the-game-changer/. Accessed 26 March 2023.
27. 'Jim Ratcliffe Talks About INEOS £2.5bn Shale Gas Offer', https://www.youtube.com/watch?v=EolzsjygaTk 0:40–0:44. Accessed 27 March 2023.
28. https://www.heraldscotland.com/news/13205767.revealed-energy-giants-plans-lovebomb-scotland-backing-fracking/. Accessed 26 March 2023.
29. https://www.ineos.com/businesses/ineos-shale/why-shale-gas/. Accessed 27 March 2023.
30. INEOS (2015), 'The Importance of Gas', https://www.youtube.com/watch?v=DVwuJEbFxec Accessed 27 March 2023.
31. Marriott and Macalister, *Crude Britannia*, p. 293.
32. https://www.heraldscotland.com/news/13421511.government-must-come-clean-fracking-ineos-boss-claims-ministers-privately-supportive/. Accessed 28 March 2023.
33. Marriott and Macalister, *Crude Britannia*, p. 275.
34. J. Armstrong, and J. MacLaren (2017), 'The Oil and Gas Sector', In: K. Gibb, D. Maclennan, D. McNulty, and M. Come (eds.), *The Scottish Economy: A Living Book*, p. 115. London: Routledge.
35. I. Wood (2014, February), 'UKCS Maximising Recovery Review: Final Report', https://www.nstauthority.co.uk/media/1014/ukcs_maximising_recovery_review.pdf. Accessed 30 March 2023.
36. Ibid., Annex A. 'Arguments for a New Arm's Length Body', p. 55.
37. https://www.nstauthority.co.uk/the-move-to-net-zero/. Accessed 30 March 2023.
38. https://www.nstauthority.co.uk/licensing-consents/. Accessed 30 March 2023.
39. https://www.offshore-mag.com/regional-reports/north-sea-europe/article/14233757/oguk-becomes-offshore-energies-uk. Accessed 31 March 2023.
40. https://oeuk.org.uk/who-we-are-offshore-energy-industry/. Accessed 31 March 2023.
41. https://oeuk.org.uk/net-zero/. Accessed 19 September 2023.
42. GMB (2021), https://www.gmb.org.uk/news/political-and-industrial-failures-will-fuel-climate-and-employment-crises. Accessed 31 March 2023.
43. GMB (2021). 'Congress 2021 CEC Special Report on Energy and the Environment', p. 5.
44. Ibid., p. 8.
45. J. Stewart (2002, 12 June), 'Merchants of Doom Give Wrong Message', *The Scotsman*, p. 4.
46. A. Grant (2022, 13 December), 'Outgoing Boss of Oil and Gas Trade Body Warns Against "Environmental Populism" Amid North Sea Debate', *The Scotsman*.
47. Climate Action Against Disinformation (2022), *The Impacts of Climate Disinformation on Public Perception*. CAAD, p. 22. https://caad.info/wp-content/uploads/2022/11/The-Impacts-of-Climate-Disinformation-on-Public-Perception.pdf
48. Ibid., p. 45.
49. https://www.heraldscotland.com/news/23773004.neil-oliver-tv-presenter-resigns-royal-society-edinburgh-fellow/.
50. https://dailysceptic.org/.
51. https://richardlyon.substack.com/.

52. R. Lyon (2023, 18 January), 'The Dangerous Fantasy of Scotland's Net Zero Energy Transition', The Daily Sceptic, https://dailysceptic.org/2023/01/18/the-dangerous-fantasy-of-scotlands-net-zero-energy-transition/
53. Climate Action Against Disinformation (2022), p. 10.
54. https://www.conservativewoman.co.uk/.
55. C. Cross (2021, 9 March), 'Twenty-Five Years of Hot Air', *The Conservative Woman*, https://www.conservativewoman.co.uk/twenty-five-years-of-hot-air/; W. Loneskie (2023, 8 February), 'Save the Planet—Stop This Net Zero Lunacy', *The Conservative Woman*, https://www.conservativewoman.co.uk/save-the-planet-stop-this-net-zero-lunacy/.

4
Climate Obstruction in Ireland

The Contested Transformation of an Agricultural Economy

ORLA KELLY, BRENDA MCNALLY, AND JENNIE C. STEPHENS

INTRODUCTION: IRELAND'S NUANCED LANDSCAPE

Contrary to its 'green' international image as the 'Emerald Isle', Ireland has a bleak environmental record. While the country has demonstrated a commitment to environmental goals by adopting several ambitious climate policies, Ireland has one of the highest rates of greenhouse gas (GHG) emissions per capita in the European Union, and it is not on track to meet its emissions reductions targets.[1]

Ireland's ambitious climate policies include the Climate Action Act of 2015, which was amended in 2021 to include legally binding, economy-wide carbon budgets and sectoral emissions ceilings. At the international level, Ireland became the first country in the world to commit to divesting from fossil fuels (in 2018), the second to declare a climate and biodiversity emergency (in 2019), and, in 2021, it became a core member of the Beyond Oil and Gas Alliance (BOGA), an international alliance of governments and stakeholders working to facilitate the managed phase-out of oil and gas production.[2]

Despite these aspirational policy commitments, Ireland consistently ranks among the lowest within the European Union (EU) across a range of environmental indicators.[3] The country is not on track to meet its commitments under the Paris Agreement, and emissions are increasing rather than decreasing in some key areas, including agriculture and transportation.[4] In addition to GHG emissions, Ireland has multiple other troubling environmental indicators. It holds the dubious record of being the country with the worst wetlands depletion of any nation in the world over the past three centuries.[5] Ireland also scores below the EU average on multiple metrics including air quality, the percentage of river water that is unpolluted, and the proportion of land that is protected.[6]

A small European island-nation with a population of just over 5 million, the Republic of Ireland[7] has a comparatively small fossil fuel industry and a strong cultural tradition of agriculture and burning high-carbon-emitting peat for home heating in rural areas. Ireland also has a long history of ecological exploitation and extraction derived from its colonial past as part of the British Empire. This legacy continued post-independence with successive national policies that incentivized draining wetlands to intensify food production and planting non-native monoculture forestry.

Within this context, climate obstruction in Ireland has emerged in a complex and dynamic policy landscape characterized by government efforts to meet the European Union's mandated environmental targets while simultaneously maintaining Ireland's position as a business-friendly, foreign-direct investment hub and subsidizing an ecologically intensive domestic meat and dairy sector.[8] As a result, the Irish landscape is a net source of, rather than a sink for, GHG emissions.

To better understand the disconnect between Ireland's climate policy ambition and its policy implementation failure, this chapter presents an overview of the institutional, sectoral, and individual interests that facilitate climate obstruction in Ireland. 'Climate obstruction' in this chapter is meant to include both outright denial of the climate crisis and intentional efforts to delay climate action. It describes how Ireland's colonial legacy, its unique economic context, its political system, and the country's historically uncritical news media have contributed to a lacklustre approach to environmental policymaking and implementation as well as scepticism in public discourse about the urgency of the climate crisis. The chapter also provides an overview of the sectoral interests that have stymied ambitious policy reform, including a case study of the tactics employed by the Irish agri-food sector. It concludes by highlighting the strong potential for this small, wealthy, and socially cohesive country to overcome climate obstruction and become a global leader in climate action and just climate policy.

BACKGROUND: THE IRISH CLIMATE CONTEXT

An extensive public survey conducted by the Irish Environmental Protection Agency in collaboration with the Yale Program on Climate Communication found that 84% of people living in Ireland are alarmed or concerned about climate change, with only 3% expressing doubt. This analysis demonstrates extremely low levels of climate scepticism in Ireland and widespread public concern and acceptance of climate science.[9]

Despite the high levels of public concern, the misalignment between Ireland's economic and environmental policies have led to GHG emissions increasing by 11.64% between 1990 and 2021 (Figure 4.1). GHG emissions also rose sharply between 1990 and 2008, and increased dramatically after the 2020 pandemic low. The former increases represent a period of economic boom, often referred to as the Celtic Tiger, during which the nation's economy grew rapidly due to EU subsidies and a rapid influx of foreign direct investment (FDI) from US companies. Since then, both the economic downturn caused by the 2008 financial crisis and the 2020 COVID-19 pandemic led to temporary emission reductions that rebounded after these crises. Ireland's emissions increases stand in contrast to emissions declines in other parts of the European Union. Notably, Ireland was the European country with the highest GHG emissions per capita and the highest growth rates of GHG emissions in the third quarter of 2022.[10]

The historical economic context

The ecologically unsustainable nature of Ireland's economy is linked to the legacy of British imperialism and the associated dispossession, commodification, mass deforestation, and plantations of colonial exploitation.[11] In the early post-colonial period (1920s–1959), the Irish economy remained largely closed, stagnant, and heavily reliant on subsistence agriculture. Throughout the 1960s, successive governments pursued policies of trade openness, foreign investment, and economic growth, leading to Ireland's admittance to the European Economic Community in 1973. For economic diversification, the country strived to integrate into the global economy throughout the 1970s and 1980s by securing FDI.[12] Leveraging close historical and linguistic ties to the United States, low corporate taxes and a minimal environmental regulatory landscape positioned Ireland to attract FDI from the global chemical industry sector in the 1970s and the pharmaceutical and computer manufacturing sectors in the 1980s.

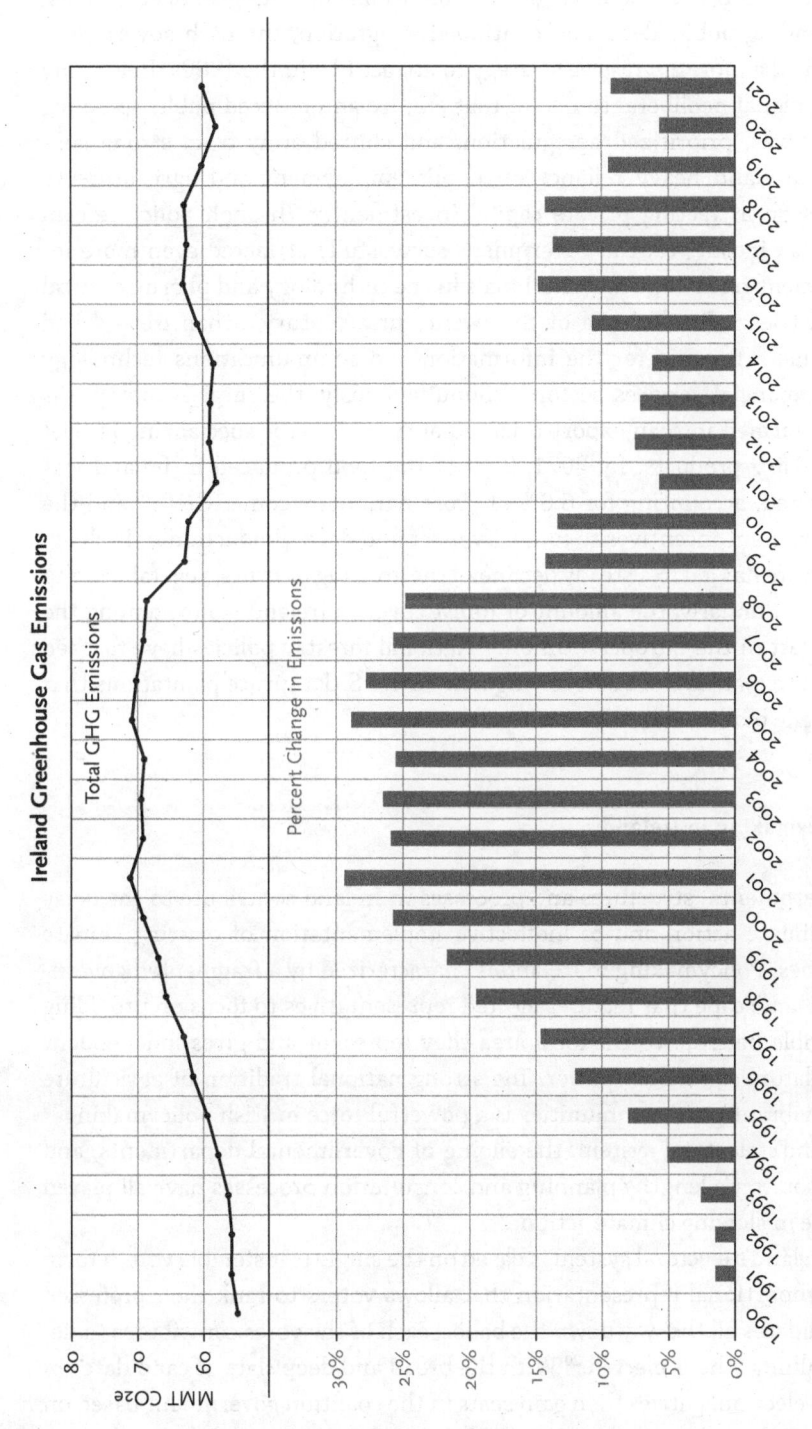

Figure 4.1. Total greenhouse gas (GHG) emissions (in MMT CO_2e) and percentage change in emissions in the Republic of Ireland between 1990 and 2021, inclusive.

Source: Total GHG emissions based on data provided by Gütschow and Pflüger (2023) for Kyoto Six Greenhouse Gas Totals.

In response to poor economic conditions including budget deficits, expanding public debt, and continued emigration, the Irish government pursued a more aggressive strategy to attract FDI in the 1980s. In keeping with global neoliberal trends of that era, Ireland reduced public spending and taxes, prioritized deregulation, and shifted away from strong support for and heavy reliance on public employment and agriculture to focus on attracting private capital investment.[13] Through policy reform and tax incentives, the government successfully attracted even more investment from large multinationals in the technology and pharmaceutical industries.[14] By the turn of the twenty-first century, urban Ireland had become a hotspot for the information and communications technology and financial services sectors.[15] Simultaneously, the rural economy was transformed into an export-oriented agri-food sector specializing in beef and dairy products. By 2021, 90% of the food produced in Ireland was exported, accounting for 6.6% of gross national income (GNI),[16] and the government incentives to intensify meat and dairy production in Irish agriculture has led to a steady decline in the growing of fruits, vegetables, and grains. Similarly, the amount of forest cover in Ireland is now among the smallest in the European Union.[17] National forestry policies have focused on the expansion of fast-growing non-native Sitka spruce plantations that degrade biodiversity.[18]

Policymaking in Ireland

Governmental structures and processes in Ireland contribute to the delay of climate action and to ineffective implementation of existing climate policies. Policymaking in Ireland is characterized by a fragmented governance landscape that requires elected representatives to focus on providing tangible benefits to the local area they represent and gives independent regulators significant power. The strong national tradition of agriculture and vibrant rural communities is a powerful force in Irish policymaking.[19] Ireland's electoral system, the siloing of governmental departments, and the country's lengthy planning and consultation processes have all played a role in slowing climate action.

Ireland's electoral system is based on the single transferable vote, a form of proportional representation that allows voters to rank their preferred candidates all the way down the ballot; each of the votes contributes to determining who is elected.[20] With the broad and deep slate of candidates in each election, parties then gain seats in the coalition government based on the number of votes cast for individuals in their party. Within this system, voters tend to focus on individual representatives rather than the party or the party's policy agenda. This arrangement means that politicians tend to focus delivering tangible benefits directly to their constituents, resulting in a parochial approach that has deprioritized national-level issues such as climate change.

Silos, administrative burdens and limited institutional capacity among governmental departments also contribute to climate delay. For example, although the government has committed to ending the sale of fossil-fuel-powered cars by 2030, the infrastructure for electric vehicles is not yet available in many places, both urban and rural. This delay is partly due to a lack of implementation capacity at the local governance level.[21] Further, Ireland's planning processes are participatory and slow, allowing long periods for community and constituent consultation, which frequently leads to contestation, mobilization, and subsequent changes to proposed infrastructure, buildings, and policies.[22] Beyond the multiple benefits of community-engaged planning, these processes have inhibited efficient policymaking and delayed policy implementation.

Evolving climate policy in Ireland

Membership in the European Union has embedded Ireland in an environmental regime that has created legal obligations and normative expectations of environmental protection. Despite the introduction of two major climate change policies in the early to mid-2000s[23] aimed at helping the state to meet its EU commitments, climate policy implementation was limited during this period, with Ireland's carbon emissions increasing by 5%. This rise was attributed, in part, to the large increase in private vehicle ownership.[24] Ireland did meet its targets set out in the Kyoto Protocol due to the sharp decline in economic activity following the banking and financial crisis in 2008. During the 2010s, emissions increased sharply again, driven by development-oriented national policy signals including the national agricultural strategies: Food Harvest 2020 and Food Wise 2025. Published in 2010 and 2015, respectively, these programmes prioritized substantial expansion of the methane intensive beef and dairy sectors.

Ireland's 2021 Climate Action Plan provides a roadmap for decisive action to halve emissions by 2050. Under this legally binding plan, the state must reduce GHG emissions by 51% compared with 2018 levels by 2030 and reach climate neutrality by no later than the end of 2050. In 2022, the government also adopted an economy-wide carbon budget that included specific emissions ceilings for seven distinct sectors: electricity, transport, commercial and public buildings, residential buildings, industry, agriculture, and other miscellaneous areas including petroleum refining, waste, and fluorinated gases used in refrigeration. Reductions in the overall carbon budget are allocated among these different sectors, so there is competition among them. For example, the July 2022 announcement included a 75% reduction target from 2018 emissions levels for electricity but only a 25% reduction target for agriculture—a controversial distribution.

The policy measures proposed to meet these targets also vary for each sector and include ensuring at least 70% of electricity demand is met from renewable sources, retrofitting 500,000 homes for energy efficiency, and increasing the number of electric vehicles on the roads to 556,000 by 2030.[25] Such a shift will be a considerable challenge given that Ireland still relies on oil and gas for about 80% of its energy needs, including transport, heat, and electricity, with renewables comprising about 13% of supply.[26] Renewables accounted for 34.8% of electricity generated in Ireland in 2021, but natural gas still accounted for 46.0%.[27]

These national targets align with Ireland's statutory obligations as a member of the European Union and a signatory to the Paris Climate Accord. According to civil society groups, this agreement 'hardwired' accountability and transparency into the public and administrative system. The accord gives Ireland's Department of the Environment, Climate and Communications more power to ensure the enforcement of emissions targets and, most significantly, the sectoral ceilings, marking a distinct shift in accountability. Despite these positive developments, Ireland is still considered a 'climate laggard', as policy implementation has not yet resulted in significant emissions reductions.[28] A report released in July 2023 by the Climate Change Advisory Council, an independent watchdog organization, warned that the government's implementation of climate policies is unacceptably slow and ineffective so far; the report pointed out that Ireland will not meet its legally binding targets unless more urgent action is taken.[29]

The Irish government has openly acknowledged the delay in implementing the country's ambitious climate goals. In the November 2022 Climate Action Plan progress report, the government explicitly addressed climate delays, identifying three primary causes: (1) lack of capacity and capability constraints across the public sector, (2) lengthy stakeholder consultation processes, and (3) the complexity of climate action delivery.[30] This report highlights mechanistic capacity challenges to explain the delay rather than identifying individuals or organizations who are intentionally slowing things down. Several of those we interviewed also informally confirmed that the structure and processes within the government are major contributors to implementation delay.

IRISH MAINSTREAM MEDIA AND CLIMATE OBSTRUCTION

Ireland is a small media territory dominated by the national television and radio broadcaster Raidió Teilifís Éireann (RTE), in addition to commercial broadcast stations (Virgin Media and Sky News) as well as local radio. Press coverage includes national dailies (*The Irish Times*, *The Irish Independent*, and *The Irish Examiner*) as well as the Sunday and Irish editions of popular UK tabloids (such as the *Irish Sun* and the *Irish Daily Mail*). Overall, Irish media coverage of climate change has been largely event-driven, focusing on the publication of reports by the International Panel on Climate Change (IPCC) and extreme weather events, rather than exploring the broader social, environmental, and economic contexts.[31] Research shows that episodic framing of climate change is associated with reduced citizen perceptions of the need for government accountability, whereas thematic and contextual coverage increases the potential for citizens to hold governments accountable for enabling climate action.[32] Over the past decade, Irish media established a legacy of accepting the science of climate change while also promoting reasons not to act, thereby contributing to the delay of climate policymaking and effective implementation. Research shows how media discourses reproduced political and elite framings that serve to maintain the status quo and, in so doing, marginalize alternative framings of transformative climate action.[33]

In the post-financial crash years (2008 onward), media narratives about national competitiveness also contributed to normalizing public discourse aimed at obstructing and delaying effective climate policy. This period saw significant political interest in leveraging 'green' solutions to aid Ireland's economic recovery.[34] In parallel, research shows that the green growth agenda and an overriding concern with protecting the economy over meeting environmental challenges became the predominant media trope in climate policy coverage and reveals that media privileged a top-down, supply-side, technological framing of climate change mitigation.[35]

Climate delay tactics

Research highlights three notable delay tactics associated with Irish media coverage of climate change: (1) failing to report critically on the topic,

(2) presenting a polarized debate, and (3) creating a political 'hot potato'. One of the first reviews of climate change communication in Ireland found a lack of critical engagement with the nature of the problem, its causes, and the need for systemic change.[36] This review also highlighted the media's tendency to focus on 'conflict frames' in climate coverage, such as rural resistance to wind farms, and to pitch agriculture against environmental protection, both of which emphasize a polarized debate.

Interestingly, Irish media also act as a platform for the creation of political 'hot potatoes'. This tactic provides a way of exerting pressure on the government in a political culture that tends to avoid contentious issues. Most recently, a government memo to develop a strategy on how to reduce private car emissions as part of the agreed-upon climate targets offers a good example of this media-driven delay tactic.[37] Discussion of the government's transport strategy to reduce car use by half through congestion charges, among other measures, was pulled from a cabinet meeting as it was deemed 'too controversial' by government ministers. This followed intense media coverage the day before that focussed on the controversial nature of congestion charges and division within government coalition parties about the strategy.

Platforming climate contrarians and sceptics

While climate denial is often considered marginal in Irish public discourse, a small number of high-profile actors have historically held sway in challenging climate science. Crucially, their positions of power and close links to media ensured their contrarian claims garnered wide public attention. An analysis of climate change coverage in Irish print media from 2007 to 2016 observed the presence of a 'contested science' frame among columnists and in the *Irish Daily Mail* (a sister publication of the UK tabloid). Key actors included well-known pundits Kevin Myers and Maurice Nelligan, who denounced concern about climate change as alarmist and 'kitchen-sinkology', as well as John Fingleton, meteorologist for the state weather service Met Éireann, who promoted natural climate variation arguments.[38] Another study analysing decarbonization discourses in print media from 2000 to 2013 also identified a 'climate denial' theme (most prominently via Irish editions of the UK tabloid press).[39] In this case, the contrarian arguments pointed to deep divisions among scientists about the causes of global warming and were employed largely by business actors to resist climate policy during debates about a carbon tax in the early 2000s.

Coverage of climate sceptic views is far more evident in Irish media than of outright denial. Sceptical beliefs can be divided into three categories: trend, attribution, and impact, which capture doubts about climate science, belief in human-made causes, and whether there will be negative impacts.[40] In the early phase of climate policymaking, sceptics' arguments focussed primarily on trend and attribution scepticism.[41] Most notably, Pat Kenny, one of RTE's high-profile current-affairs broadcasters (and an engineer by training), regularly included climate sceptics such as David Bellamy on programmes about global warming and infamously argued that rising GHGs were not a problem.[42] Sceptic voices also focussed on response scepticism, questioning the desired level of government regulation of industry as well as the efficacy of climate taxes and policies.[43] Notably, Michael O'Leary, the head of Ryanair, a major Irish airline, did not publicly accept that climate change was real until as recently as 2017.[44] He continues to be given a platform to question the government's ability to deal with the crisis and to engage in public media campaigns against mandatory emissions reductions for the aviation industry.[45]

While climate sceptic views have evolved with the changing policy context, research indicates a media focus on 'dismissive' voices, anti-environmentalists who deride those advocating climate action or attack environmentalist stances for being overly earnest or sanctimonious. Examples include references (often in headlines) to 'environmental nutters', 'lunatic environmentalists', 'headbangers', and 'Luddites marching us back to the 18th century'. Rather than denying the science, these actors dismiss environmental protection based on the view that the economic project is more urgent than tackling climate change.

Another prominent discursive strategy involves the use of religious metaphors. A study of media discourses about the low-carbon transition identified the presence of a 'Church of Green' discourse used by sceptics to challenge perceived 'green authoritarianism'.[46] The analysis found that this discourse was antagonistic toward the perceived 'moralizing' of those advocating carbon-reduction activities. It mobilized an Irish sense of humour to ridicule the imposition of a green orthodoxy with references to a 'tax on fun', 'green sins', a 'carbon confession box', and 'guilt and finger-waving from the environmentalists' response.[47]

KEY ACTORS DRIVING CLIMATE OBSTRUCTION IN IRELAND

Climate obstruction has been advanced through individual and organizational efforts as well as through governmental processes and coordinated lobbying by sectoral interests. While some key Irish actors are actively delaying climate policy in support of their personal or professional interests, others are inadvertently causing delay because they are focused on non-climate-related priorities. Although this distinction may seem clear in theory, in practice it is often challenging to discern why different people and institutions advocate against climate action.

Fringe academics and think tanks

Within the scientific and academic communities in Ireland, a few individuals have been outspoken, claiming that the science of climate change is not settled. These outliers tend to be networked with international climate denial groups.

Ray Bates, a meteorologist, member of the Royal Irish Academy, and retired Adjunct Professor at University College Dublin, is among the most controversial and internationally recognized of such figures. Bates has leveraged his scientific credentials to advocate against climate action by claiming that the science is not settled and has become politicized. The impact of a sole climate denier was highlighted when, in December 2015, RTE invited Bates to participate in a prime-time discussion with climate policy experts and the minister for the environment on the costs of climate action. In response to the programme, An Taisce (the National Trust for Ireland, focused on environmental conservation) filed a complaint against RTE for failing to provide fair, objective, and impartial current affairs content, which was a violation of Broadcasting Authority of Ireland rules.[48] Bates also wrote a report 'Deficiencies in the IPCC's Special Report on 1.5 Degrees', published in 2018 by the UK-based climate obstruction organization Global Warming Policy Foundation (GWPF),[49] which was heavily excoriated by climate scientists around the world.[50] The report's foreword was written by Edward Walsh, the President of the University of Limerick, who had served as chair of Ireland's National Council for Science, Technology and Innovation, providing additional legitimacy to this effort. Most recently, in 2021, Bates was appointed to the GWPF's academic advisory board.[51]

The Irish Climate Science Forum (ICSF), co-founded by Bates and led by Jim O'Brien, an energy consultant and retired engineer, is one of the most well-known climate-denying organizations in Ireland. According to their website, the ICSF is a voluntary organization composed of scientists, engineers, and other professionals dedicated to disseminating 'objective science' and to providing 'the good news on climate'.[52] The organization's stated aim is to promote 'realism' in climate science and 'prudence in climate and energy policy'. Their main activity involves holding public lectures on national climate and energy policy, often by high-profile climate deniers, as well as submissions to relevant public consultations. In 2023, the lecture series included presentations by several well-known climate deniers including David Horgan, head of Petrel Resources, one of Ireland's most prominent oil and gas exploration companies, who argued that Ireland's current energy policy was 'tantamount to economic suicide'. Other recent speakers included Marcel Crok, co-founder of Climate Intelligence (CLINTEL), a Dutch foundation aimed at obstructing climate policy, who spoke about why the IPCC needs to be reformed; Christopher Monckton, one of the most cited and widely published climate sceptics; and Professor William van Wijgaarden, a member of the CO_2 Coalition (a US think-tank) who argued that GHG emissions are insignificant.

The ICSF also published a critique of the IPCC AR6 Synthesis Report, arguing that the report was 'seriously flawed' based on the view that 'real world observations point to only a modest 1 degree warming up to 2100' and that 'the IPCC should be disbanded'.[53] While the organization's influence on climate obstruction is difficult to assess, the ICSF provides an important platform for international climate-denying voices that seek to challenge the prevailing scientific consensus on climate change and the need for ambitious climate policies. The organization has links, through its members and lecture series, to the GWPF[54] and to CLINTEL[55] as well as the denialist groups EIKE in Germany and the Stockholm Initiative in Sweden.

Sectoral lobbyists

Although outright denial of climate change is increasingly rare in Ireland, many industry actors are actively engaged in climate policy discussions, trying to slow change. Lobbying groups representing various constituents within multiple large sectors including agriculture, energy and transportation, delay action by highlighting a broad array of social, economic, and cultural costs of implementing changes.

Evidence of environmental lobbying and counter-lobbying activities can be uncovered through a review of public records held on the public database Lobbying.ie, a web-based register of lobbying of designated public officials on policies, legislative matters, or prospective decisions, which is mandated under the Regulation of Lobbying Act 2015.[56] A preliminary keyword search of records using the subject 'climate' found more than 4,000 records filed on this subject during the period September 2015–December 2022. These public records show that the Irish Business and Employers Confederation (IBEC) and the Irish Farmers Association (IFA) have engaged in the highest volume of lobbying of public officials on this issue during this period. The third most frequent lobbying group was Wind Energy Ireland (WEI), a renewable energy lobbying group.

An initial review of this database shows a range of 'intended results' from the lobbying efforts. Table 4.1 includes samples of actors' stated intended results sampled from January 2016. Notably, IBEC often lobbied to governments to consider national competitiveness alongside climate action targets. Similarly, the IFA sought to protect the economic interests of the farming sector in the context of discussion on environmental policies. In contrast, WEI sought to highlight the importance of indigenous renewable energy sources. It is important to note that these records

Table 4.1 TOP THREE LOBBYING ACTIVITIES ON 'CLIMATE' BY ORGANIZATIONS, SEPTEMBER 2015–DECEMBER 2022

Rank	Organization	Total returns filed on the subject 'climate'	Sample of organization's stated 'intended results'
1	Irish Business and Employers Confederation (IBEC)	262	Effective mitigation of greenhouse gas emissions in a manner that enhances rather than damages Ireland's prosperity Date published: 20 January 2016
2	The Irish Farmers' Association (IFA)	212	Agreement on a common position on Climate Change Fair deal for Ireland in International Agriculture Trade Negotiations Support for IFA request for increased competition in EU on Inputs Date published: 21 January 2016
3	Wind Energy Ireland	113	Awareness and possible support for 'The Power to Power Ourselves' communications campaign, highlighting Ireland's 85% dependency on imported energy, and promoting increased attention on the use of indigenous renewable energy sources. Date published: 21 January 2016

do not capture, or reflect, the extent or effectiveness of lobbying activity conducted for each subject. However, the number of records provide an indication of the frequency of actors' engagement with public officials on climate action. Further research could be helpful to better capture the effectiveness of these lobbying efforts.

The energy sector

With a low share of energy-intensive industry, Ireland's carbon intensity relative to its gross domestic product (GDP) is among the lowest in the European Union.[57] Furthermore, the Irish energy sector publicly conveys strong support for and a deep commitment to climate action and the energy transition away from fossil fuels. Although the government has outlined a path to an eventual elimination of fossil fuels from the country's energy systems,[58] the country remains heavily reliant on fossil fuels and is ranked lowest in Europe for renewable energy readiness.[59] The national 2030 target of a 34% renewable energy share is focused mainly on harnessing wind, with some solar and biomass, with a renewable energy in electricity target of 70% by 2030.[60]

The Electricity Supply Board (ESB) is the nationally owned company charged with delivering the country's electricity and maintaining its grid. The ESB has committed to achieving net zero emissions by 2040 by increasing renewable generation, investing in electric grid infrastructure, and empowering consumers to electrify. Its website claims an 'unwavering commitment to tackling some of the biggest challenges we face as a society, including climate change'.[61] Despite these vague public messages, the ESB has been accused of slowing the transition to renewable energy by not making the infrastructure investments needed and using its dominance to push new actors[62] out of the energy generation market.

Despite the European Union's encouragement and specific recommendations from Ireland's Citizens' Assembly on Climate Change[63] for cooperative or community-owned, distributed renewable energy in Ireland, this resource has been slow to deploy. One notable exception is led by Community Power, the country's first community-owned renewable electricity utility company.[64] Despite its success in selling and distributing local renewable electricity, the organization has faced difficulties in accessing the grid, and their model has not yet been widely replicated. Complex factors have contributed to the delay in expanding community-owned renewable energy, including a lack of capacity for innovation in the public sector. Community-driven energy initiatives also face significant competition from international investments funds, which have identified Ireland as a key market. The government, too, has been criticized for creating administrative bottlenecks.[65]

Meanwhile, the ongoing proliferation of data centres in Ireland represents a significant challenge to Ireland's efforts to reduce emissions from electricity generation.[66] A recent investigation revealed that onsite carbon emissions from data centres are more than 35 times higher than during the previous decade.[67] By 2021, data centres consumed 14% of Ireland's total electricity, more than rural dwellings combined. Although the electricity for data centres could be renewably generated, Ireland's renewable energy capacity is not yet sufficient to cover the amount of energy required for the growing demand. Activists have highlighted that such trends are misaligned with climate goals, but addressing these concerns represents a significant challenge to government because of the economic benefits these centres offer. While the centres themselves do not provide much employment, their parent companies are large sources of urban employment for highly skilled information technology workers.[68]

The transport sector: Reinforcing car culture

Decarbonizing the transport system is a major focus of the current government. For example, €35 billion has been earmarked for active travel under the latest climate action plan.[69] Progress in decarbonizing this sector has been slow to date. Ireland's transport sector has reduced its GHG emissions by just 7.5% since 2005. Emissions reductions have stagnated in recent years[70] due in part due to continual reinforcement of the nation's car-dependent transport system. Car-dependent transport systems are a critical component of 'carbon lock-in' in national energy systems,[71] and the Irish government has to date been largely ineffective in reducing reliance on automobiles. Car dependency can become entrenched through several factors including (1) advocacy from the automotive industry; (2) the proliferation of car infrastructure; (3) the political economy of urban sprawl; (4) the lack of alternative modes of transport, including public transport and bicycle infrastructure; and (5) strong cultural norms that promote car use.[72] All of these factors are present in Ireland and undermine efforts to transform Ireland's transport system. Transportation is particularly challenging in rural Ireland, where car dependency is among the highest in Europe due largely to minimal public transport particularly outside of major cities.[73] Evidence of the entrenchment of car dependency can be seen in the recent resistance to the government's efforts to redesign roadways to provide more space for walking and biking as well as public transport.[74]

While Ireland does not have its own domestic auto manufacturing company (since Ireland joined the European Union, all cars are imported), it does have a strong automotive industry that sells and maintains the nation's more than 2.5 million cars.[75] The network of automobile suppliers selling European, Japanese, and American cars is extensive, and the motor industry promotes electric vehicles but resists efforts to reduce car dependency.[76] Car sales and electric car infrastructure are accelerating quickly in Ireland, per the goals of the national Climate Action Plan, although local authorities have struggled to build a network of charging stations,[77] reflecting the government's ongoing capacity challenges in implementing decarbonization strategies. Ongoing efforts to overcome car dependency contentious, as demonstrated by widespread political controversy in response to the July 2023 release of the first All-Island Strategic Rail Review, which included recommendations for developing an electrified regional rail network.[78]

The agri-food sector: A case study in obstruction in the Irish context

The agriculture sector's historical importance and its ongoing role as a key rural employer give it strong influence in Irish policymaking circles. The broader agri-food sector includes those involved in primary production in farming, fishing, and forestry and those engaged in the production and processing of food, beverages, and wood. There are both indigenous and export-oriented dimensions to the sector. Dairy is the largest component of Irish food and drink exports, followed by meat and livestock. The sector accounts for 7% of the total Irish workforce and is critically important for many rural areas. A key source of the sector's sway on these matters is that it represents the interests of a politically active rural minority on which the two main centrist parties rely for votes.

The agri-food sector represents a major and entrenched stumbling block in Ireland's efforts to reach its emission targets because agriculture is the single largest contributor to Irish GHG emissions, accounting for 37.5% of the national total in 2021. The source of these emissions is mainly methane from livestock and nitrous oxide from the use of nitrogen fertilizer and manure.[79] In addition to its climate impacts, agriculture is also the predominant cause of Ireland's water pollution, ammonia air pollution, and biodiversity loss.[80] The sector is particularly environmentally destructive compared with other European nations, emitting three times more pollution than the sectoral EU average.[81] Notably, only 1.3% of Irish agricultural land is farmed organically, the second lowest area in the European Union.[82] Moreover, while other traditionally agricultural nations within the European Union, including France, have reduced pollution associated with agriculture in recent years,[83] Ireland is among those that have seen a significant increase, with emissions rising 9.3% between 2011 and 2021.[84] This increase is linked to a 50% rise in agri-food exports during this same period, a strategy explicitly supported by the Irish government over the past two decades.

The political influence of the agri-food sector

Many have argued that Ireland's beef and dairy farmers and their corporate partners have had disproportionate influence on the nation's agricultural policymaking.[85] As shown earlier in Table 4.1, the IFA has engaged in extensive political lobbying and public engagement activities around climate-related matters. Much of this activity has been aimed at limiting the impact of environmental legislation on the existing business model.[86]

A source of the sector's influence is their close ties with the Department of Agriculture. For example, observers note that Teagasc—the state agency providing research, advice, and education in agriculture, horticulture, food, and rural development in Ireland—is heavily influenced by representatives from the dairy industry, with five of the eleven members of this advisory council identifying as dairy farmers.

Evidence of this strong industry influence can be found in Ireland's national agricultural strategies communicated in reports published in 2010 and 2015 (Food Harvest 2020 and Food Wise 2025, respectively), which explicitly called for the expansion of methane-intensive meat and dairy production. The agricultural strategy report published in 2020, which was titled Ag Climatise, proposed climate neutrality by 2050, but still assumed intensive meat and dairy production; this report was deemed 'not fit for purpose' by expert analysts and environmental groups because of the level of emissions it allowed.[87]

As of 2023, the Irish government, which had a Green Party minority, has also been accused of pandering to the interests of the agri-food sector.[88] For example, livestock farming, particularly the raising of cattle and sheep, continues to be heavily subsidized by the state, with only 27% of all cattle farms classified as economically viable.[89] The dairy industry, while more economically lucrative, is environmentally problematic, with GHG emissions per hectare on dairy farms two to four times higher than on other farm systems. The environmental intensity of the dairy sector has grown steadily since the 2010s in response to government policies that pressured and incentivized dairy farmers to increase the size of their farms.[90]

The misalignment of the country's agricultural strategy with its climate commitments is becoming increasingly divisive. Notably, the Environmental Pillar, a non-profit organization that represents Irish environmental civil-society groups, withdrew from the drafting process of the latest agri-food strategy (Food Vision 2030), claiming that the process was too industry-dominated; did not seriously integrate consideration of climate, biodiversity, and water and air quality; and relied too heavily on future action, yet-to-materialize innovations, and potential abatement technologies.[91] The Irish dairy industry, on the other hand, welcomed the final recommendations, noting that 'it allows Irish dairy further to enhance its competitive advantage'. Seven of the thirty-three members of the Food Vision 2030 stakeholder committee were leaders directly engaged in the global food industry.[92] The current 'roadmap' for the dairy sector allows for continued expansion in dairy output until at least 2027.[93]

Extensive lobbying has also had significant influence in reducing the agricultural sector's legally binding emissions targets under the Climate Action Amendment Bill (2022). A sectoral target originally proposing a 30% emissions reduction by 2030 was successfully resisted and reduced to 25%. The agricultural sector is among the most active in lobbying in Ireland (Table 4.1), and our analysis of the lobbying register revealed that members of the agri-food sector, including the IFA, lobbied government representatives at the EU and national levels to negotiate lower emission targets on the basis that meeting the proposed reductions would (1) devastate the sector, (2) compromise global food systems, (3) result in 'carbon leakage', and (4) allow insufficient time for technologies to be implemented.[94]

There is also doubt about whether these lower targets will even be achieved. An assessment of three scenarios for emission reductions in agriculture found that even in the scenario with the most state support (in which a mandatory 4% emissions reduction would be enforced and subsidies provided to farmers), emissions would decrease by only 6.78% relative to 2005 levels.[95] Furthermore, some of the mitigation approaches these strategies propose remain contested within the scientific literature.[96] An additional concern is whether the government will apply the necessary oversight to ensure that emissions reductions and pollution strategies are enforced. To date, the government has been reprimanded at the EU level for failing to enforce such regulations.[97]

The misalignment of Ireland's climate commitments and the country's agricultural policies (including incentives for farmers) is creating increasing frustration throughout the country among both urban and rural communities concerned about climate, food production, and the Irish economy. While the Irish media often portray a rural–urban divide in Ireland regarding support for climate policy, 2022 research shows that concern about climate change is just as strong in rural communities as in is in urban ones,[98] providing the government with a growing mandate for change.

The agricultural sector and discursive tactics of delay

In their efforts to minimize the financial and regulatory impacts of climate action on their members, the agricultural sector has frequently employed a range of denial and delay discourses engaged in a range of discursive tactics of delay (Table 4.2). The association's flagship publication, *Irish Farmers Journal* (IFJ), has been criticized for giving a platform to debunked climate science. Such discourses have also infiltrated sections of the education system. For example, Agri Aware, a charitable trust controlled and funded by a consortium of agricultural industry players, distributed a series of four workbooks under the title 'Dig In' to more than 3,200 primary schools, misrepresenting Ireland's carbon footprint by underplaying the biodiversity loss and methane emissions attributable to agriculture.[99]

In the context of policy obstruction, sectoral representatives often deploy three clearly identified discourses of delay[100] to resist climate policies: (1) redirecting responsibility, (2) pushing non-transformative solutions, and (3) emphasizing the downsides of climate policy. Notably, the Irish agri-food sector employed the services of a well-known communications consulting agency, Red Flag,[101] which used similar tactics when representing the interests of British American Tobacco, Monsanto, and other agri-chemical companies in the European Union.

Lobby groups regularly use the term 'carbon leakage' to describe the unfairness of the 'free rider' problem; that is, unless all individuals, industries, or countries undertake emissions reductions, some will benefit from the actions of the others. In the Irish context, carbon leakage is frequently used to describe a scenario whereby emissions savings from livestock cuts would be reversed by increased production elsewhere.[102]

Interestingly, many of the discursive frames used by lobbying groups such as the IFA are echoed in policy documents and repeated by politicians representing rural constituents. In this way, the non-transformative

Table 4.2 DISCOURSES OF DELAY FROM AGRICULTURAL SECTOR AND RURAL POLITICIANS IN RESPONSE TO PROPOSED EMISSIONS-REDUCTIONS TARGETS

Discourse of delay		Example
Redirect responsibility	The 'free rider' excuse	'While agriculture has a responsibility to protect the environment, the imposition of a target without accounting for global carbon leakage arising from food production is ill-informed and more likely to lead to a rise in global GHG emissions. Food production must be encouraged in areas where it is the most carbon efficient to do so'. —Irish Farmers Association Annual Report 2022, p. 13[a]
Push non-transformative solutions	Technological optimism; holding that technological progress will rapidly bring about emissions reductions in the future	'Solutions are in development, we are seeing great progress in relation to feed additives, particularly the 3NOP additive which is under research at Teagasc. This additive was designed for continual feeding in indoor systems, the challenge now is to develop options for our pasture-based animals'. —Minister Martin Heydon Statement on Emissions Reductions 28 July 2022[b] 'Irish farmers are embracing measures to reduce emissions and there are significant scientific developments on feed additives and other technologies, but it will take time'. —Irish Farmers Association return recoded to lobbying (i.e. September 2022)[c]
Emphasize downsides	Appeal to social justice: claims that the cost of climate action will reduce social justice	'Reducing the national herd to meet emissions targets would be paramount to "ethnic cleansing" of the agricultural community'. —Michael Fitzmaurice, Independent member of Dáil Eireann (Irish Parliament) for Roscommon-Galway, 12 February 2023, in response to the release of a commissioned EPA research report which recommended reducing the national herd to meet emissions reductions targets[d] 'Who will supply this food? There is a real risk that we will create a global food emergency trying to solve the climate emergency'. —Speech by IFA President Tim Cullinan, 27 January 2022[e]

[a] Irish Farmers Association (2023, January), 'Irish Farmers' Association Annual Report 2022', https://www.ifa.ie/wp-content/uploads/2023/01/IFA-Annual-Report-2022-Published-Jan-23.pdf.
[b] Department of Agriculture Food and Marine, 'Pathway to 51% Reduction'.
[c] Lobbying.ie, 'The Irish Farmers' Association'; Lobbying.ie, 'Regulation of Lobbying Act'.
[d] M. Maguire (2023, February), 'Government Using Farmers as "Scapegoats" for Increase in Emissions – McNamara'. *NewsTalk*, https://www.newstalk.com/news/government-using-farmers-as-scapegoats-for-increase-in-emissions-mcnamara-1436598
[e] IFA (2022, April), 'Agriculture Sectoral Emissions Ceiling Is a Potentially Devastating Blow for Irish Farming', https://www.ifa.ie/farm-sectors/agriculture-sectoral-emissions-ceiling-is-a-potentially-devastating-blow-for-irish-farming/.

discourse of delay supporting a 'green economy' is purveyed not only by lobby groups but also by its government representatives. For example, the national agricultural strategy published in 2010 declared : 'The modern use of "green" to identify concern for the natural environment has, for some time, been recognized as representing a natural marketing opportunity for Irish agri-food to build on'.[103] Similar narratives continue to be perpetuated by semi-state bodies such as the International Development Authority (IDA) and An Bord Bia (The Food Board) as well as ministerial trade missions.[104]

CONCLUSION

Climate obstruction in Ireland is complicated and nuanced, primarily taking the form of delay and inertia rather than promoting climate denial. Ireland has ambitious climate goals and policies, and most Irish people are alarmed or concerned about the climate crisis.[105] Yet policy implementation has been largely ineffective so far. This review of climate obstruction in Ireland suggests that transformative change is stymied by the country's long colonial history of economic and ecological exploitation, its reliance on foreign direct investment, the political and cultural power of the agri-food sector, inertia resulting from limited institutional capacity for change, and a slow planning process within the public sector.

Research on Irish news media coverage of climate change shows how mainstream media have normalized climate denial and delay in public discourse and provided a platform for climate contrarians and sceptic viewpoints. However, recent developments by media organizations, such as more frequent coverage of climate and biodiversity issues, an increase in environmental correspondents, and dedicated climate sections in the press as well as the promotion of climate literacy training by Coimisiún na Meán (the new Irish media regulator) suggest that opportunities for media-driven climate misinformation could decline. Nonetheless, given the entrenched resistance to change among high-carbon sectors, the sophistication of climate obstruction tactics, and the significance of media coverage for democratic debate about the radical social transformations required to address the climate crisis, ongoing research to identify and counter climate denial and delay narratives in public discourse is essential.

Understanding climate obstruction in Ireland requires consideration of the unique Irish context. As a small, English-speaking, post-colonial islanded country, Ireland has an often contradictory economic and environmental agenda. Although the energy sector has ambitious decarbonization targets, the pace of change is slow and the scale of investments required to phase out fossil fuel reliance in heating, transport, and electricity have not yet been prioritized. Planning regulations, the dominance of the public sector provider, and the conflicting demands of the economically important technology sector for energy-intensive data centres have also stymied efforts to reduce emissions.

Similarly, the government continues to subsidize environmentally intensive agricultural production due to the political influence of the sector. After two decades of supporting beef and dairy expansion, there is now pressure on the agriculture sector to reduce its emissions under the Climate Action Plan. The agri-food lobby is resisting such efforts to protect its economic interests. Many rural communities are also feeling increasingly threatened by or mistrustful of the government's climate policies due to inconsistent and misaligned approaches. For example, the government's climate policies to incentivize the forestry industry to increase carbon sinks throughout the Irish landscape has resulted in the proliferation of industrial, non-native monoculture forests that are often owned and managed by foreign companies, offering no economic benefit to rural communities and harming local biodiversity.[106] To tackle this issue, the government is now investigating how to incentivize radical shifts in land use and forestry through research and stakeholder engagement.[107] Transforming toward a low-emission economy will also require strong political leadership and new coalitions to collectively tackle powerful actors within the sector whose economic interests lie in maintaining the status quo.

More research is needed to better understand how climate obstruction in Ireland is changing over time and how the media, government, civil society, and interest groups are adapting their strategies, especially as pressure mounts to make more drastic changes. Universities in Ireland are increasingly engaged with creating and expanding multiple innovative climate-related programs and research centres that have had broad social impact around the country.[108] But as university administrations are increasingly driven to seek alternative forms of funding such as industry partnerships, caution is warranted to ensure that they are not influenced by corporate interests to strategically resist climate policy as universities in the United States, Canada, and the United Kingdom have been.[109]

The CEO of Friends of the Earth Ireland characterized the mainstream Irish response to taking transformative climate action with the phrase 'Not us, Not yet, Not this. . . '. Despite the slow pace of change, anti-fossil fuel norms are expanding,[110] and Irish elected officials have supported the 2023 European Union Nature Restoration Law that commits member nations to restoring ecological health by 2050. Furthermore, as the climate crisis escalates, advocacy and appetite for larger transformative change is growing.[111] So, too, is government investment in community engagement on climate issues.[112]

As a small, wealthy country the potential for Ireland to become a global climate leader is high. Ireland has a recent history of making major social changes such as the overwhelming support for same-sex marriage, the its Gender Recognition Act, allowing trans people to apply to have their preferred gender legally recognised by the state, and the legalization of abortion. During the COVID pandemic, Ireland was among the countries with the highest vaccine uptake, additional evidence of the Irish people's collective sense of social responsibility, justice, and accountability. This strong sense of fairness and social justice can be harnessed to further resist climate obstruction in Ireland and leverage the country's potential for climate justice leadership.

On the other hand, the resounding rejection of two proposed amendments to Ireland's constitution in 2024 regarding women's role, caregiving, and family structure highlights the need for strong public engagement on social change initiatives. This outcome serves as a reminder that government-led change cannot be successful without public understanding and support. This lesson must be heeded by the government if it hopes to overcome obstrucutionism and secure public backing for crucial climate policies.

NOTES

1. Environmental Protection Agency (2023, June), 'Ireland Projected to Fall Well Short of Climate Targets', https://www.epa.ie/news-releases/news-releases-2023/ireland-projected-to-fall-well-short-of-climate-targets-says-epa.php.
2. L. M. Fitzgerald (2023), 'Tracing the Development of Anti-Fossil Fuel Norms: Insights from the Republic of Ireland'. *Climate Policy*, 1–14.
3. Climate Change Performance Index (2023), 'Ireland: Climate Performance Ranking 2023', https://ccpi.org/country/irl/.
4. Environmental Protection Agency (2023, June).
5. Etienne Fluet-Chouinard, Benjamin D. Stocker, Zhen Zhong, et al. (2023), 'Extensive Global Wetland Loss over the Past Three Centuries'. *Nature*, 614, 7947: 281–286.
6. Central Statistics Office (CSO) (2021, September), 'Environmental Indicators Ireland 2021', Central Statistics Office, Ireland, https://www.cso.ie/en/releases andpublications/ep/p-eii/environmentalindicatorsireland2021/.2021.
7. In this chapter, we focus on the Republic of Ireland. The six counties in the northeast of the island of Ireland share an ecological identity with the Republic but are under the administration of the United Kingdom.

8. S. Dekker and D. Torney (2021, January), 'Evaluating Ireland's Climate Policy Performance', Environmental Protection Agency, Ireland Report, https://www.epa.ie/publications/research/climate-change/Research_Report_362.pdf Climate Change Performance Index, 'Ireland'.
9. A. Leiserowitz, J. Carman, S. Rosenthal, et al. (2021, December), 'Climate Change in the Irish Mind', Yale Program on Climate Change Communication, https://climatecommunication.yale.edu/publications/climate-change-in-the-irish-mind/.
10. Eurostat (2022), 'Performance of the Agricultural Sector', https://ec.europa.eu/eurostat/statistics-explained/index.php?title=Performance_of_the_agricultural_sector.
11. R. Crotty (2001), *When Histories Collide: The Development and Impact of Individualistic Capitalism*. Lanham, MD: AltaMira Press.
12. Ibid.
13. Ibid.
14. Ibid.
15. S. Deckard, S. (2016), 'World-Ecology and Ireland: The Neoliberal Ecological Regime', *Journal of World-Systems Research*, 22, 1: 145–176.
16. Department of Agriculture, Food and Marine (2022, July), 'Pathway to 51% Reduction in Economy-Wide Emissions Agreed: McConalogue Confirms 25% Reduction in Agricultural Emissions', https://www.gov.ie/en/press-release/40b39-pathway-to-51-reduction-in-economy-wide-emissions-agreed-mcconalogue-confirms-25-reduction-in-agricultural-emissions/.
17. Department of Agriculture, Food and Marine (2023), 'Forestry Facts and News', https://www.gov.ie/en/publication/57d2a-forestry-facts-and-news/.
18. R. Carroll (2019, July), 'The Wrong Kind of Trees: Ireland's Afforestation Meets Resistance', *The Guardian*, https://www.theguardian.com/world/2019/jul/07/the-wrong-kind-of-trees-irelands-afforestation-meets-resistance.
19. M. Lockwood, C. Kuzemko, C. Mitchell, and R. Hoggett (2017), 'Historical Institutionalism and the Politics of Sustainable Energy Transitions: A Research Agenda', *Environment and Planning C: Politics and Space*, 35, 2: 312–333.
20. M. Gallagher (1986), 'The Political Consequences of the Electoral System in the Republic of Ireland', *Electoral Studies*, 5, 3: 253–275.
21. S. Burns (2022, May), 'Local Authorities Have Not Installed Any Electric Vehicle Chargers for Public, Says Bruton', *The Irish Times*, https://www.irishtimes.com/politics/oireachtas/2022/05/25/local-authorities-have-not-installed-any-electric-vehicle-chargers-for-public-says-bruton/; and Government of Ireland (2023), 'Climate Action Plan 2023: Changing Ireland for the Better', https://www.gov.ie/pdf/?file=https://assets.gov.ie/243585/9942d689-2490-4ccf-9dc8-f50166bab0e7.pdf#page=null.
22. M. Scott (2006), 'Strategic Spatial Planning and Contested Ruralities: Insights from the Republic of Ireland', *European Planning Studies*, 14, 6: 811–829.
23. Department of Environment and Local Government (2002), 'The National Climate Change Strategy 2002'; Department of Environment, Heritage and Local Government, 2007. *The National Climate Change Strategy (2007–2012)*.
24. I. Conway (2000, November), 'Celtic Tiger Economy Gets Blame for Failure to Cut Pollution', *Irish Independent*, https://www.independent.ie/irish-news/celtic-tiger-economy-gets-blame-for-failure-to-cut-pollution-26104751.html.
25. Government of Ireland, 'Climate Action Plan'.

26. G. Lee (2022, September), 'Where Does Ireland's Energy Come From?' *RTE*, https://www.rte.ie/news/environment/2022/0907/1320733-where-does-irelands-energy-come-from.
27. Environmental Protection Agency (2022), *Energy*, https://www.epa.ie/our-services/monitoring--assessment/climate-change/ghg/energy-/.
28. Climate Change Performance Index, 'Ireland'.
29. K. O'Sullivan (2023, July), 'Climate Change: Ireland's Pace of Action "Not Acceptable Given Existential Threat", Watchdog Warns', *The Irish Times*, https://www.irishtimes.com/environment/climate-crisis/2023/07/25/irelands-pace-of-climate-policy-implementation-not-acceptable-says-watchdog.
30. Government of Ireland, 'Climate Action Plan'.
31. T. Morgan (2020), 'Challenges and Potentials for Socio-Ecological Transformation: Considering Structural Aspects of Change'. In: T. Robbins, D. Torney, and P. Brereton (eds.), *Ireland and the Climate Crisis*, 149–168. Basingstoke: Palgrave.
32. Frameworks Institute (2002), *Framing Public Issues: A Toolkit*. Baltimore, MD: The Annie E. Casey Foundation.
33. E. Fox and H. Rau (2017), 'Climate Change Communication in Ireland'. In: M. C. Nisbet, S. S. Ho, E. Markowitz, et al. (eds.), *Oxford Research Encyclopedia of Climate Change Communication*. New York: Oxford University Press.
34. Department of Jobs, Enterprise and Innovation (2012, November), 'Potential for Over 10,000 Extra Jobs in Green Economy by 2015—Minister Bruton', http://www.djei.ie/press/2012/20121123.htm.
35. B. McNally (2015), 'Media and Carbon Literacy: Shaping Opportunities for Cognitive Engagement with Low Carbon Transition in Irish Media, 2000–2013', *RAZÓN Y PALABRA*, http://www.razonypalabra.org.mx/N/N91/Monotematico/05_McNally_V91.pdf
36. Fox and Rau, 'Climate Change Communication'.
37. Friends of the Earth (2023, March), 'Running Scared on Transport No Way to Tackle Congestion and Pollution', https://www.friendsoftheearth.ie/news/running-scared-on-transport-no-way-to-tackle-congestion-and/.
38. D. Robbins (2019), *Climate Change, Politics and the Press in Ireland*. Routledge: London.
39. B. McNally (2017), *Media and Low Carbon Transition: A Multi-Modal Analysis of Print Media Themes and Their Implications for Broader Public Engagement*. PhD Thesis, Dublin City University.
40. S. Rahmstorf (2004), *The Climate Sceptics*. Potsdam: Potsdam Institute for Climate Impact Research, http://www.pik-potsdam.de/~stefan/Publications/Other/rahmstorf_climate_sceptics_2004.pdf.
41. Robbins, *Climate Change*.
42. J. Gibbons (2009, November), 'Kenny Stirs up Bogus Climate Change Debate', *The Irish Times*, https://www.irishtimes.com/opinion/kenny-stirs-up-bogus-climate-change-debate-1.774639.
43. A. Pringle and D. Robbins (2022), 'From Denial to Delay: Climate Change Discourses in Ireland', *Administration*, 70, 3: 59–84; and Robbins, *Climate Change*.
44. Reuters Fact Check (2022, August), 'Fact Check-Quote from Ryanair's Chief Executive Denying Man-Made Climate Change Is from 2010; Michael O'Leary Has Since Changed His View', *Reuters*, https://www.reuters.com/article/factcheck-ryanair-climate-idUSL1N2ZU1E3.

45. G. Ni Aodha (November 2022), 'O'Leary Accuses Government of "Squandering" Climate Funds on School Bus Scheme', *BreakingNews.Ie*, https://www.breakingnews.ie/ireland/oleary-accuses-government-of-squandering-climate-funds-on-school-bus-scheme-1399458.html.
46. McNally, *Media and Low Carbon Transition*.
47. Ibid.
48. P. McGarry (2016, April), 'An Taisce Accuses RTÉ of 'False Balance' in Climate Change Debate', *The Irish Times*, https://www.irishtimes.com/news/social-affairs/an-taisce-accuses-rte-of-false-balance-in-climate-change-debate-1.2483769.
49. R. Bates (2018, January), 'Deficiencies in the IPCC's Special Report on 1.5 Degrees', The Global Warming Policy Foundation, https://www.thegwpf.org/content/uploads/2019/01/Bates-2018b.pdf.
50. P. Thorne (2019, December), 'Addressing Stated Concerns from Ray Bates around the SR1.5', Irish Climate Analysis and Research Units, http://icarus-maynooth.blogspot.com/2018/12/addressing-stated-concerns-from-ray.html.
51. P. Cooke (2021), 'UK Climate Denial Group "Slides Further into Obscurity" with Latest Appointment, Say Academics', *Desmog*, https://www.desmog.com/2021/09/28/uk-climate-denial-group-slides-further-into-obscurity-with-latest-appointment-say-academics/.
52. Irish Climate Science Forum (ICSF) (2023), Welcome to the Irish Climate Science Forum 'The Good News On Climate'. https://www.icsf.ie/. Accessed June 2023.
53. Irish Climate Science Forum (ICSF) (2023, March), *ICSF Critique of the IPCC Sixth Assessment Synthesis*. Report, Summary for Policymakers (AR6 SYR SPM), and of the Irish Government's Climate Action Plan 2023 (CAP23), https://static1.squarespace.com/static/579892791b631b681e076a21/t/641c7353f5d0fc2ff444e1f2/1679586131462/ICSF+Critique+of+AR6+SYR+%26+CAP23+-+Mar21FF.pdf
54. DeSmog. 'Irish Climate Science Forum', https://www.desmog.com/irish-climate-science-forum/#s10. Accessed June 2023.
55. Clintel Foundation (2023), 'Home Page', https://clintel.org/. Accessed June 2023.
56. Lobbying.ie. 'Regulation of Lobbying Act', https://www.lobbying.ie/about-us/legislation/regulation-of-lobbying-act/'. Accessed 11th March, 2015.
57. Ibid. Accessed 11th March, 2015.
58. A. Halligan, and D. Lawlor (2018, December), 'Ireland's Energy Transition – Challenges and Opportunities', Oireachtas Library & Research Service, https://data.oireachtas.ie/ie/oireachtas/libraryResearch/2019/2019-01-02_ireland-s-energy-transition-challenges-and-opportunities_en.pdf.
59. V. Gain (2022, November), 'Ireland Ranked Lowest in Europe for Renewable Energy Readiness', *Silicon Republic*, https:// https://www.siliconrepublic.com/innovation/ireland-renewable-energy-readiness-index-evs.
60. L. Jensen (2021, April), 'Climate Action in Ireland: Latest State of Play', European Parliament Climate Action Research and Tracking Service, Members' Research Service, https://www.europarl.europa.eu/RegData/etudes/BRIE/2021/690580/EPRS_BRI(2021)690580_EN.pdf.
61. ESB, 'Sustainability', https://esb.ie/sustainability.
62. Business Plus, 'Cowen: Sell Off ESB and Spend the Money on Green Energy', https://businessplus.ie/news/cowen-esb-green-energy/.
63. The Citizens' Assembly (2018, April), 'Third Report and Recommendations of the Citizens' Assembly How the State Can Make Ireland a Leader in Tackling Climate Change', https://2016-2018.citizensassembly.ie/en/How-the-State-can-make-Ireland-a-leader-in-tackling-climate-change/Final-Rep

ort-on-how-the-State-can-make-Ireland-a-leader-in-tackling-climate-change/ Climate-Change-Report-Final.pdf.

64. Community Power, 2023. 'Community Power | Ireland's First Community Owned Electricity Supplier', Website accessed March 20, 2023. https://communitypower.ie/.
65. K. O'Sullivan (2022, April), 'Government Inaction Stalling Scale-up of Wind Energy—McDonald', *The Irish Times*, https://www.irishtimes.com/business/energy-and-resources/government-inaction-stalling-scale-up-of-wind-energy-mcdonald-1.4851882.
66. D. Robinson (2022, August), 'More Irish Data Centers Coming Despite Concerns over Energy', *The Register*, https://www.theregister.com/2022/08/15/irish_data center_ban_reversed/
67. N. Sargent (2022, December), 'Drilling for Data', *The Journal*, https://www.thejournal.ie/drilling-for-data/news/.
68. V. Clark (2023, July), 'Data Centres Are "Key to Ireland's Economic Model", Says IDA Head', *The Irish Times*, https://www.irishtimes.com/business/2023/07/07/data-centres-are-key-to-irelands-economic-model-says-ida-head/.
69. Department of Transport (2022), 'Transport Vision in Climate Action Plan Will Transform How We Travel Over the Coming 7 Years', https://www.gov.ie/en/press-release/d3341-transport-vision-in-climate-action-plan-will-transform-how-we-travel-over-the-coming-7-years/#:~:text=Government%20will%20provide%20E2%82%AC35,the%20sustainable%20mobility%20option%20first.
70. Jensen, 'Climate Action in Ireland'.
71. G. Mattioli, C. Roberts, J. K. Steinberger, and A. Brown (2020), 'The Political Economy of Car Dependence: A Systems of Provision Approach', *Energy Research & Social Science*, 66: 101486.
72. Ibid.
73. P. Carroll, R. Rodolfo Benevenuto, and B. Caulfield (2021), 'Identifying Hotspots of Transport Disadvantage and Car Dependency in Rural Ireland', *Transport Policy*, 101: 46–56.
74. S. Murray (2021, December), 'Green Light for Galway Ring Road – But Roads Projects Have Many More Miles to Cover', *The Examiner*, https://www.irishexaminer.com/news/spotlight/arid-40763547.html; M. Hilliard (2022, October), 'It's as If They Are Trying to Alter My Daughter's Grave', *The Irish Times*, https://www.irishtimes.com/ireland/dublin/2022/10/06/its-as-if-they-are-trying-to-alter-her-grave-families-oppose-deansgrange-cycle-path-plan/.
75. N. Fulham (2022, June), 'Car Ownership In Ireland: Private Fleet Size Hits Record High', *Cartell*, https://www.cartell.ie/2022/06/car-ownership-in-ireland/.
76. Irish Examiner (2023, 22 July), 'Banning Advertising for Cars Would Be Hard to Sell', *The Irish Examiner*, https://www.irishexaminer.com/opinion/ourview/arid-41188870.html; and J. Wickham (2022, 14 December), 'Ireland's Climate Policies Will Make Us More Car Dependent Than Ever', *The Irish Times*, https://www.irishtimes.com/opinion/2022/12/14/irelands-climate-policies-will-make-us-more-car-dependent-than-ever/.
77. C. Dalby (2021, May), 'Unable to Roll-Out and Run a Network of Electric Charging Points, Council Seeks Help', *The Dublin Inquirer*, https://dublininquirer.com/2021/05/12/unable-to-roll-out-and-run-a-network-of-electric-charging-points-council-seeks-help/.
78. Department of Transport (2023, July), 'Putting Communities on Track for a New Age of Rail', https://www.gov.ie/en/press-release/df8dd-putting-communities-on-track-for-a-new-age-of-rail/.

79. Environmental Protection Agency (2023, June), 'Agriculture', https://www.epa.ie/our-services/monitoring--assessment/climate-change/ghg/agriculture/.
80. Government of Ireland (2022, November), 'Joint Committee on Environment and Climate Action Report on Biodiversity', https://data.oireachtas.ie/ie/oireachtas/committee/dail/33/joint_committee_on_environment_and_climate_action/reports/2022/2022-11-17_report-on-biodiversity-november-2022_en.pdf.
81. Jensen, 'Climate Action in Ireland'.
82. Eurostat, 'Performance of the Agricultural Sector', https://ec.europa.eu/eurostat/statistics-explained/index.php?title=Performance_of_the_agricultural_sector.
83. European Environmental Agency (2022, October), 'Green House Gas Emissions from Agriculture', https://www.eea.europa.eu/ims/greenhouse-gas-emissions-from-agriculture.
84. Environmental Protection Agency (2023, June).
85. H. Daly (2023, February), 'Beef Is Not Sustainable, So Why Are We Subsidising It for Export?' *The Irish Times*, https://www.irishtimes.com/science/2023/02/02/beef-is-not-sustainable-so-why-are-we-subsidising-it-for-export/m; J. Gibbons, (2022, July), 'In Climate Policy, Agri-Lobbyists Are the Tail Wagging the Government's Dog', *The Journal*, https://www.thejournal.ie/readme/lobbying-government-climate-5823956-Jul2022/.
86. Government of Ireland (2023); Torney D. (2020), 'The Politics of Emergency? Ireland's Response to Climate Change', *Irish Studies in International Affairs*, 31, 1: 13–26.
87. K. O'Sullivan (2020, December), 'Analyst Says Climate Plan Incompatible with Government's Environmental Policy', *The Irish Times*, https://www.irishtimes.com/news/politics/analyst-says-climate-plan-incompatible-with-government-s-environmental-policy-1.4442329.
88. Gibbons, 'In Climate Policy'.
89. E. Dillon, T. Donnellan, B. Moran, and J. Lennon (2022, September), 'Teagasc National Farm Survey 2021 Final Results', https://www.teagasc.ie/media/website/publications/2022/Teagasc-National-Farm-Survey-2021.pdf.
90. C. Buckley and T. Donnellan (2022, October), 'Teagasc National Farm Survey 2021 Sustainability Report', https://www.teagasc.ie/media/website/publications/2022/2021-Sustainability-Report.pdf.
91. F. McNulty (2021, February), 'Government Agri-Food Strategy Dealt Blow', *RTE News*, https://www.rte.ie/news/2021/0225/1199312-agriculture/.
92. F. Convery (2022, November), 'Climate Policy for Ruminant Agriculture In Ireland Blog 6 | UCD Earth Institute', https://www.ucd.ie/earth/blog/climate-policy-agriculture-ireland-blog/climatepolicyforruminantagricultureinirelandblog6/.
93. J. Gibbons (2022, July), 'Are the Government's Compromise Emissions Targets Even Achievable?' *The Irish Examiner*, https://www.irishexaminer.com/opinion/commentanalysis/arid-40930436.html.
94. Lobbying.ie (2022, 1 May–31 August), 'The Irish Farmers' Association—IFA, https://www.lobbying.ie/return/92586/the-irish-farmers-association---ifa; Lobbying.ie, 'Regulation of Lobbying Act'.
95. Lucie Adenaeuer, James Breen, Peter Witzke, et al. (2020, April), 'The Potential Impacts of an EU-Wide Agricultural Mitigation Target on the Irish Agriculture Sector', https://t-stor.teagasc.ie/handle/11019/2643.
96. C. M. Richardson, TTT Nguyen, M Abdelsayed, et al. (2021), 'Genetic Parameters for Methane Emission Traits in Australian Dairy Cows', *Journal of Dairy Science*,

104, 1: 539–549; S. Van Engelen, et al. (2018), 'Genetic Background of Methane Emission by Dutch Holstein Friesian Cows Measured with Infrared Sensors in Automatic Milking Systems', *Journal of Dairy Science*, 101, 3: 2226–2234.
97. J. Power and V. Clark (2022, November), 'EPA Criticises Low Level of Water Quality Inspections by Councils', *The Irish Times*, https://www.irishtimes.com/environment/2022/11/15/epa-criticises-low-level-of-water-quality-inspections-by-councils/; K. O'Sullivan (2021, January), 'European Commission Brings Ireland to Court over EU Water Directive "Failure"', *The Irish Times*, https://www.irishtimes.com/environment/2023/01/26/european-commission-brings-ireland-to-court-over-eu-water-directive-failure/.
98. A. Leiserowitz, J. Carman, S. Rosenthal, et al. 'Climate Change in the Irish Mind'.
99. J. Gibbons (2020, December), 'How Big Ag Is Influencing What Irish Students Learn About Climate Change', DeSmog (blog), https://www.desmog.com/2020/12/15/big-ag-influencing-irish-students-climate-change/.
100. William F. Lamb, G. Mattioli, S. Levi, et al. (2020), 'Discourses of Climate Delay', *Global Sustainability*, 3: 1–5, doi:10.1017/SUS.2020.13.
101. M. Donnelly and C. Fox (2019, September), 'Agri-Food Alliance Brings in Big Guns to Fight Negative Press', *The Irish Independent*, https://www.independent.ie/farming/agri-business/agri-food-alliance-brings-in-big-guns-to-fight-negative-press/38502886.html.
102. Daly, 'Beef Is Not Sustainable'.
103. S. Deckard (2016), 'World-Ecology and Ireland: The Neoliberal Ecological Regime', *Journal of World-Systems Research*, 22, 1: 145–176.
104. I. Curran (2022, September), 'Irish Beef and Lamb to Be Sold in Singapore Supermarkets for the First Time', *The Irish Times*, https://www.irishtimes.com/business/2022/09/05/irish-beef-and-lamb-to-be-sold-in-singapore-supermarkets-for-the-first-time/.
105. A. Leiserowitz, J. Carman, S. Rosenthal, et al. 'Climate Change in the Irish Mind'.
106. L. M. Fitzgerald (2023, February), 'What's the Impact of Large-Scale Forestry on Irish Communities?' *RTE News*, https://www.rte.ie/brainstorm/2023/0201/1352939-ireland-forestry-plantations-coillte-sitka-spruce-local-communities/.
107. National Economic and Social Council (NESC) (2023, June), 'Just Transition in Agriculture and Land Use Council Report', https://www.nesc.ie/app/uploads/2023/06/162_just_transition_in_ag_land_use-1.pdf.
108. Kelly, O., Illingworth, S., Butera, F., Dawson, V., White, P., Blaise, M., . . . & Cowman, S. (2022). 'Education in a warming world: Trends, opportunities and pitfalls for institutes of higher education,' *Frontiers in Sustainability*, 3, 920375.
109. S. Hiltner, E. Eaton, N. Healy, et al. (forthcoming), 'Fossil Fuel Industry Influence in Academia: A Research Agenda', *WIRES Wiley Interdisciplinary Reviews*.
110. Fitzgerald, 'What's the Impact'.
111. A. Leiserowitz, J. Carman, S. Rosenthal, et al. 'Climate Change in the Irish Mind'.
112. Department of the Environment, Climate and Communications (2022), 'Climate Conversations 2022 Report', https://www.gov.ie/en/consultation/0de7f-climate-conversation-climate-action-plan-2022/. Published 6 September 2023; L. Devaney, D. Torney, P. Brereton, and M. Coleman (2020), 'Deepening Public Engagement on Climate Change: Lessons from the Citizens' Assembly. Environmental Protection Agency'.

5

Climate Obstruction in Sweden

The Green Welfare State—Both Progressive and Obstructionist

KJELL VOWLES, KRISTOFFER EKBERG, AND MARTIN HULTMAN

INTRODUCTION: A MEDIOCRE ENVIRONMENTAL MIDDLE WAY

In the spring of 1972, in Stockholm, just months before the United Nations Conference on the Human Environment was held, leading experts gathered in the house of the Worker's Educational Foundation (ABF) for a conference. The Club of Rome's *Limits to Growth* report had just been released and environmental issues had recently become a subject of public debate.[1] The meteorologist Bert Bolin, Sweden's foremost climate scientist who would later become the first chair of the International Panel on Climate Change (IPCC), and then-Prime Minister Olof Palme were among the experts who had come together to discuss the theme 'Is the future possible?' Palme opened the conference by reflecting on how Sweden could find a balance between those arguing that 'Man´s ingenuity is unlimited' and those who believed 'the coming catastrophe is inevitable if not the present society is completely overthrown'.[2]

Palme contended that, between what he considered an idealist and defeatist reaction and a cornucopian illusion, there was a compromise: a planned social democratic society which limited the ill side effects of modern society while expanding the wellbeing of its citizens. It was a continuation of a form of compromise with corporatist tendencies, which in this chapter we call 'middle-way politics'. This was a type of politics that Sweden had promoted and for which it had become internationally renowned and was based firmly in the Swedish Social Democratic Party. The Social Democrats had dominated Swedish politics, holding government in the consecutive years between World War II and 1976; since then, it has been an opposition party for a total of seventeen years, until 2022. This middle way included military neutrality, a labour–industry compromise negotiated through collective agreements, and the idea of 'the people's home' (*folkhemmet*), adopted in the 1930s as a metaphor for the inclusive welfare state.[3] Palme argued that the middle way would improve society through a successful compromise between radical reform and business-as-usual, which also applied to environmental politics. In hindsight, the policy resonates with the argument of historian Kasimierz Musiał, who claimed that 'in Scandinavia there exists a certain frame of mind, a mental capacity by virtue of which a change for the better comes to be regarded as inevitable'.[4] In a similar vein, historian Melina Antonio Buns and sociologist Dominic Hinde have recently argued that 'this [Nordic environmental model] allowed for the creation of an image of a green modernity, one that not only incorporated environmental protection into welfare but made environmental protection itself the catalyst for technological innovation, political progressiveness, and economic growth'.[5]

In this chapter, we use the term 'Swedish middle way' to signify a political compromise that, simply by virtue of being Swedish, would lead to a brighter, low-carbon future.[6] What the main political parties, labour unions, and corporate associations all tacitly agreed upon was that environmental concerns and climate change were important but could be fixed incrementally with technical solutions, challenging neither economic growth nor contemporary lifestyles. It was a compromise aiming for an energy transition rather than a social transformation. At the same time, it was a compromise inherently devoid of internal coherence, a void that was filled by different actors according to their political preferences. For the Social Democrats in the 1970s, the promised future would be achieved through state-led investments, while the conservative party and corporate interests promoted mainly market mechanisms and consumer choice, especially after the neoliberal trends of the 1990s.[7] The middle way, we argue, became a hegemonic discourse that dictated from above what constituted reasonable actions to deal with climate change. It marked a kind of common-sense position between outright denial and an urgent push for transformation to a low-carbon society.

Middle-way policies managed to slowly reduce territorial greenhouse gas (GHG) emissions. They peaked in 1970, but declined over the next decade and a half as oil for household heating and industry was replaced, mainly by nuclear power and biomass, and industrial production was outsourced. Since the mid-1980s, the production of electricity has been dominated by nuclear and hydropower.[8] Since 1990, territorial GHG emissions have continued to fall (Figure 5.1), from 72 million metric tonnes of carbon dioxide equivalents (MMT CO_2e) to 44.1 MMT CO_2e, but consumption-based emissions have not declined at the same rate.

For a long time, the slow downward trend in emissions of roughly 2% per year[9] was enough to curtail the influence of those who argued that Sweden needed to do more on climate. Toward the end of the 2010s, however, it became obvious that the track record was compliant with neither the climate pledges the parliament had made when ratifying the Paris Agreement in 2016 nor the rules of the Swedish Climate Act, passed in 2017. The Swedish Climate Council, which has the mandate to evaluate whether the Swedish government is doing enough to meet the climate law goal of reaching net zero emissions in 2045, argued in its 2019 report that 'the pace [of emissions reductions] is way too slow to be in line with the climate-policy goals'.[10] A year later, professor of energy and climate change Kevin Anderson and colleagues published a study arguing that the climate law goal itself is 'less than half of what is the absolute minimum necessary to deliver on the Paris Agreement'.[11] The unchanged policy culture, weak governance of transport and consumption, and unrecognized potential of the forest as a natural carbon sink have since been highlighted as examples of the lack of transformative climate policies in Sweden.[12] The latest calculations regarding Sweden's carbon budget show that emission reductions need to increase nearly tenfold to 20% per year (from 1 January 2022) to be in line with the 1.5° threshold in the Paris Agreement, or 12% per year to contribute to limiting warming to 2°C.[13] Similarly, decoupling rates would need to quadruple by 2025 if Swedish policy is to be compliant with the Paris Agreement while still pursuing economic growth.[14]

Analyses such as these lend weight to the position of activists such as Greta Thunberg, who started her school strike in 2018 by saying that the Swedish parliament is not treating the climate crisis as a crisis. At the same time, there has been an increasingly vocal opposition to more ambitious climate policies, such as that of then-conservative opposition leader (and later prime minister) Ulf Kristersson, who said in 2019 that 'I don't believe that you can say that we have a specific time to act. I am scared of the alarmistic'.[15] Hence, the end of the 2010s marked a period when the climate compromise ultimately broke down.

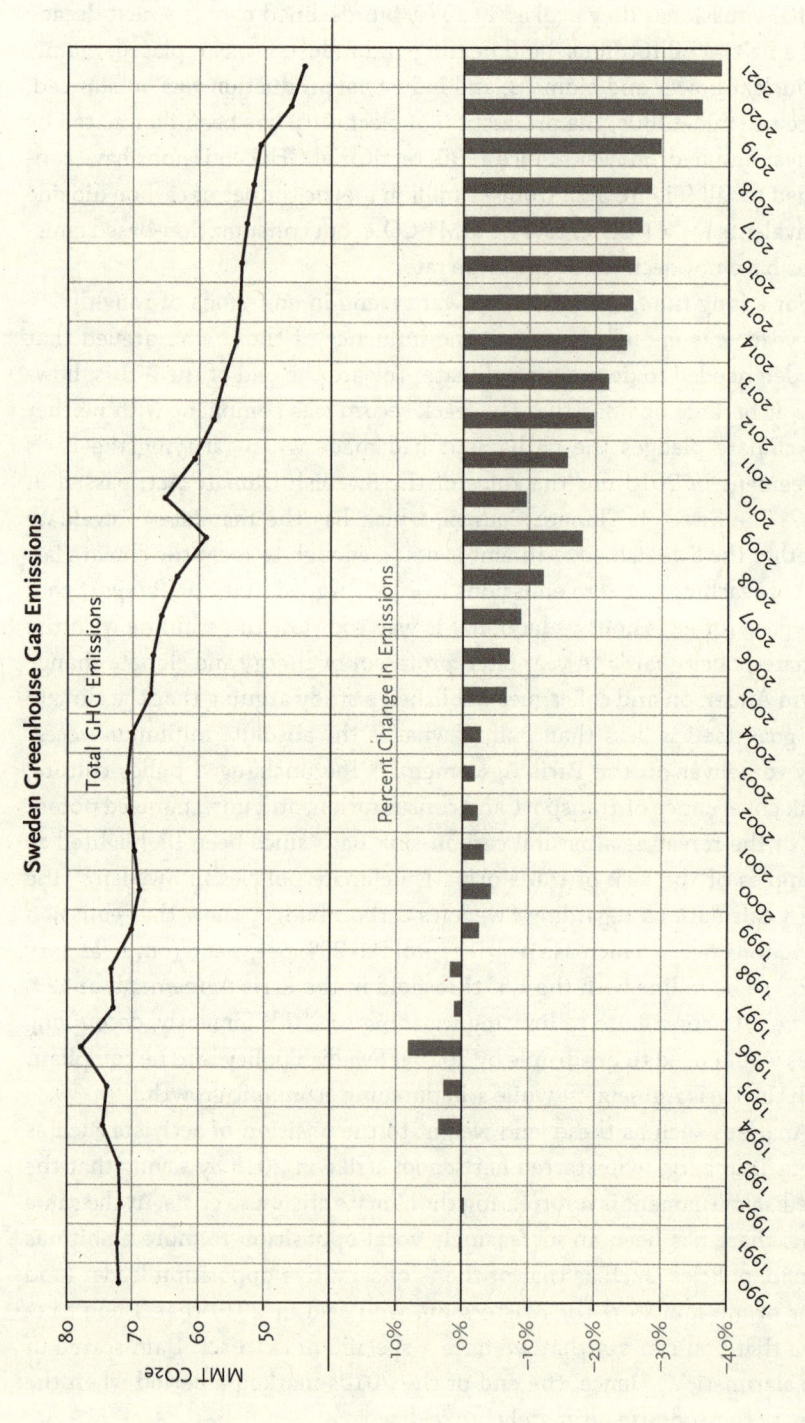

Figure 5.1 Total greenhouse gas (GHG) emissions (in MMT CO$_2$e) and percentage change in emissions in Sweden between 1990 and 2021, inclusive.

Source: Total GHG emissions based on data provided by Gütschow and Pflüger (2023) for Kyoto Six Greenhouse Gas Totals.

THE MIDDLE WAY AS SECONDARY OBSTRUCTION

This chapter conceptualizes climate obstruction using a three-part typology two of us (Ekberg and Hultman) developed with colleagues. Here, *primary obstruction* denotes the '[d]enial of the scientific evidence of human-induced climate change, and consequently, actions which undermine climate policy'. In *secondary obstruction*, '[s]cience is at least tacitly accepted but meaningful climate action is delayed because of for example ideological, economic or political reasons'. *Tertiary obstruction* denotes '[c]ultures, hierarchies and values, as well as for example infrastructures that stand in the way of necessary action'.[16] It is important to emphasize the notion of delay in secondary obstruction as actors often claim to be content with current policies as a delaying tactic to oppose additional reform.[17] This means that policies that can be seen as progressive when implemented, such as the Swedish carbon tax, can later be used to obstruct further action by arguing that Sweden has already done enough. To exemplify tertiary obstruction, this chapter discusses the roles of gender and industrial/breadwinner masculinities enacted by those (mainly men) who have gained the most from extractivist policies.[18] These tertiary obstruction identities are also part of, and shape, secondary and primary obstruction.

Using this classification system, we make two arguments. The first is that certain aspects of the Swedish middle way can be seen as secondary obstruction. By displacing actions in time and space (e.g. relocating policies from the national to the international arena and limiting the space available for socially transformative politics and more radical climate movements), it has provided the public with a comfortable sense that the problem is being addressed.[19] In this way, secondary obstruction policies have helped to create cultures of tertiary obstruction, and vice versa. To put it bluntly: Sweden's incremental emissions reductions have allowed the nation to claim to be a frontrunner by pouring a little less fuel on the burning planet compared with most other wealthy, Western industrial nations. The second argument is that whenever primary obstruction narratives have appeared, they have usually been directed toward those who have argued for climate policies more ambitious than the middle way. In this way, primary obstruction often takes the form of a countermovement.

While we acknowledge that it makes sense to talk about climate obstruction in Sweden from the late 1980s onward (the period when the IPCC was formed and carbon tax discussions began), we start our story in the 1970s for two reasons. The first is to show how the idea of Social Democratic middle-way politics was expanded to apply to the environment and later influenced climate policy. The second is because contemporary climate debates evolved from contestations of the middle way that have existed from the start, especially by certain actors close to the Swedish Employer's Confederation, Svenska Arbetsgivarföreningen (SAF, later the Confederation of Swedish Enterprise), which organizes all major businesses and industries in Sweden. Indeed, Swedish corporate interests and the high concentration of wealth and capital among a few actors stand out in comparison with many other nations.[20] This means that industry opposition to strong environmental policies is generally to be found in centralized business organizations such as SAF and its affiliated actors. It is important to note that Sweden has been an export-oriented country. Large portions of this small nation's gross domestic product (GDP) still stem from energy-intensive industries: manufacturing, mining, and forestry.[21] Therefore, especially after the 1970s, Sweden's economy has been highly dependent on international competitiveness.

From oil to nuclear: Early contestations of Swedish environmental policies

As a country without viable fossil fuel reserves, Sweden has been entirely dependent on imports. When the oil embargo of 1973 increased pressure on the Swedish energy system, the state-led response was decarbonization. Nuclear expansion had been planned since the early 1950s and was in progress. The rise in the price of oil gave the project further impetus, which in turned spurred an anti-nuclear movement. The concern over energy issues also incentivized energy-saving measures and triggered refurbishments of the existing building stock, while industries such as pulp-and-paper and forestry reduced their carbon emissions by substituting biofuels for oil.[22]

This early transition coincided with Sweden's attempt to showcase itself as a frontrunner in environmental policy leadership. In the early 1970s, Sweden and its Nordic neighbours were pushing for international agreements on transboundary pollution.[23] While anthropogenic global warming was not a top priority at that time, it had been recognized (due in part to climate scientist Bolin, then a government advisor) and utilized to promote the expansion of nuclear energy.[24] The opening paragraphs of the government's 1975 statement outlining future energy policies stated that 'according to some scientists, this [fossil fuel combustion] could lead to climatic change that in time could bring about catastrophic consequences for our way of life'.[25] Historians have pointed out that the business community in Sweden was generally not as antagonistic to environmental legislation and regulation as its US counterpart until the 1980s. Instead, corporatist structures emerged, such as the public–private research institute IVL (formerly known as the Institute for Water and Air Quality research, now the Swedish Environmental Research Institute), in which the state and industry shared knowledge and costs.[26]

Simultaneously, mobilization against the strong state had been mounting among industry and affiliated think tanks. By 1971, SAF had begun advocating direct engagement with public opinion, and, in 1978, it founded the free market-oriented think tank and publisher Timbro as part of this push.[27] Researchers have described the creation of Timbro as an 'undisguised attempt to pursue the interests of the capitalist class in opposition to the Swedish labour movement and to counter any ideas connected with socialist economic planning and the rapid expansion of the welfare state'.[28]

Proponents of nuclear energy, including the Social Democrats, argued in the 1970s that newly built and planned reactors would increase energy use and living standards while phasing out oil and its polluting emissions.[29] Industry-affiliated thinkers even pushed the sort of cornucopian narratives that then-Prime Minister Palme had brushed off as an excuse for inaction. Most prominent among these pundits were physicist Tor Ragnar Gerholm and PR firm Kreab; together with SAF and the Confederation of Swedish Industry (Industriförbundet, since 2001 part of the Confederation of Swedish Enterprise), they initiated early efforts to counter a growing environmental and anti-nuclear movement that contested the idea that economic growth and environmentalism could go hand in hand.[30] The response from economists and business-affiliated experts signified the first anti-environmental opposition and shaped the debate in the following decades.[31]

In 1980, a national referendum on the future of Swedish nuclear power was held. A narrow majority voted for a controlled phase-out, allowing nuclear power to be used until 2010. A year later, parts of the anti-nuclear movement were consolidated into the Green Party, which promoted itself as an ecological alternative to the left–right political divide. Environmental themes grew in importance among voters, and, in 1988, the party gained seats in the Swedish parliament for the first time.[32] At the same time, the environmental discussion became more pronounced within the Social Democrats, with some members of the party leaning toward the Greens' position and others remaining closely tied to the industrial unions and promoting continued economic growth and expansion of the welfare state.[33] There was joint opposition to the environmental movement from some of the unions and business actors, primarily in export-oriented industries.[34] Simultaneously throughout the 1980s, Swedish businesses promoted themselves as part of the solution to climate change, in line with the rise of sustainable development discourse internationally.[35]

Sustaining the unsustainable

As climate change became more prominent in the public debate, Swedish voices echoed some of primary obstructionist tactics and arguments prevalent in the United States. The Swedish business organizations and their allies, which in the mid-1980s had been successful in countering wage earner funds that would guarantee a degree of union ownership in companies, now turned their focus to the environment. In the 1980s, Gerholm joined forces with physicist and climate sceptic Fred Singer, becoming scientific advisor to Singer's Science and Environmental Policy Project (SEPP) as well as its transatlantic counterpart, the European Science and Environment Forum (ESEF).[36] The two organizations aimed to relativize and question the science of environmental and medical hazards such as climate change and tobacco.[37] In 1992, Gerholm was one of the authors of a SEPP report challenging the work of the IPCC.[38] Pushing scientific uncertainty to the Swedish public, Gerholm wrote in 1990 that 'We know too little about the workings of carbon in the biosphere. Nature—predominantly the oceans—seems to manage the increased amount of carbon dioxide in the atmosphere. Without facts every effort to strike international agreements is pointless'.[39] Similar arguments were made by prominent resource economist, Marian Radetzki, who, like Gerholm, had been a vocal member of the pro-nuclear camp in the 1970s.[40] As early as 1987, Radetzki portrayed climate change as a potential blessing and, in the early 1990s, authored the book *Growth and Environment* (*Tillväxt och miljö*) promoting growth as compatible with, or even a prerequisite for, environmental protection.[41] The book was published by SNS Energy, a part of SNS (Centre for Business and Policy Studies), which was led by Radetzki and founded in collaboration with the publicly owned utility company Vattenfall. Later, SNS Energy was funded by major energy companies, among them the Swedish branches of Preem and Shell.[42]

While some voices were trying to fend off regulation through pushing scientific uncertainty, or what we define as primary obstruction, other industry actors, including representatives from fossil fuel companies such as Shell, were promoting sustainability through business self-regulation and consumer citizenship, a strategy we call secondary obstruction.[43] In Sweden, such efforts were visible in reports like 'The Citizens Environmental Manifest' (*Medborgarnas miljömanifest*) part of a series called MOU, or

Medborgarnas Offentliga Utredningar, published by Timbro in collaboration with the new think tank New Welfare (*Den nya välfärden*).[44] The abbreviation mimicked SOU, the letters used for official government reports. The MOU report promoted a 'green business' model wherein consumer choice would steer production in a more sustainable direction. The lead author, Lars Bern, was an engineer who had worked with Volvo's ethanol projects in the 1970s and had been CEO of the IVL, the joint industry- and state-run environmental research institute. The promotion of consumer preferences was part of a wider shift toward depoliticization of environmental issues in the 1990s, following neoliberal government reforms and the implementation of Agenda 21 after the 1992 UN conference on environment and development in Rio de Janeiro. Using the discourse of 'sustainable development', Agenda 21 was framed as an initiative to promote citizen participation but was pushed from above. Emphasizing cooperation rather than conflict, it fit well with the idea of the Swedish middle way.

The idea of a middle way in environmental politics has continued during the era of the United Nations Framework Convention on Climate Change (UNFCC) and international climate negotiations. In 1991, the Swedish carbon tax came into effect. The origin of the proposal is unclear, but it has been argued that its implementation was the result of a compromise between business interests (evident in the mobilization described earlier) and environmental concerns during an era of tax reform.[45] In the late 1980s, the planned nuclear power phase-out, which had been decided in the 1980 referendum, became a pressing issue, at a time when nuclear power amounted to around 45% of Sweden's total electricity production. While certain actors argued that it was impossible to address climate change, maintain welfare, and shut down nuclear power, others claimed that energy efficiency and renewable energy should be the way forward.[46] The carbon tax thus became a compromise that would lower emissions while keeping the possibility of a nuclear phase-out alive. In the words of political scientists Roger Hildingsson and Åsa Knaggård: 'Although no party got exactly what they wanted, the proposal was balanced enough to prevent any stronger opposition'.[47] A reduction in the tax rate and other deductions were initially granted to heavy industry and, in 1994, were replaced by a general discount implemented by the conservative government. This arrangement allowed exceptions for heavy industry up until the implementation of the European Union's Emissions Trading System (EU ETS) and the subsequent phasing out of discounts between 2011 and 2018. The carbon tax has been raised continuously since its inception, and per capita territorial emissions from fossil fuels and industry have declined, but consumption-based emissions are still high.[48]

SWEDISH CLIMATE DEBATES IN THE UNFCCC ERA

In 1991, the Social Democrats lost the election, and a weak liberal/conservative government took power. The same year a key feature of the middle-way compromise waned as SAF withdrew their representatives from the boards of government agencies.[49] As the Swedish economy entered a crisis during the 1990s recession, environmental issues were not a priority.[50] Nonetheless, the Climate Delegation (Klimatdelegationen) was formed in 1993 and, in 1994, was given the task of producing an official government report that would guide Sweden´s position on the UNFCCC.[51] The following year Bolin, who had become the IPCC's chair, and his co-authors stated in the panel's second assessment report that 'the balance of evidence suggests a discernible human influence on global climate'.[52]

The formal obligations of the Climate Delegation and the IPCC's alarming scientific statements intensified the conflict between pro-business actors who wanted to stave off all environmental considerations and pro-market supporters who argued for consumer power rather than state-enforced regulations on business. In 1995, a public rift emerged between the green growth and market-friendly Bern and the corporate greening consultancy The Natural Step (Det naturliga steget) on one side and Gerholm and the think tank Timbro on the other. It was essentially a battle between primary and secondary obstruction. Gerholm, who refuted the IPCC consensus on climate change, got support from the political editor of SvD, a newspaper closely affiliated with SAF that has a history of giving space to contrarian voices on its opinion pages.[53] Following the conflict, Bern, who had published his *Citizens Environmental Manifesto* at Timbro, slowly adopted a more sceptical attitude toward environmental issues. As we will show, he was later key in launching a more clearly defined movement of primary obstruction in the 2000s.

In 1996, the think tank Timbro sharpened its focus on environmental issues. With Gerholm's attack on The Natural Step and the promotion and distribution of the book *The True State of the Planet*, Timbro claimed to 'describe the actual state of the world and push back prophecies of doom'.[54] According to later accounts, the environmental movement was seen as the latest iteration of an anti-intellectualism that Gerholm fiercely resisted.[55] Part of Timbro's campaign was to launch an attack on the Social Democrats, who had returned to government and were now led by Göran Persson. Persson, while no classical Keynesian, envisioned using the transition toward sustainability to take Sweden out of the early 1990s financial crisis. This plan included removing part of the heavy-industry reductions in the carbon tax. Drawing on the Social Democratic welfare project of the 1930s, Persson's idea was quickly labelled 'the green people's home' (resembling later international calls for a Green New Deal). The state-led environmental agenda, according to Timbro, was a religion, privileging nature before humans. The think tank used this argument to question both specific subsidies and local investments as well as Sweden's goals for the forthcoming COP 3 meeting in Kyoto.[56]

After the meeting, Gerholm and colleagues made a concerted effort to oppose the Kyoto Protocol, the international treaty to limit GHGs, by gathering leading contrarians in Sweden and international actors such as atmospheric physicist Richard Lindzen and Shell-funded Frits Böttcher to write the edited volume *Climate Policy after Kyoto* (*Klimatpolitik efter Kyotomötet*), published in 1998.[57] The same year, the neoliberal Atlas network proposed to Exxon that Timbro could be an important European ally in promoting market-friendly policies and engagement with environmental issues.[58] However, these efforts from Swedish businesses and other actors failed to gain political party support for primary obstruction.

After the Kyoto Protocol was signed, Swedish actors who rejected climate science and opposed mitigation policies concentrated on minimizing Swedish domestic efforts, thereby shifting from Gerholm and colleagues' strategy of primary obstruction to secondary obstruction. While few voices argued that no action should be taken, a fear of free-riding and the comparative disadvantages to the Swedish export-oriented industry were often highlighted by opponents of climate action. According to EU ETS, the cap and trade emissions trading system then being developed within the European Union, Sweden would be permitted to increase its emissions by 4%, but the Swedish government argued that the country should be an environmental frontrunner and instead proposed a target of *decreasing* emissions by that amount. During the early 2000s, the carbon tax was also raised substantially. Industry actors who voiced their opposition argued that it was important that Sweden follow the same pace as others and, perhaps even more importantly, that it was more effective to finance mitigation abroad rather than at home.[59]

A PUBLIC CLIMATE CHANGE COUNTERMOVEMENT

During the second half of the 2000s, climate change rose on the political agenda in both Sweden and internationally. Following the release of Al Gore's film 'An Inconvenient Truth' and the *Stern Review* on the economics of climate change in the autumn of 2006, along with the publication of IPCC's fourth assessment report the following year, climate change became a central political issue. One example of this trend is that the conservative prime minister and leader of the centre-right coalition, Fredrik Reinfeldt, who had hardly spoken about climate change during the 2006 election, soon thereafter began to argue that Sweden could and should be an environmental leader. In this way, Reinfeldt adopted the idea of a compromise between economic growth and environmental protection, now focused mainly on company-led innovation and consumer power rather than regulation. Once again, nuclear power was seen as the core technology that would lead to environmentally sustainable economic growth. The Liberal and Conservative parties, both part of the government coalition, and the Swedish Trade Union Confederation argued that Sweden should invest in nuclear in the name of fighting climate change.[60] A new energy plan was developed by the centre-right coalition, and, for the first time since the referendum in 1980, new reactors would be allowed to be built to replace retiring ones.

During the years when the climate issue was high on the political and media agendas, there was activity on both sides of the carbon compromise. While several of the big environmental nongovernmental organizations (NGOs), which had been institutionalized since the 1990s, were holding regular meetings with the Swedish government to discuss climate strategies, newer activist organizations were taking climate campaigning to the streets, demanding much more rapid emissions reductions than the incremental steps of middle-way politics.[61] At the same time, a more concerted and open Swedish climate change countermovement revived doubt about the science, thereby shifting their strategy from secondary to primary obstruction.

Several conservative think tanks spread arguments and materials previously distributed in the United States. For example, in 2007, the think tank Eudoxa (now defunct) translated and published the Competitive Enterprise Institute's report 'What Every European Should Know about Global Warming'.[62] The think tanks Timbro and Captus, both financed by the Confederation of Swedish Enterprise, now argued that the climate had always been changing and that the science regarding human influence was not settled.[63] In 2008, Sweden's foremost climate denialist network, the Climate Realists (formerly known as The Stockholm Initiative) had their first public event, a seminar titled 'Time for Reason/Common Sense Regarding the Climate Issue'. The seminar was organized by the PR firm Kreab, mentioned earlier. Several of the people involved in the Climate Realists network held prominent positions within media, academia, and industry. One was a former board member of the car manufacturer Volvo and president of the large industrial component manufacturer Sandvik;

another was a well-known TV presenter. They were mainly men who had enjoyed careers closely connected to the modern industrial Swedish welfare state that had expanded in the post-World War II era. Their influential positions also helped them attract attention in national media through op-eds and a series of programs on Swedish public radio.[64]

Shortly after the Climate Realists' first public seminar, its leading figures published an opinion piece titled 'Don´t Throw Money on the Climate Scam' while Bern co-authored the book *Chill Out*, arguing that human influence on the climate was negligible and increased carbon dioxide levels in the atmosphere were positive.[65] In the book, Gerholm, with whom Bern had been in a public dispute a decade earlier, was now praised for his fight 'against unscientific opinions'.[66] The network was also given media space in connection with the event labelled Climategate: representatives were asked to comment on public radio and in local media on the stolen e-mails and the state of the science as part of the run-up to the COP 15 in Copenhagen.[67] As they were based in a country lacking major fossil fuel interests, the Climate Realists' primary obstruction campaign was not solely about economics. Their counterclaims about climate science appeared to be rooted in deep-seated values connected to industrial modernity, which are in turn connected to rationality, economic growth, patriarchy, and industrial progress. In this way, the men of Climate Realists were enacting industrial/breadwinner masculinities and, through these gendered identities, upholding white, male, patriarchal privilege.[68]

Together with conservative think tanks, another network called Klimatsans (Climate Sense), and a few independent opinion makers, the Climate Realists formed an organized Swedish climate change countermovement that took shape as a response to the heightened public awareness of and increased activism around the issue during the second half of the 2000s. This Swedish countermovement contained some contrarian scientists who, like Gerholm, had international connections. For example, the retired geophysicist Nils-Axel Mörner was the former head of the paleogeophysics and geodynamics department at Stockholm University. Mörner claimed to be an expert in sea-level rise and argued against the IPCC's conclusion that climate change contributes to sea level change. In 2011, he published a cover story in the UK magazine *The Spectator* titled 'The Sea Level Scam: The Rise and Rise of a Global Scare Story', in which he insinuated that the then-president of the Maldives was not truly concerned about climate change as '[the president] has authorized the building of many large waterside hotels and 11 new airports. Or could it perhaps be that he wants to take a cut of the $30 billion fund agreed at an accord in Copenhagen for the poorest nations hit by "global warming"?' Mörner's use of ironic quotation marks—or scare-quotes—around *global warming* is noteworthy, as this tactic would later become commonplace in Swedish far-right media.[69] Mörner had been a speaker at events organized by US countermovement organizations such as the Cooler Heads Coalition and The Heartland Institute. In 2017, he conducted research paid for by a US contrarian advocacy organization, the CO2 Coalition, which he later published in journals with little or no peer review.[70]

The far right as a countermovement ally

Despite the media visibility of the Swedish climate change countermovement at the end of the 2000s, their views and arguments were not adopted by any of the seven parliamentary parties. But, in the autumn of 2010, the far-right Sweden Democrats, an anti-immigration party with roots in the neo-Nazi milieu of the late 1980s, entered the parliament. Influenced by opinion makers connected to the Climate Realists, most notably Bern, party representatives started spreading denialist arguments within the Swedish government. The Sweden Democrats thus became the political ally the countermovement needed to be heard.[71] The party saw itself as the only opposition party in parliament, claiming all the others were part of the political establishment. Hence the Sweden Democrats could use its populist, anti-establishment rhetoric in the climate debate, where it argued that middle-way politics were alarmist. In January 2013, the party's environmental spokesperson, Josef Fransson, used the well-known and thoroughly debunked 'hiatus' argument (the claim that global warming stopped in 1998) to argue that the 'apocalyptic future scenario' would not happen, something which should be good news 'unless you are one of plenty who have built a lucrative career in warning humanity about the doom of the planet'.[72] He thereby repeated the anti-environmentalist trope that concern for climate or the environment was nothing more than a thinly veiled project of 'the new class' to gain power.[73] This argument—that environmental issues were pushed by an educated middle class—was also present in Timbro publications of the late 1980s and early 1990s.[74]

During the coming years, representatives of the Sweden Democrats would mix primary and secondary obstruction, arguing both that there was no anthropogenic global warming and that Sweden's carbon emissions were too low to matter. In 2016, the party also used ironic quotations marks around 'climate' in its proposed budget bill.[75] During the 2010s, the Sweden Democrats gained support in every election. After the election of 2014, the party held the balance of power in parliament and would have become a kingmaker were it not for a parliamentary agreement between the other parties to minimize its influence. But the agreement lasted only a year, after which conservative parties started talking and negotiating with the Sweden Democrats.

Deploying a far-right media ecosystem

Aiding the Sweden Democrats in its popularity was an influential, far-right, alternative digital media ecosystem. The ecosystem consists of a plethora of news sites and video channels, often with personal and organizational ties to both the Sweden Democrats and the extreme-right party Alternative for Sweden. Just as movements and countermovements need political allies, they can also be advanced via social and partisan media, where traditional gatekeepers have been removed.[76] In the summer of 2018, Sweden and large parts of Europe experienced a record-breaking heat wave and drought, which had become more likely due to human-induced climate change.[77] Forest fires swept the country and the total area burnt, approximately 25,000 hectares, was an anomaly during the era of the modern fire defence.[78] The country was also on the verge of a national election scheduled for early September. Three weeks before the election, a fifteen-year-old girl named Greta Thunberg sat down outside the parliament building bearing a sign saying: 'School strike for climate'. In line with scientific assessments, she demanded policies that went beyond the middle way.

During the same period, SwebbTV, a nationalist and conspiracist online video channel, aired an interview with Bern under the headline 'The Environmental Movement's Scare-Mongering'. The channel, with personal ties to both the Sweden Democrats and the extreme-right Alternative for Sweden, had been launched as a YouTube channel three years earlier but was later expelled from the platform because of broadcasting disinformation about the COVID virus. From the start, the channel was focused mainly on immigration, but, after the summer of 2018, climate change became a prominent issue. Bern, who in 1990 had said that '[t]he emerging environmental commitment is really nothing more than another step in the long civilizational process of humanity', was now enlisted as the channel's political and scientific commentator. As such, he continuously spread doubt about climate science. During 2018–2019, the channel hosted several prominent members of the Swedish climate change countermovement, such as the former president of Sandvik, as well as the contrarian scientist Mörner. Indeed, many prominent industry leaders and academics with highly distinguished careers have been guests on a far right, conspiracist video channel.[79] One recurring guest was a former employee of the Ministry of Enterprise and Innovation, Elsa Widding. As a writer for the Climate Realist blog and through videos on her own YouTube channel, Widding had just emerged as a leading voice in the climate change countermovement, standing out as the female exception in a culture of older males. On these shows, SwebbTV often discussed the science of climate change, using contrarian graphs which used cherry-picked or obsolete data to deny the trend, attribution, and negative consequences of climate change.[80]

Through these interviews, SwebbTV became a nexus of primary obstruction, connecting the organized Swedish climate change countermovement with the far-right alternative media ecosystem. Other news sites, such as Samhällsnytt and Fria Tider, which at the time reached roughly 10% of the Swedish online population, started reporting widely on climate change in the autumn of 2018, and, during 2019, the issue became prominent across such media. But compared with SwebbTV, these news sites didn't discuss the science per se; instead, they occasionally referenced SwebbTV or some international contrarian source to create an anti-establishment discourse in which it was tacit knowledge that climate change was a hoax. By scare-quoting climate and related words, they signalled to the reader that this was a non-existent problem and thereby attacked anyone who was talking about it—usually those they marked as belonging to a globalist elite.

The fact that primary obstruction, led by the far right, has lately become conspicuous in Sweden should not be taken as a sign that secondary obstruction has disappeared. Rather, these forms are often advanced simultaneously, sometimes by the same actors. Industry actors who publicly accept the reality of climate change have continued to use secondary obstruction to promote emissions reductions abroad, which they deem 'efficient climate policy'. This argument has also been the most important among the conservative and far-right think tanks and parties' critiques against domestic climate policy.[81] Similarly, the research institute Ratio, closely affiliated to Timbro and the Confederation of Swedish Enterprise, has disputed a growing interest in planning and entrepreneurial state thinking.[82] In so doing, these actors continue to promote a neoliberal critique that in recent years has included attacks on the state-led HYBRIT-project, an ambitious plan to use (vast amounts of) electricity to produce steel using hydrogen instead of coal. Another argument that has appeared since the European Union adopted its new climate framework, 'Fit for 55', is that national climate policy is superfluous, ineffective, and expensive and that Sweden will be doings its part simply by remaining a member of the European Union.[83]

FAR-RIGHT NATIONALISM DEFENDS SWEDISH INDUSTRIAL MODERNITY

Discourses of the far right are based on a nostalgic longing for a lost patriarchal and homogenous national community, a community now perceived as decaying and further threatened by immigration, feminism, and a corrupt elite. This idealized community is also an industrially prosperous one. The reactionary nostalgia of the Sweden Democrats is for a world before globalization, a world built around nation-states, nuclear families, and industrial capitalism where the male breadwinner benefitted and gained security through his work in successful industries.[84] It was also a supposedly rational world where scientific progress and innovation improved living conditions and material welfare for everyone in the Swedish 'people's home', but where racism and discriminatory policies ensured that that home was ethnically homogenous.[85]

The far right's claim to rationality also ties into an argument that they stand for the common-sense position while everyone else is alarmist. The Sweden Democrats often claim to be truth tellers, arguing that other parties were wilfully blind to reality until their recent adoption of more restrictive immigration policies. The party leader, Jimmie Åkesson, has thus compared immigration with climate, using the familiar trope of climate change as a religion: '[the climate change debate] is very reminiscent of how the immigration debate sounded some years ago. You can't question or lay down different perspective because then you are called a climate denier'.[86]

The Swedish climate change countermovement and the Swedish far right meet in defending the values of patriarchal, industrial capitalism. Their national industrial project is based on domination of nature and extraction of resources. It is also an industrial/breadwinner masculinities project, in which those men who have earned the most from the burning of fossil fuels are the ones holding prominent positions within industry and working in high-emitting, high-resource-use sectors.[87] In far-right media discourses, masculine rationality is often pitted against feminine emotionality, with the latter now portrayed as destroying the nation. One example comes from the digital media site Samhällsnytt, which claimed that the social democratic government was leading a destructive cultural process by 'moving from a rational patriarchy to an emotional feminism'.[88] Feminine, irrational climate 'hysteria' is seen as a threat to an industrial world built by generations of hard-working, white, Swedish men.[89] Similar sentiments have also been found in Norway and several other countries.[90]

The collaboration between the far right and the climate change countermovement has recently resulted in the Swedish government's backtracking on its climate policies. In the 2022 national election, the Sweden Democrats received 20.5% of the vote and became the biggest party in the winning nationalist-conservative block. While not part of the government, the Sweden Democrats hold direct influence over its policies, which include reducing taxes on petrol, lowering standards on the amount of biofuels required in diesel fuel, dismantling subsidies for electric vehicles, withdrawing state support for connecting offshore wind power to the electricity grid, and initiating a massive investment in building new nuclear reactors. The Swedish Climate Council has already asserted that the new policies will increase Sweden's emissions and make it even harder to meet existing climate goals.[91] The conservative minister for finance, Elisabeth Svantesson, however, has shrugged off the consequences of not meeting these targets, saying that 'if we don't do it, we don't do it'.[92]

Probably the most explicit example of how the far right and the climate change countermovement have joined hands is the election, in September 2022, of Elsa Widding as an MP for the Sweden Democrats, making her one of the party's main voices in the climate and energy debate.[93] Apart from writing on the Climate Realists blog and frequently appearing on SwebbTV, Widding has also been named a member of the Norwegian Climate Realists' scientific board.[94] She was also a signatory to the international CLINTEL declaration, which stated that there is no climate emergency and that 'The Little Ice Age ended as recently as 1850. Therefore, it is no surprise that we now are experiencing a period of warming'.[95] In her 2022 book *Common Sense about Energy and Climate*, Widding argued that common sense in the Swedish climate debate had disappeared because the politicians and media often talked about the problem as a crisis, a view she strongly opposed. In Widding, the Swedish climate change countermovement gained its first member of parliament.[96]

Several surveys have shown there is an electoral base for far-right obstructionist policies in Sweden.[97] A recent study showed that 6% of Sweden's population doubt that climate change is anthropogenic, a majority of whom sympathize with the Sweden Democrats.[98] There is also a clear trend of a growing left–right divide in climate change public opinion, with voters to the right recently becoming less concerned.[99] This is another sign that Swedish climate politics has moved further away from the compromise of incremental change and middle-way politics, instead becoming fiercely contested.

CONCLUSION: NATIONAL PROTECTIONISM AND THE ALLURE OF THE MIDDLE WAY

If we are to take seriously the question of why countries are failing on climate mitigation, we need to understand the different forms of obstruction.[100] In this chapter, we have seen how the Swedish 'middle way' of 'balancing' environmental and economic concerns has led to incremental reductions of territorial GHG emissions that were more substantial than those of many other countries of the Global North, but far short of Sweden's commitments in the Paris Agreement. Therefore, we argue, these climate policies can in part be seen as secondary obstruction: through being concerned primarily with technical solutions and economic growth, they have limited the space available for discussion and implementation of more ambitious policies.

While the strategies and tactics of those opposing climate and environmental policies in Sweden have shifted over time, there is also continuity. From the 1970s onward obstruction has been expressed and organized as a countermovement in response to environmentalists' demand for more radical and transformative policies. Central throughout the period was the countermovement's positioning of neoclassical and neoliberal economic doctrine and industrial practice as the main sources of knowledge and containers of reasonable action. During the 2010s, the far right became the political ally the Swedish climate countermovement needed to be heard, and the latter has, to a large degree, seamlessly merged with the former. The countermovement did so first by giving advice to far-right politicians and later through appearing in far-right media. Today, the far right is the driving force of primary obstruction in Sweden, which has led the current government to leave behind the middle way of incremental, but insufficient, reductions in carbon emissions. Instead, the government has adopted policies that will increase Sweden's territorial carbon emissions for the first time in two decades (if we exclude the rebound years after the financial crises and the pandemic). In the crosshairs of both conservatives and the far right are not only policies but also activists, often deemed 'alarmists'. Efforts to portray climate activist groups as extremists have sought to conflate the position of business interests and the protection of domestic exporting companies with common sense. The concept of common sense plays to the allure of a middle way, prominent in Swedish cultural self-understanding. But as we have shown, it has more recently been used to defend a primary obstructionist position.

Moving forward, it is important to further study secondary obstruction. We have argued that certain aspects of Sweden's climate policies, which have often been hailed as progressive, can also be seen as obstructionist in that discussion of the deep societal transformation needed to reach the Paris Agreement targets remains strictly off limits. This reality suggests that consensus in climate politics can sometimes be problematic, as it usually means that incumbent interests gaining the most from the status quo remain unchallenged. More research is needed to distinguish how, why, and when certain policies can become both progressive and obstructionist, and what can be done to overcome such obstruction. Studies examining certain industries in Sweden (e.g. forestry, pulp-and-paper, and the auto industry) could be helpful to determine which sectors have been most active in pushing obstructionist perspectives.

Finally, further analysis is needed to understand how to facilitate policies that go beyond ecomodernism, green growth, and technological change. The current state of the planet requires Sweden to reduce its emissions by more than 12% per year to deliver on its commitment to a 2° warmer future, as stated in the Paris Agreement, and by 20% per year if the country intends to help limit warming to 1.5°C.[101] The middle-way politics of the Swedish welfare state has mainly served to obstruct discussion of such levels of mitigation.

The few times this obstruction has been overcome and more concerned climate voices have been heard, it was often regarding the science on mitigation and the need for rapid carbon phaseout. The most obvious example is the activist campaign of Greta Thunberg who, during the early days of her school strike, argued that it made no sense for her to be in school because 'Facts don't matter any more, politicians aren't listening to the scientists, so why should I learn?'[102] Another example is the Stay on the Ground movement, which helped create a widespread debate about the climate impact of aviation.[103] One thing that both Thunberg and Stay on the Ground had in common was the appeal to morality and to adjusting lifestyles accordingly. Leading by example, they managed to break through the noise of obstruction.

NOTES

1. David Larsson Heidenblad (2021), *Den Gröna Vändningen: En Ny Kunskapshistoria Om Miljöfrågornas Genombrott under Efterkrigstiden*, Checkpoint. Lund: Nordic Academic Press.
2. 'Är Framtiden Möjlig? Några Vetenskapliga Förutsägelser', Konferens i Z-Salen, ABF-Huset, Audio recording, Rullband (Stockholm, 1972), Swedish Movement's Archives and Library. https://smdb.kb.se/catalog/id/002585615.

3. This has later become a national trope for both progressives and conservatives to signal a secure and solidaric society; see Jenny Andersson (2009, September), 'Nordic Nostalgia and Nordic Light: The Swedish Model as Utopia 1930–2007', *Scandinavian Journal of History*, 34, 3: 229–245, doi:10.1080/03468750903134699.
4. Kazimierz Musiał (2002), 'Roots of the Scandinavian Model: Images of Progress in the Era of Modernisation', Die Kulturelle Konstruktion von Gemeinschaften Im Modernisierungsprozess, Band 8. Nomos Verl, as quoted in Andersson, 'Nordic Nostalgia and Nordic Light', 10.
5. Melina Antonia Buns and Dominic Hinde (2023), 'Green States in a Dirty World: 1975 and the Performance of Nordic Green Modern'. In: Fredrik Norén, Emil Stjernholm, and C. Claire Thomson (eds.), *Nordic Media Histories of Propaganda and Persuasion*, p. 244. Cham: Springer International Publishing, doi:10.1007/978-3-031-05171-5; Erland Mårald and Christel Nordlund (2020, November) even argue that this idea of green modernity stretches all the way back to the beginning of the twentieth century in 'Modern Nature for a Modern Nation: An Intellectual History of Environmental Dissonances in the Swedish Welfare State', *Environment and History*, 26, 4: 495–520, doi:10.3197/096734019X15463432086883.
6. As a term, the Swedish middle way was first used by the American journalist Marquis Childs in the 1930s to signify a compromise between Soviet-style communism and US-style capitalism. Since then the term has often been used as a synonym to concepts such as 'the Swedish model', 'Swedish exceptionalism', or even the 'Swedish welfare miracle' to portray, in essence, a modern, progressive, green, welfare state. There is, however, an ongoing scholarly debate about the exact meaning—and meaningfulness—of all of those terms.
7. The two ecomodern discourses 'industrial fatalism' and 'green Keynesianism', which Jonas Anselm and Martin Hultman (2014) identify as dominant in the years 2006–2009, can both be seen as part of this middle way. See *Discourses of Global Climate Change: Apocalyptic Framing and Political Antagonisms*.Abingdon, Oxon: Routledge, doi:10.4324/9781315769998.
8. Energimyndigheten (2022), *Energiläget 2022 – En översikt*, ET 2022:02. Stockholm, Energimyndigheten.
9. Klimatpolitiska rådet, 'Årsrapport 2023' (Stockholm: Klimatpolitiska rådet, 2023).
10. Klimatpolitiska rådet, 'Årsrapport 2019' (Stockholm: Klimatpolitiska rådet, 2019), 6.
11. Kevin Anderson, John F. Broderick, and Isak Stoddard (2020, May), 'A Factor of Two: How the Mitigation Plans of "Climate Progressive" Nations Fall Far Short of Paris-Compliant Pathways', *Climate Policy*, 12, doi:10.1080/14693062.2020.1728209. The analysis is not based on the national determined contribution (NDC), where Sweden is following the European Union's NDC. Instead it is based on the commitment in the Paris Agreement to limit global warming to well below 2°C based on common but differentiated responsibilities, but acknowledging feasibility when dividing the global carbon budget.
12. Göran Sundqvist (2021), *Vem Bryr Sig? Om Klimatforskning Och Klimatpolitik*. Göteborg: Daidalos.
13. Isak Stoddard and Kevin Anderson (2022), *A New Set of Paris Compliant CO2-Budgets for Sweden*. Carbon Budget Briefing Note 1. Uppsala, Sweden: CEMUS

14. Jefim Vogel and Jason Hickel (2023, September), 'Is Green Growth Happening? An Empirical Analysis of Achieved versus Paris-Compliant CO2–GDP Decoupling in High-Income Countries', *The Lancet Planetary Health*, 7, 9: e759–e769, doi:10.1016/S2542-5196(23)00174-2.
15. Filippa Rogvall (2019, 17 November), 'Ulf Kristersson ställs till svars om klimatet: Jag är rädd för alarmisterna', https://www.expressen.se/nyheter/qs/klimat/alarmisterna-sager-at-oss-att-sluta-leva/.
16. Kristoffer Ekberg, Bernhard Forchtner, and Martin Hultman (2023), *Climate Obstruction: How Denial, Delay and Inaction Are Heating the Planet*, 1st ed. New York: Routledge, p. 12.
17. Secondary obstruction has been recognized by scholars from different fields using the terminology of for example post-politics, ecomodern utopias, sustained unsustainability and discourses of delay. See Erik wyngedouw, Erik (2010), 'Apocalypse Forever?,' *Theory, Culture & Society*, 27, 2–3: 213–232; Martin Hultman and Christer Nordlund (2013), 'Energizing Technology: Expectations of Fuel Cells and the Hydrogen Economy, 1990–2005', *History and Technology*, 29, 1: 33–53; I. Blühdorn (2013), 'The Governance of Unsustainability: Ecology and Democracy after the Post-Democratic Turn', *Environmental Politics*, 22, 1: 16–36, doi:10.1080/09644016.2013.755005
18. Martin Hultman and Paul M. Pulé (2018), *Ecological Masculinities: Theoretical Foundations and Practical Guidance*, Routledge Studies in Gender and Environments. New York: Routledge.
19. Regarding displacement, see Kristoffer Ekberg and Martin Hultman (2021), 'A Question of Utter Importance: The Early History of Climate Change and Energy Policy in Sweden, 1974–1983', *Environment and History*, doi:10.3197/096734021X16245313030028; regarding comforting the public, see the affirmation techniques discussed by Nathan Young and Aline Coutinho (2013, May), 'Government, Anti-Reflexivity, and the Construction of Public Ignorance about Climate Change: Australia and Canada Compared', *Global Environmental Politics*, 13, 2: 89–108, doi:10.1162/GLEP_a_00168.
20. Majsa Allelin, Markus Kallifatides, Stefan Sjöberg, and Viktor Skyrman (2018). *Ägande- och förmögenhetsstrukturen och dess förändring sedan 1980*. Katalys rapport, n 55.
21. Statistics Sweden, 'Swedish Exports', https://www.scb.se/hitta-statistik/sverige-i-siffror/samhallets-ekonomi/sveriges-export/ Accessed 26 June 2023.
22. Ann-Kristin Bergquist and Kristina Söderholm (2014), 'Industry Strategies for Energy Transition in the Wake of the Oil Crisis', *Business and Economic History On-Line*, 12.. We note that switching oil to biofuel reduces official carbon emissions but that there is an ongoing debate about the actual climate impact of biofuels due to the time lag between releasing carbon when burning biofuel and sequestering it in growing new crops or trees.
23. Eric Paglia (2021, January), 'The Swedish Initiative and the 1972 Stockholm Conference: The Decisive Role of Science Diplomacy in the Emergence of Global Environmental Governance', *Humanities and Social Sciences Communications*, 8, 1: 2, doi:10.1057/s41599-020-00681-x; Melina Antonia Buns (2020), 'Green Internationalists: Nordic Environmental Cooperation, 1967–1988', PhD Dissertation, University of Oslo.
24. Ekberg and Hultman, 'A Question of Utter Importance'.

25. Riksdagsförvaltningen, 'Regeringens proposition om enerighushållningen m.m. Proposition 1975:30', https://www.riksdagen.se/sv/dokument-lagar/dokument/proposition/regeringens-proposition-om-enerighushallningen-m_FY0330. Accessed 8 February 2023.
26. Ann-Kristin Bergquist and Kristina Söderholm (2017), 'Business and Green Knowledge Production in Sweden 1960s–1980s', *SSRN Electronic Journal*, doi:10.2139/ssrn.3086774.
27. Linderström, 'Industrimoderniteten Och Miljöfrågans Utmaningar: En Analys Av LO, SAF, Industriförbundet Och Miljöpolitiken 1965–2000'; Rikard Westerberg (2020), *Socialists at the Gate: Swedish Business and the Defense of Free Enterprise, 1940–1985*. Stockholm: Stockholm School of Economics; Which shows similar developments as in the United States; Melissa Aronczyk and Maria I. Espinoza (2022), *A Strategic Nature: Public Relations and the Politics of American Environmentalism*. New York: Oxford University Press.
28. Sigurd Allern and Ester Pollack (2020), 'The Role of Think Tanks in the Swedish Political Landscape', *Scandinavian Political Studies*, 43, 3, https://onlinelibrary.wiley.com/doi/abs/10.1111/1467-9477.12180. Inspiration for the creation of came from actors such as Mobil Oil, Heritage Foundation, Institute of Economic Affairs and Hudson institute. See Kristina Boréus (1994), *Högervåg: nyliberalismen och kampen om språket i svensk debatt 1969–1989*, Stockholm Studies in Politics 51. Stockholm: Tiden; Mats Svegfors, Johan Hjertqvist, and Gunnar Fröroth, eds. (1998), *Timbro.20.Nu: Tre Inlägg Om En 20-Årig Tankesmedjas Insatser Och Uppgifter*. Stockholm: Timbro.
29. Jonas Anshelm (2010, February), 'Among Demons and Wizards: The Nuclear Energy Discourse in Sweden and the Re-Enchantment of the World', *Bulletin of Science, Technology & Society*, 30, 1: 43–53, doi:10.1177/0270467609355054.
30. Such examples could be found in, for example, Tor Ragnar Gerholm's publications. See *Futurum Exaktum: Den Tekniska Utmaningen* (Stockholm: Aldus/Bonnier, 1972); and *What about the Future? [En Bättre Framtid]* (Stockholm: Kreab, 1984), partly funded by Swedish industry.
31. Linderström, 'Industrimoderniteten Och Miljöfrågans Utmaningar: En Analys Av LO, SAF, Industriförbundet Och Miljöpolitiken 1965–2000', 137.
32. Martin Hultman, Ann-Sofie Kall, and Jonas Anshelm (2021), *Att ställa frågan – att våga omställning: Birgitta Hambraeus och Birgitta Dahl i den svenska energi- och miljöpolitiken 1971-1991*, Pandoraserien 29. Lund: Arkiv förlag.
33. Ibid.
34. Linderström, 'Industrimoderniteten Och Miljöfrågans Utmaningar: En Analys Av LO, SAF, Industriförbundet Och Miljöpolitiken 1965–2000'.
35. Ann-Kristin Bergquist and Thomas David (2022), 'Collaboration Between the International Business and United Nations in Shaping Global Environmental Governance', *Les Cahiers de l'IEP*, 80.
36. 'Tor Ragnar Gerholm – SourceWatch', https://www.sourcewatch.org/index.php?title=Tor_Ragnar_Gerholm. Accessed 9 February 2023,
37. Naomi Oreskes and Erik M. Conway (2010), *Merchants of Doubt: How a Handful of Scientists Obscured the Truth on Issues from Tobacco Smoke to Global Warming*. London: Bloomsbury, pp. 129–130.
38. Science and Environmental Policy Project (1992, 3 June), 'International Scientific Panel Completes Analysis of IPCC Reports: Finds No Support for

Greenhouse Warming or Global Climate Treaty', Atlas B246, F07 Global Warming. Hoover Institution.
39. As quoted in Eric Dyring (1990, 31 May), 'Norden Bra Plats För Atomsopor', *Dagens Nyheter*. Thanks to Supermiljöbloggen for pointing us to this and the statement by Radetzki in 1987.
40. Eva Friman (2002), *No Limits: The 20th Century Discourse of Economic Growth*. Umeå: Umeå University.
41. Marian Radetzki (1987, 4 March), 'Växthuseffekten Kan Bli En Välsignelse', *Dagens Nyheter*; Marian Radetzki (1990), *Tillväxt Och Miljö*. Stockholm: SNS.
42. 'SNS Energy'(n.d.), https://web.archive.org/web/20000301062706/http://www.sns.se/. Accessed 20 June 2023.
43. Bergquist and David, 'Collaboration Between the International Business and United Nations'. A Swedish example of such efforts to portray industry as leading the way was the fair and conference Ecology 89 initiated by Pehr G. Gyllenhammar in Gothenburg, 1989.
44. See Kristoffer Ekberg and Victor Pressfeldt (2022, November), 'A Road to Denial: Climate Change and Neoliberal Thought in Sweden, 1988–2000', *Contemporary European History*, 31, 4: 627–644, doi:10.1017/S096077732200025X.
45. Roger Hildingsson and Åsa Knaggård (2022), 'The Swedish Carbon Tax'. In: Caroline de la Porte, et al. (eds.), *Successful Public Policy in the Nordic Countries: Cases, Lessons, Challenges*, 1st ed., p. 250. Oxford: Oxford University Press, doi:10.1093/oso/9780192856296.001.0001.
46. Magnus Linderström (2001), *Industrimoderniteten Och Miljöfrågans Utmaningar: En Analys Av Lo, Saf, Industriförbundet Och Miljöpolitiken 1965–2000*. Linköping: Linköping University.
47. Hildingsson and Knaggård, 'The Swedish Carbon Tax', p. 250.
48. Ibid.
49. Sigurd Allern and Ester Pollack (2020), 'The Role of Think Tanks in the Swedish Political Landscape', *Scandinavian Political Studies*, 43, 3: 145–69.
50. Åsa Knaggård (2009), *Vetenskaplig Osäkerhet i Policyprocessen: En Studie Av Svensk Klimatpolitik*, Lund Political Studies 156. Lund: University Statsvetenskapliga Inst. p. 191.
51. The report was published as Klimatdelegationen (1995), *Jordens Klimat Förändras: En Analys Av Hotbild Och Globala Åtgärdsstrategier*. Stockholm: Fritze.
52. IPCC (1995), 'IPCC Second Assessment – Climate Change 1995', p. 22, https://www.ipcc.ch/site/assets/uploads/2018/05/2nd-assessment-en-1.pdf.
53. See Ekberg and Pressfeldt, 'A Road to Denial'.
54. Dick Erixon (1996), 'Timbro Satsar På Miljön', *Timbro Idag* (Stockholm), 6: 1. The publication was edited by Ronald Bailey while he was a Warren T. Brookes Journalism Fellow at Competative Enterprise Institute. Thank you, Martin Arnsten, for making us aware of this document.
55. Svegfors, Hjertqvist, and Fröroth, *Timbro.20.Nu: Tre Inlägg Om En 20-Årig Tankesmedjas Insatser Och Uppgifter*.
56. Stefan Kraft (1997), *Människoartens Herrar: En Kritisk Granskning Av Statsminister Göran Perssons Miljötal Den 11 April 1997*. Stockholm: Timbro.
57. Tor Ragnar Gerholm, ed. (1998), *Klimatpolitik Efter Kyotomötet*. Stockholm: SNS (Studieförb. Näringsliv och samhälle. The book included a summarizing chapter by Bert Bolin which in general questioned the conclusions of the other authors.

58. 'PROPOSAL FOR FUNDING, Exxon Corporation, Dallas, Texas January 1998 ENERGY AND THE ENVIRONMENT Market-Based Solutions' page 4, Hoover's Atlas collection, Box 77, Folder 1 'Exxon 1998'.
59. Knaggård, *Vetenskaplig Osäkerhet i Policyprocessen*, 228; Ekberg and Pressfeldt, 'A Road to Denial', 642.
60. Anshelm and Hultman, *Discourses of Global Climate Change*. Abingdon, Oxon: Routledge.
61. Ibid.; Håkan Thörn and Sebastian Svenberg (2017), 'The Swedish Environmental Movement'. In: Carl Cassegård et al. (eds.), *Climate Action in a Globalizing World* 1st ed., Routledge Research in Environmental Politics; 27, pp. 193–216. New York: Routledge, doi:10.4324/9781315618975-11.
62. Iain Murray (2007), 'Vad Varje Europé Bör Veta Om Global Uppvärmning', Eudoxa, https://web.archive.org/web/20070110100915/http:/www.eudoxa.se/content/archives/klimatet.pdf.
63. Torbjörn Carlbom (2006 November 29), 'Skeptikerna rasar', *Veckans Affärer*.
64. Anshelm and Hultman, *Discourses of Global Climate Change*.
65. 'Om Oss – Stockholmsinitiativet – Klimatupplysningen', Klimatupplysningen, https://web.archive.org/web/20130928121556/http://www.klimatupplysningen.se/om-oss/. Accessed 9 February 2023.
66. Lars Bern and Maggie Thauersköld (2009), *Chill-out: Sanningen Om Klimatbubblan*. Stockholm: Kalla kulor, p. 232.
67. Johan Sievers (2009, 8 December), 'FN Sprider Myten Om Klimathotet', *Norrköpings Tidningar*.
68. Jonas Anshelm and Martin Hultman (2014, April), 'A Green Fatwā? Climate Change as a Threat to the Masculinity of Industrial Modernity', *NORMA* 9, 2: 84–96, doi:10.1080/18902138.2014.908627.
69. Nils-Axel Mörner (2011, 1 December), 'Rising Credulty', *The Spectator*.
70. Grahamn Readfearn (2018, 18 January), 'Climate Denial Group With Trump Admin Ties Is Funding Sea Level Research in Questionable Journals', DeSmog, https://www.desmogblog.com/2018/01/18/climate-denial-co2-coalition-trump-morner-funding-sea-level-research-dodgy-journals.
71. For more on countermovements and political allies, see Suzanne Staggenborg and David S. Meyer (2022), 'Understanding Countermovements'. In: D. B. Tindall, Mark C. J. Stoddart, and Riley E. Dunlap (eds.), *Handbook of Anti-Environmentalism*, pp. 23–42. Cheltenham, UK: Edward Elgar Publishing; Ekberg and Pressfeldt, 'A Road to Denial'; David Baas (2016, 19 October), 'SD-Politik Styrs Dolt Av Klimatförnekare', *Expressen*, https://www.expressen.se/nyheter/sd-politik-styrs-dolt-av-klimatfornekare/.
72. 'Riksdagens Protokoll 2012/13:59' (Riksdagen, January 2013), 9.
73. Alex John Boynton (2015), 'Confronting the Environmental Crisis?: Anti-Environmentalism and the Transformation of Conservative Thought in the 1970s', PhD Thesis, University of Kansas, p. 140.
74. Kurt Wickman and Viveka Wickman (1989), *Det Gröna: Varning För Miljölarmen!* Stockholm: Timbro.
75. Martin Hultman, Anna Björk, and Tanya Viinikka, 'Far-Right and Climate Change Denial. Denouncing Environmental Challenges via Anti-Establishment Rhetoric, Marketing of Doubts, Industrial/Breadwinner Masculinities, Enactments and Ethno-Nationalism'. In: Bernhard Forchtner (ed.), *The Far Right and the Environment*, pp. 122–136. Abingdon, UK: Routledge; Sverigedemokraterna, 'Budgetpropositionen för 2017', 2016.

76. Suzanne Staggenborg and David S. Meyer, 'Understanding Countermovements'.
77. Renate Anna Irma Wilcke, et al. (2020, December), 'The Extremely Warm Summer of 2018 in Sweden – Set in a Historical Context', *Earth System Dynamics* 11,4: 1107–1121, doi:10.5194/esd-11-1107-2020.
78. Johan Sjöström and Anders Granström (2020), 'Skogsbränder Och Gräsbränder i Sverige – Trender Och Mönster under Senare Decennier' (Myndigheten för samhällsskydd och beredskap).
79. An example of the conspiracist nature of SwedbbTV is that one of its most watched programmes concerns a theory that Olof Palme, who was killed in 1986, is alive; see SwebbTV (2020, 24 December), 'Våra Populäraste Program Del 1 – Om Palmemordet', SwebbTV, https://swebbtv.se/w/1JJyCRoJwJ2rXWF ESBbWSj.
80. Kjell Vowles (2023), 'Talking Heads and Contrarian Graphs: Televising the Swedish Far Right's Climate Change'. In: Bernhard Forchtner (ed.), *Visualising Far-Right Environments Communication and the Politics of Nature*, pp. 253–273. Manchester: Manchester University Press; Birgit Schneider, Thomas Nocke, and Georg Feulner (2014), 'Twist and Shout: Images and Graphs in Skeptical Climate Media' In: Birgit Schneider and Thomas Nocke (eds.), *Image Politics of Climate Change*, pp. 153–186. Berlin: Verlag, doi:10.14361/transcript.9783839426104.153; 'KÄLLA' (n.d.); Painter et al. 2023.
81. Ellen Gustafsson (2022), 'Effektiv Klimatpolitik? En Granskning Av Partiernas Klimatpolitik 2022'. Timbro, https://timbro.se/app/uploads/2022/08/effektiv-klimatpolitik.pdf; Ellen Gustafsson (2018), *Effektiv Klimatpolitik? En Granskning Av Partiernas Förslag* Stockholm: Timbro; Martin Kinnunen, Mats Nordberg, Runar Filper, Staffan Eklöf, Yasmine Eriksson (2018), *Motion 2018/19:2820 av Martin Kinnunen m. fl. (SD) En effektiv klimatpolitik*. Swedish Parliamentary Records (https://www.riksdagen.se/sv/dokument-och-lagar/dokument/motion/en-effektiv-klimatpolitik_h6022820/).
82. Karl Wennberg and Christian Sandström, eds. (2022), *Questioning the Entrepreneurial State: Status-Quo, Pitfalls, and the Need for Credible Innovation Policy*. New York: Springer.
83. Tidningen Näringslivet (2023, February), 'Professorn: Därför är svensk klimatpolitik överspelad', https://www.tn.se/naringsliv/26192/professorn-darfor-ar-svensk-klimatpolitik-overspelad/.
84. Using the language of Ulrich Beck, it could be argued that the far right is looking back toward first modernity. Regarding far-right nostalgia, see Ruth Wodak (2019), 'The Trajectory of Far-Right Populism: A Discourse-Analytical Perspective' In: Bernhard Forchtner (ed.), *The Far Right and the Environment: Politics, Discourse and Communication*, pp. 21–37. Abingdon, UK: Routledge.
85. Mårald and Nordlund, 'Modern Nature for a Modern Nation'; Andersson, 'Nordic Nostalgia and Nordic Light'.
86. Golster (2022, 10 November), 'Flera SD-ledamöter tror inte på pågående klimatkris', *SVT Nyheter*, sec. Inrikes, https://www.svt.se/nyheter/inrikes/flera-sd-ledamoter-tror-inte-pa-pagaende-klimatkris.
87. Hultman and Pulé, *Ecological Masculinities*.
88. Jan Tullberg (2019, 16 August), 'En destruktiv kulturrevolution', *Samhällsnytt*, https://samnytt.se/en-destruktiv-kulturrevolution/.
89. Kjell Vowles and Martin Hultman (2022, April), 'Dead White Men vs. Greta Thunberg: Nationalism, Misogyny, and Climate Change Denial in Swedish

Far-Right Digital Media', *Australian Feminist Studies*, 1–18, doi:10.1080/08164649.2022.2062669.
90. Krange, Olve, Bjørn P. Kaltenborn, and Martin Hultman (2019). 'Cool Dudes in Norway: Climate Change Denial among Conservative Norwegian Men', *Environmental Sociology*, 5,1: 1–11.
91. Klimatpolitiska rådet, 'Årsrapport 2023'.
92. Erik Nilsson (2022, 12 November), 'Finansministern: Svårt att nå klimatmålet nu', *Svenska Dagbladet*, https://www.svd.se/a/15MxpM/elisabeth-svantesson-om-kritiken-mot-tidopartiernas-budget.
93. Widding has since left the party but remains a member of parliament.
94. Klimarealistenes (2023), 'Klimarealistenes Vitenskapelige Råd – Klimarealistene', https://web.archive.org/web/20230129195746/https://klimarealistene.com/om-oss/klimarealistenes-vitenskapelige-rad/. Accessed 14 March 2023.
95. Clintel – Global Climate Intelligence Group (2022), 'World Climate Declaration – There Is No Climate Emergency', https://clintel.org/wp-content/uploads/2022/10/WCD-version-100122.pdf.
96. Elsa Widding (2023), *Sunt Förnuft Om Energi Och Klimat: Fakta Och Funderingar*. Uppsala: Elsa Widiing AB.
97. Kirsti Jylhä, Jens Rydgren, and Pontus Strimling (2018), 'Sverigedemokraternas Väljare. Vilka Är de, Var Kommer de Ifrån Och Vart Är de På Väg?', Forskningsrapport. Stockholm: Institutet för framtidsstudier.
98. Henrik Oscarsson, Jesper Strömbäck, and Erik Jönsson (2021), 'Svenska klimatförnekare'. In: Ulrika Andersson et al. (eds.), *Ingen anledning till oro (?)*, pp. 235–250. Göteborg: Göteborgs universitet, SOM-institutet.
99. N. Newman et al. (2020), 'Reuters Institute Digital News Report 2020', Reuters Institute for the Study of Journalism, https://reutersinstitute.politics.ox.ac.uk/sites/default/files/2020-06/DNR_2020_FINAL.pdf; Björn Rönnerstrand (2023), 'Kärnkraft och teknikoptimism när klimatfrågorna förändras'. Göteborg: SOM-institutet.
100. Isak Stoddard et al. (2021, October), 'Three Decades of Climate Mitigation: Why Haven't We Bent the Global Emissions Curve?', *Annual Review of Environment and Resources*, 46, 1: 653–689, doi:10.1146/annurev-environ-012220-011104.
101. Stoddard and Anderson, 'A New Set of Paris Compliant CO2-Budgets for Sweden'.
102. David Crouch (2018, September), 'The Swedish 15-Year-Old Who's Cutting Class to Fight the Climate Crisis', *The Guardian*, sec. Science, https://www.theguardian.com/science/2018/sep/01/swedish-15-year-old-cutting-class-to-fight-the-climate-crisis.
103. Sara Ullström, Johannes Stripple, and Kimberly A. Nicholas (2023, March), 'From Aspirational Luxury to Hypermobility to Staying on the Ground: Changing Discourses of Holiday Air Travel in Sweden', *Journal of Sustainable Tourism*, 31, 3: 688–705, doi:10.1080/09669582.2021.1998079.

6
Climate Obstruction in Germany

Hidden in Plain Sight?

ACHIM BRUNNENGRÄBER, MORITZ NEUJEFFSKI, AND DIETER PLEHWE

INTRODUCTION: GREEN GROWTH AND THE LIMITED MITIGATION COALITION

Germany is unique in the realm of climate change in Europe as it has been on a self-imposed path of energy transition, or *Energiewende*, for about fifty years. Yet, in 2020, while 65% of Germans said they regarded climate change as a very important issue,[1] environmental protection and climate policy in particular remain highly contested issues. Battles have centred on the implementation of prominent pieces of legislation—particularly the Renewable Energy Sources Act (RESA)—that attempt to redirect not just the energy system but all economic sectors to meet national and international climate goals. To better understand the energy status quo, we need to more thoroughly examine the efforts of both environmental social movements and obstructionist forces. As we will show, these tensions reveal an ambivalence (and sometimes hostility) toward the transition that is hindering progress, fed by powerful incumbents and reactionary forces that are mostly 'hiding in plain sight'.

A short history of the energy transition

The energy transition in Germany has had an eventful history. The country's journey from a centralized, 'hard' energy path dependent on large fossil and nuclear power plants toward a decentralized, 'soft' path relying on various renewable energy sources was first influenced by the work of Americans Amory and Hunter Lovins in the 1970s.[2] The first reference to an energy transition (*Energiewende*) appeared in a 1981 publication by the Institute for Applied Ecology (Öko-Institut)[3] partly in response to Europe's dependence on oil imports, which became problematic during the 1970s oil crises, and the ongoing debate about 'limits to growth'.[4] With anti-nuclear futurologist Robert Jungk's mid-1980s plea for a German soft path,[5] the energy debate had officially arrived in Germany.

The international development of alternative energy perspectives strengthened the German environmental and peace movements, from which strong anti-fossil/nuclear nongovernmental organizations (NGOs) and, ultimately, the Green Party (established as a national party in 1993) emerged. Germany's Federal Environmental Agency (UBA) had been created in 1974, following the 1972 United Nations environment conference in Stockholm. After the Chernobyl disaster in 1986, environmental policy responsibilities once distributed across various ministries were concentrated in the Federal Ministry of the Environment (BMU). The new ministry was largely responsible for the growing importance of climate change in government policy that followed the country's reunification in 1990. However, the BMU frequently had to fight an uphill battle within the government against other ministries, notably economics, transport, and finance.

Opponents and supporters of the energy transition had been openly confronting each other in various political arenas since the early 1980s. The Greens and the Social Democrats (SPD) were first to form left-leaning coalition governments. It took longer for the centre-right-leaning Christian Democratic Union (CDU) and its sister party the Christian Social Union (CSU) to embrace environmental policymaking. The party of Konrad Adenauer, Germanys first chancellor after World War II, the CDU had ruled most of the time since then and worked closely with Germany's industrial business sector. However, in the wake of the severe ecological crises of the 1980s and 1990s (e.g., rapid forest decline due to acid rain) the party's conservative wing joined avantgarde business leaders and the green-leaning political parties in integrating ecological considerations into Germany's social market economy model. It was now to be redesigned to enhance environmental responsibility in the production process and along supply chains.

The point of departure for the many climate policy debates that eventually emerged in Germany began with the multilateral United Nations climate conference, COP 1, in 1995, in Berlin. Since then, Germany has prided itself as a climate policy leader. Before Angela Merkel, a CDU member, became chancellor (2005–2021), she served as federal minister of the environment under Helmut Kohl from 1994 to 1998. Merkel contributed to the increasing attention to climate change in Germany and supported the establishment of climate research facilities, including the Potsdam Institute for Climate Impact Research (PIK) and, much later, the Institute for Advanced Sustainability Studies (IASS; since 2023 Research Institute for Sustainability, or RIFS). Indeed, although many other centre-right parties have demonstrated more ambivalent attitudes, Germany's majority conservatives have supported high-level climate science and demonstrated support for climate action.

Backed by a cross-party coalition behind the think tank Green Budget Germany and the SPD-Green coalition, since the late 1990s, Germany has embraced a new paradigm, the 'ecological social market economy'.[6] This model seeks to integrate environmental and social concerns into the principles of a market-based economy, aiming to achieve sustainable development by promoting the efficient allocation of resources, social welfare, and ecological balance. Following the implementation of some of these principles during the Social Democrat and Green coalition governments (1998–2005), the climate and energy political landscape changed drastically. Measures such as the ecological tax reform (a tradeoff of higher taxes on fossil energy for a reduction in social wage contributions) in 1999 aimed at a larger social and ecological transformation of the economy. The RESA of 2000 provided financial stability for the influx of electricity from renewable sources into the public grid to promote energy conversion from fossil fuels to renewables. The act spurred the rapid growth of renewables by providing a secure investment via a guaranteed feed-in tariff for twenty years. Amended several times, the most recent version of the RESA, as of 1 January 2023, set a goal of 80% of electricity supply from renewable energy sources by 2030.

In parallel, the share of nuclear energy in the electricity mix had been falling steadily. The German government had already moved to phase out nuclear power in 2001. This commitment was amended by the Merkel government, which extended the deadline for reactor phaseout in 2010. However, these extensions were revoked again in 2011 following the Fukushima power plant disaster in Japan. Nuclear production peaked in

1997 at around 31% of the energy mix and fell to zero after the last plants were shut down in April 2023. Against this, renewables increased continuously and, by 2022, accounted for 48.3% of Germany's gross electricity generation.

Germany's climate policy at a crossroads

Since the approval of the first draft of the RESA in 2000,[7] the landscape of actors has changed considerably. Germany's government since 2021—the 'traffic light coalition' of Social Democrats (red), Free Democrats (yellow), and Greens—has further elevated climate protection as a guiding principle in national and international politics. New groups of civil society actors have emerged since 2018, with a vocal climate movement now including Fridays for Future (FfF), Extinction Rebellion (XR), and Last Generation. This activism in Germany has intensified significantly and has once again led to a stronger public debate in the climate policy field.

These trends, however, have not meant that decarbonization is already well on its way across all relevant sectors, not least due to persistent opposition to ambitious climate action. Although Germany has experienced growing conflicts around climate policy, outright denialism has played a subordinate role.[8] Rather, the 'traffic light' coalition has repeatedly failed to turn ambition into reality. Within the government, the right-leaning liberal Free Democratic Party (FDP) has been the most vocal opponent of ambitious climate policymaking. Key climate protection measures, including the phasing out of coal and nuclear production, the 'mobility transition' toward widespread sustainable transportation, and the replacement of fossil heating devices, have been subject to numerous delay strategies to accommodate the preferences of fossil interest groups and individuals with close ties to the major German political parties (e.g. the Wirtschaftsunion lobby group in the CDU, the SPD's business-friendly subgroup Seeheimer Kreis, and the fundamentalist neoliberal wing of the FDP's Member of German Parliament, Frank Schäffler).

The rise of a new right-wing populist party, Alternative für Deutschland (AfD), Germany's only party that openly features climate denial positions, has added additional weight to obstructionist efforts against the more ambitious climate policies promoted by the Green Party or the left-wing opposition party Die Linke (The Left). The country's official climate goal is to achieve climate neutrality ('net zero') no later than 2045, but political backsliding and 'horse trading' to meet the demands of the FDP in the traffic light coalition have continued to undermine the implementation of necessary measures. Thus, the time frame for a slated phaseout of the combustion engine in road transport and fossil gas-dependent heating in buildings has been continuously postponed, most recently in 2023.

Germany at a crossroads

Due to efforts to undermine ambitious mitigation efforts, Germany is expected to fall short of its pledges (nationally determined contributions [NDCs]) under the Paris Agreement, which are designed to keep global warming below a threshold of 1.5°C. Although the country has decreased greenhouse gas (GHG) emissions continuously between the 1990s and the present, it remains the largest GHG emitter of the European Union (Figure 6.1).

Substantial efforts will be needed to turn the tide in the coming years. As the German Environment Agency (*Umweltbundesamt*) stated in 2022, after Germany managed to cut its emissions by 1.6%, 'We need a rate of six percent reduction per year from now until 2030'.[9]

As Figure 6.2 shows, most emission cuts in Germany so far were made within the electricity production sector. In contrast, emissions in the transport sector have remained almost unchanged since 2010, as have those of most other sectors.

Thus, the industry has failed to reach the sector-specific climate goals stipulated under the RESA in 2021 and 2022. Rather than increasing political pressure, in 2023, the German government abandoned the concept of mandatory, sector-specific goals and now focusses solely on the overall reduction of emissions nationwide. In a recent ministerial report, experts concluded that Germany will most likely not meet its national climate goal of reducing GHG emissions by 65% compared with 1990 levels as planned[10] and would actually need to reduce GHG emissions by 70%.[11]

How can this situation be explained in this alleged 'climate pioneer' country? First, we must distinguish between primary and secondary obstruction. Primary obstruction, according to scholars Ekberg and colleagues, refers to the denial of climate science and the very existence or relevance of global warming. Secondary obstruction 'includes all those calls which do not deny the human-induced nature of the climate crisis (science), but nevertheless delay or forestall meaningful climate action'.[12] Such efforts to delay (1) question the measures required to tackle climate change in general, (2) emphasize the downside of climate policies, and/or (3) present allegedly better, alternative, and market-oriented solutions for transition.[13]

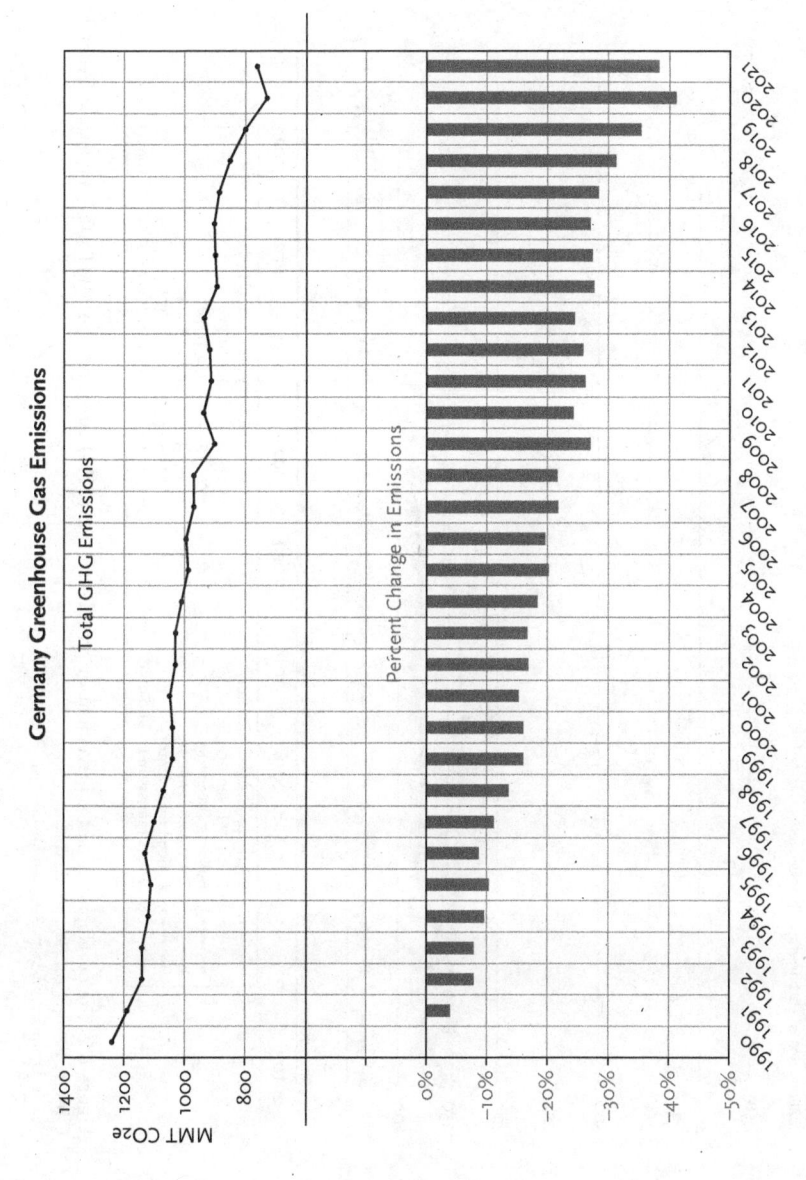

Figure 6.1 Total greenhouse gas (GHG) emissions (in MMT CO_2e) and percentage change in emissions in Germany between 1990 and 2021, inclusive.

Source: Total GHG emissions based on data provided by Gütschow and Pflüger (2023) for Kyoto Six Greenhouse Gas Totals.

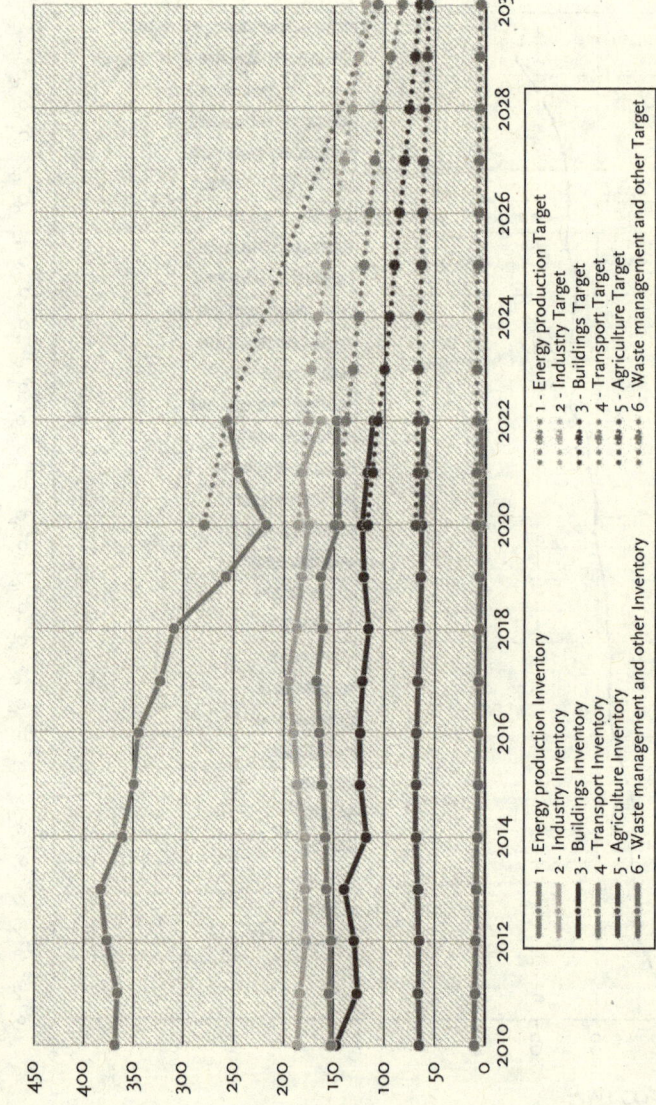

Figure 6.2 The development and achievement of greenhouse gas (GHG) emissions (in MMT CO_2e) under the Federal Climate Protection Act, by sector.

Source: Umweltbundesamt (German Environment Agency) press office.

In Germany, both types of obstructionism have played a role in maintaining the status quo, especially the latter. When the 'traffic light coalition' took leadership of the government in 2021, climate protection was transferred to the Ministry of Economic Affairs and Climate Action (BMWK) under the leadership of Green Party Vice Chancellor Robert Habeck. The merger of two traditionally hostile ministries under the Greens represented a new strategy to align economic and climate policy goals. This trend was also seen at the state level (the Länder), where earlier antagonism between the pro-business parties of the centre-right and the Greens had progressively given way to 'conservative-green' coalition governments.

Against this background, and unlike in the United States or United Kingdom, the voices of climate deniers—the first form of obstructionism—had been marginalized in Germany. But they had become institutionalized in the second decade of the new millennium with the rise of AfD in 2013 (noted earlier) and emerging networks of climate-sceptic civil society actors.[14] In opposition to the mainstream parties, AfD—much like other right-wing populist parties in neighbouring European countries—has recently gained strong support in public polls. Despite the fringe character of German denialism, there remains other significant opposition to ambitious climate policy, particularly command-and-control regulatory instruments, from fossil interest groups and in neoliberal policy expert circles.[15] Indeed, between 2010 and the Russian invasion of Ukraine, the earlier push for energy transition instead encountered significant resistance, culminating in the 2014 amendments to the RESA, which replaced the successful feed-in tariff incentive for the expansion of renewable energy capacity with an auction system.

Thus, in terms of political strategies and policy instruments for climate action, Germany can hardly be called progressive or pioneering. Instead, the country stands at the crossroads between energy regimes: one based on conventional fossil fuels and the other on more sustainable renewables.

THE OPPONENTS OF CLIMATE ACTION

One useful way to obtain an overview of the relevant actor landscape with regard to climate policy is to focus on the major sources of CO_2 emissions in Germany, which in 2016 were energy generation (37.8%), industrial production (20.7%), transport (18.2%), and households (10.2%).[16] Agriculture (7.8%) also played a role, but large, energy-consuming and emissions-intensive animal farming partly benefits from the transition to renewable energy (the use of, e.g. biogas and biomass; for sectoral drivers of CO_2 emissions, see Note 21).

Once the RESA went into force in 2000, interest groups representing these GHG sources went on the defensive. Due to the rapid expansion of the share of renewable energy used in electricity production, various fossil interest groups interested in the preservation of the traditional production system mobilized. Germany's car industry, with 800,000 employees, its influential lobby association, the Association of the German Automotive Industry (VDA), and allies in industry and politics, was quite successful in slowing the transition to renewable energy in private (road) transport. The speed of transformation in heating has also been slow.[17] For example, a law passed in 2023 to push for a fast replacement of fossil fuel-based heating was first diluted by the smallest party of the government coalition, the market liberal FDP, and then blocked by a legal challenge. Finally, it was adopted in September 2023.

The pressure on utilities and customers in energy-intensive industries due to the renewable policy was high, which set the stage for sometimes furious campaigns against the feed-in tariff (noted above) and the energy transition in general.[18] The resilience of the fossil interest groups also became evident through their efforts to maintain Russian gas supplies in spite of Russia's annexation of Crimea in 2014, and the growing Russian pressure on the Ukraine, 'weaponizing' fossil fuel dependency in Europe.[19]

However, the full range of supporters (green alliance) and opponents (grey alliance) of ambitious climate policy in Germany is more diverse.[20] The two groups comprise a variety of actors including companies, business associations, academic and partisan think tanks, and civil society actors with various ties to the progressive and conservative political party spectrum. Following is a summary of the most powerful actors in these groups, emphasizing the obstructionist (grey) camp, comprising mainly those who want to preserve Germany's centralized fossil fuel energy infrastructure and the traditional industrial production system, along with a less influential cluster of climate deniers.

Major German grey companies

Companies from the energy production sector, including Germany's four major utilities (E.ON, RWE, Vattenfall, and EnBW), belong to the traditional, structural conservative grey coalition. They account for the bulk of nuclear, fossil, and some renewable energy production and distribution, although local grids are often wholly or jointly owned by municipal governments. Whereas E. ON and EnBW were directly involved in slowing down the energy transition (discussed later), many smaller firms supplying the car manufacturing or chemical industries have also been players, sharing a vested interest in the fossil fuels sector (e.g. reliance on plastic parts). Despite the phaseout of nuclear power in Germany, nuclear energy producers (typically owned by major energy companies and competitors with renewables for energy market share) can also be considered part of the grey coalition. Following the 2022 Russian invasion and in line with the opposition parties AfD and CDU, the FDP have called for a renaissance of nuclear energy in Germany, emphasizing the need to maintain energy security and to protect the climate.[21] Evidently, German producers of nuclear technology have not given up on their home market.

Large customers of electricity and heating fuels, including the German car manufacturers (VW, Mercedes Benz, and BMW) and foreign car producers in Germany represent another key industry group in the grey coalition. Airbus and many suppliers of auto and aircraft products (especially traditional motor part producers) also still depend on the fossil fuel regime, as do gas station chains, which usually belong to the oil majors; airports; and most tourism-related services.

Most major industrial corporations in energy-intensive industries such as aluminum, steel, and processed chemicals are also part of the grey group. For example, Aurubis AG elected a leading German climate science denier, Fritz Vahrenholt (discussed later), as chair of its supervisory board in 2018.[22]

The major firms and business associations of the grey energy coalition, with their vested interests in fossil industries, have mobilized against the recent advance of renewables. In spite of the companies' official endorsement of the Paris treaty goals, they have made numerous attempts to slow or dilute ambitious climate policies, maintaining close relationships with both the German centre-right and centre-left political parties. Of Germany's largest CO_2-emitting firms, only the utilities have taken climate policy positions substantially aligned with the Paris targets in several policy areas, according to an analysis of official company documents by the NGO InfluenceMap (see Table 6.1).

Financial firms also need to be considered part of the grey coalition. For example, Germany's largest investment fund, DWS, a subsidiary of Deutsche Bank, has been accused of 'greenwashing' for making advertising promises that are untenable given its continuing investments in coal, natural gas, and oil.[23] Both the Deutsche Bank and Commerzbank belong to the Net Zero Banking Alliance.[24] Germany's GLS Bank, a founding member of the alliance, recently dropped out due to continued investment

Table 6.1 MAJOR GERMAN GREY (NUCLEAR/FOSSIL) FIRMS THE RANKING RATES LOBBYING TRANSPARENCY AND POSITIONS TAKEN ON THE PARIS AGREEMENT GOALS ON A SCALE FROM *A* TO *F*.

Sector	Firm	Employment	Paris treaty ranking
Utility	E.ON	72,169	B-
Utility	EnBW	26,064	B-
Industrials	Siemens	311.000	C+
Utility	RWE	18,246	C
Energy	Siemens Energy	88,000	C
Automobiles	VW Group	672,800	C
Industrials	Airbus Group	143,358	C-
Automobiles	Mercedes Benz	172,425	C-
Metals & Mining	Thyssenkrupp	103,598	D+
Automobiles	BMW	118.909	D+
Chemicals	BASF	111,047	D+
Transportation	Lufthansa	107,643	D-

Source: InfluenceMap (https://europe/influencemap.org), the ranking takes lobbying transparency and positions taken with regard to the Paris goals into account on a scale of A-F; on methodology see: https://lobbymap.org/page/Our-Methodology

in fossil industries by its members.[25] The Nuclear Waste Management Fund (KENFO), the first sovereign state fund in Germany, has the political task of ensuring that its investments in the financial markets meet sustainability criteria and the Paris climate targets. Nevertheless, in 2020, the fund invested €757.9 million (3.2% of its assets) in oil and gas companies[26] and has also been criticized for its investments in Russian financial and energy companies such as Sberbank and the oil company Lukoil.

German business associations

Looking at the major business associations (Table 6.2) we can also see that individual firms seem to be somewhat better aligned with the Paris treaty goals than the associations to which they belong, revealing inconsistencies in their public affairs strategies. While certain auto manufacturers have moved to embrace the transition to electric cars, for example, the VDA has continued to oppose car sector-related climate regulations.[27] Possibly the biggest success of the VDA and the German car producers was recorded in 2013, when, following aggressive interventions and policy-drafting activities by the German industry lobby, German luxury car producers saw their

Table 6.2 MAJOR GERMANY GREY (NUCLEAR/FOSSIL) BUSINESS ASSOCIATIONS
THE RANKING RATES LOBBYING TRANSPARENCY AND POSITIONS TAKEN ON THE PARIS AGREEMENT GOALS ON A SCALE FROM A TO F.

Sector	Business association	Paris treaty ranking
All sector	Federation of German Industries (BDI)	D
Chemicals	German Chemical Industry Association (VCI)	D
Automobiles	German Association of the Automotive Industry (VDA)	D-

Source: InfluenceMap (https://europe/influencemap.org), the ranking takes lobbying transparency and positions taken with regard to the Paris goals into account on a scale of A-F; on methodology see: https://lobbymap.org/page/Our-Methodology

interests accommodated through changes to the European fuel efficiency label and a related EU directive relating 'efficiency standards' to the weight of cars.[28]

The complexity of association lobbying can be further illustrated with a case from the gas industry. While the Bundesverband der Energie und Wasserwirtschaft (BDEW) is the largest energy business association, a PR-lobby alliance, Zukunft Gas (Future Gas), was founded in 2013 to support product marketing. One hundred thirty-five firms across the gas production and distribution chain (including former Gazprom gas station subsidiaries NGV and Wingas) backed this effort to promote narratives of gas as an allegedly efficient and cheap energy source that is also climate friendly. A study by the German NGO LobbyControl identified additional lobby groups working for specific segments of the fossil gas business and noted the role of cross-sectoral and consumer business organizations,[29] which allow the gas industry to work across multiple channels. LobbyControl has shown how these and other associations from the gas industry played a key role in vilifying and weakening the law mandating decarbonization of heating devices in 2023, which aimed at gradually replacing oil and gas heating systems in Germany.[30] In another study, LobbyControl revealed the multiple connections between promotors from foreign gas-producing states, such as Russia and Azerbaijan, and German politicians and businessmen close to the SPD and CDU. Politicians from both parties held key positions on supervisory boards of companies and forums such as the Deutsch-Russisches Rohstoffforum (Michael Kretschmer, CDU), the Nord Stream 2 pipeline (Gerhard Schröder, SPD), gas company VNG (Edmund Stoiber, CSU), and the Germany–Azerbaijan Forum (Thomas Bareiss, CDU). According to the study, these close ties have increased Germany's dependence on Russian gas markets and prevented the timely switch to renewable forms of energy.

Similarly, in 2020, a cross-sectoral coalition of 180 (as of May 2023) companies and groups from seventeen countries formed the eFuel Alliance; members include big oil and gas firms, car and truck manufacturers including Porsche, and technology companies such as Siemens and Bosch.[31] Although e-fuels are nominally carbon-neutral because electricity generated from renewables is used in their production and only as much CO_2 is emitted during use as was bound during production, e-fuels release other forms of exhaust, similar to fossil fuels. They also enable the continuing production of cars that can also run on traditional fuels. Another major cross-sectoral player is the family business association Die Familienunternehmer e.V. While voicing support for climate protection, the lobby group wants to reach climate policy goals without state support for renewable energy or a single price for CO_2, advocating stronger competition in the energy sector rather than taxes on certain fuels or prices set through emissions trading.

The political influence of companies and business associations is sustained through frequent use of 'revolving doors' through which former politicians and government officials find employment in the business sector after their political careers have ended. The hiring of former Chancellor Gerhard Schröder by the Russian oil company Rosnef and the Nord Stream 2 AG consortium marks the most prominent example. In 2021, State Secretary for Energy and Digital Andreas Feicht, under Minister Altmaier (CDU), became chairman of the board of RheinEnergy. Thorsten Herdan, from 2014 until 2022 head of Department II Energy Policy – Heat and Efficiency in the Federal Ministry of Economics, later became CEO of the global eFuels company HIF EMEA. The revolving door can also swing the other way, as when the economics minister of the first Social Democrat–Green coalition government, Werner Müller (no party affiliation), entered government after a career working for German energy firms RWE and VEBA.

Academic and partisan think tanks

Numerous academic research institutes in Germany have supported the continuation of the fossil energy system. Partly funded or supported (via research contracts) by major utilities like RWE and E.ON, the Energiewirtschaftliche Institut at the University of Köln (EWI) and the Leibniz-Institut für Wirtschaftsforschung (RWI) in Essen are prominent examples. Both organizations attacked the funding of Germany's energy transition through feed-in tariffs from the beginning. They also supported the extension of nuclear energy production when the Social Democrat–Green coalition government negotiated the phasing out of nuclear power. Their pro-fossil fuel positions have been widely publicized in the conservative media and business press (e.g. *Frankfurter Allgemeine Zeitung*, *Die Welt*, and *Handelsblatt*).

Several prominent think tanks and campaign organizations are also part of the grey alliance. The main think tank of the top German employer organization BDI is the Institut der Wirtschaft (IW), with offices in Köln and Berlin. IW oversees the Initiative for a New Social Market Economy (INSM), a lobby organization funded in 2000 by the German metal industry association (*Gesamtmetall*), an organization dominated by the major car and steel manufacturing firms. INSM led several campaigns against the RESA and the broader energy transition, all of which emphasized economic efficiency, energy efficiency, and security and prioritized market principles and technological openness. The Centre for European Policy in Freiburg is the latest addition to an already large number of German neoliberal think tanks (e.g. Eucken Institut, Stiftung Marktwirtschaft, and the Röpke Institut) opposed to the state-led energy transition and 'non-market instruments' such as price regulation and subsidies favouring renewable forms of energy.[32]

The realm of climate change policy denial

The only political party in Germany officially opposed to climate action is the right-wing AfD. This singular position offers the party a unique selling point in the German political landscape, catering to a significant minority of the electorate. The main focus of the party and the AfD-aligned Desiderius Erasmus Foundation is resisting the energy transition, which allegedly threatens the prosperity of German society. The AfD seeks to end the decarbonization project *Energiewende* at large and to repeal the German government's Climate Protection Plan 2050.[33]

Closely aligned with the AfD is the EIKE think tank (Europäisches Institut für Klima- und Energieforschung) in Jena, which claims to be the leading European 'institute' advocating 'climate realism' and spreads the largest number of denial and obstruction messages of all European denial think tanks.[34] It is closely connected to the climate countermovement in English-speaking nations, whose prominent members include the US-based Committee for a Constructive Tomorrow (CFACT) and the Heartland Institute. EIKE has organized German denial conferences modelled after

Heartland's (and with the participation of experts featured at Heartland conferences) together with the Institut für Unternehmerische Freiheit (IUF), a small neoliberal think tank in Berlin. Social media influencer Naomi Seibt has appeared at these conferences, giving talks and presenting videos in which, among other things, she denies the reality of climate change. Seibt is sometimes referred to as the 'anti-Greta' in contrast with Swedish climate protection activist Greta Thunberg.[35]

While AfD's and EIKE's positions do not frequently enter the mainstream media, a dedicated group of AfD party and right-wing media outlets such as *Freie Welt* feature denialist arguments and authors. These publications, in addition to their social media channels, help these groups to sustain 'varieties of right-wing populist climate politics'.[36]

STRATEGIES AND TACTICS UTILIZED

German companies and allied interest groups engage in a number of delay strategies to preserve fossil fuel dependency. Beyond traditional business associations, the public campaigns of fossil interest coalitions rely on think tanks and NGOs to influence public opinion. Medium- and long-term campaigns have been key to the considerable efforts undertaken to slow and shape Germany's energy transition following the approval of the RESA in 2000. Apart from the fringe right-wing groups that continue to deny the existence or relevance of man-made climate change, most grey energy groups officially endorse international climate policy commitments. However, many firms and associations fail to live up to their official positions and frequently lobby to lower ambitions, engage in greenwashing, and attempt to shift the burden of change to others to protect traditional business.[37]

Championed by a coalition of Social Democratic and Green MEPs, the RESA of 2000 was unusual as it did not originate in the ministries but resulted from a parliamentarian initiative. It was built on the aforementioned grid-opening 1990 Electricity Feed-In Act, which allowed small renewable electricity producers to sell to the utilities. In addition to grid access, the RESA provided additional support for the development of renewable electricity production along the entire production chain, with long-term stable prices provided by the feed-in tariff.[38]

Utility companies opposed energy liberalization. The Hannover-based large utility PreußenElektra (later merged into E.ON) in particular fought hard against the rise of renewables in northern Germany, where it had held the regional grid monopoly. The company even pressed its case before the European Court of Justice, but lost that legal battle.[39] Incumbent fossil energy producers and large industrial customers opposed the emerging support for renewables but fought an uphill battle until the conservative liberal coalition government led by Angela Merkel took office in 2009. Their traditional influence in the Ministry of Economics, which had been in charge of energy policy, no longer sufficed during the first decade of the new millennium. Responsibility for renewable energy had been moved from the Ministry of Economics to the Environmental Ministry (until 2005 headed by Jürgen Trittin, a member of the Green Party, until 2009 by Sigmar Gabriel, member of SPD) in the early 2000s. Under the Christian Democratic and liberal leadership of the Ministry of Economics and the Environmental ministry, respectively, access for industry groups once again improved.

Fossil industry supporters originally were also ill-prepared for the challenge of energy conversion politics. Most experts were surprised by the rapid expansion of the share of decentralized electricity produced by wind and solar energy after the grid opening. At the beginning of the new millennium, incumbent fossil producers and the large electricity customers, unlike the utilities, were not yet alarmed by the development. Most experts (including Angela Merkel at the time) expected only a low-single-digit share of renewable energy production to result from the legislation. In the course of the 2000s, dedicated actors from industry, academia, and the think tank world aimed to ensure such limits by undermining the incumbent renewable and climate regime through a variety of strategies and tactics.

Scientific studies, lobbying, and media campaigns

Shortly after the passage of the RESA, academic and think tank opponents of the state-led effort to increase wind, solar, and biomass sources of electricity generation advanced arguments against the feed-in tariff-based incentive for renewable investment. Institutions involved included the academic council of the Federal Ministry of Economics, the RWI, and the industry co-financed EWI. In 2004, a group of three research institutes published a study contracted through the academic council of the Federal Ministry of Economics on the general economic, sectoral, and ecological impact of the renewable energy act.[40] The authors claimed the system in place would not be an efficient way to proceed in the long run and emphasized the emergence of unnecessarily high consumer prices as a result of the guaranteed tariff then in place. To mend this problem, the study proposed incentives to increase innovation efforts and move toward competition between different types of renewable energy. This endeavour was in marked contrast to the political effort to develop all renewable sources together to take advantage of their complementarity (the wind blows when the sun is down, solar works whether or not the wind is blowing, and so on). Instead of the feed-in tariff, the study proposed a quota system (which already existed in the United Kingdom and Sweden and compared poorly with the feed-in tariff in Germany in terms of expanding the share of renewable energy). Besides raising the spectre of incompatibility with EU law, the main concerns of the study were efficiency and cost.

Based on that study, the academic council of the Federal Ministry of Economics (headed by Wolfgang Clement, a conservative Social Democrat from the coal and steel state of Nordrhein-Westfalen) demanded the RESA be cancelled. Subsequent studies published by RWI's energy department repeated the core messages of the early expert document: the feed-in tariff is inefficient, alternative solutions based on competition are superior, and German law may not be compatible with EU law (despite the European Court of Justice's favourable ruling in 2001). Ultimately, various academic and partisan think tanks, including RWI and the employer-funded IW, as well as government expert commissions such the German Council of Economic Advisors (SVR) and the Monopoly Commission, converged on proposing a quota system as an alternative to the feed-in tariff. This alignment on an alternative policy instrument was unsurprising due to the interlocking positions of key academics involved simultaneously in academic research, government commissions, and industry-financed think tank and campaign efforts.[41]

In the meantime, additional arguments had been developed by the range of research institutes also opposed to the feed-in tariff. They focused on the growing cost of financing the fixed tariff for renewable energy. Although the figures provided in industry-funded studies were inflated (up to an 'unnecessary' €52 billion in additional expenses[42]), and, taken out of context, they served to feed an extended public media campaign against the tariff.

During the 2000s, criticism from RWI intensified. RWI researcher Manuel Frondel provided a study on the supposed high cost of German renewable energy to a US think tank, the Institute of Energy Research (IER), renewing the claim that the RESA was ineffective.[43] According to Frondel, the EU ETS, a market-based approach to reducing GHGs that sets a cap on emissions and allows allowance trading, undermined the ecological impact of Germany's own renewables policy. However, the claim did not take into account the practice of reducing the number of certificates traded according to the effectiveness of the feed-in tariff.[44] Although the European emissions trading scheme failed to live up to its CO_2- reduction promises (see Chapter 13, on the European Union), the German Innovation Council—an expert commission composed of economists and management scholars—also demanded the elimination of public support for renewables. While mentioning one study that recognized innovation in wind energy, the council's report relied on studies claiming the opposite and summarily denied 'measurable' innovation effects.[45]

Opposition also came from Germany's fringe climate-denial camp. For example, between 2013 and 2018, EIKE placed criticism of RESA at the centre of its social media activities.[46] In addition, the neoliberal INSM ran a dedicated media campaign demanding the abolition of the feed-in tariff, part of a sustained effort to mobilize the public against the RESA. Relying on RWI-contracted research and operating with a budget of up to €8 million per year, the 2012 campaign focused attention across the spectrum of mass and social media, helping to pave the way for the elimination of the feed-in tariff in 2014 (we provide a more detailed analysis of narratives deployed in this highly successful campaign in the final section of this chapter).

These various academic, legal, and media strategies ultimately contributed to the major revisions to the RESA in 2014, ending the fast tracking of renewable energy conversion in Germany—at least until the Russian invasion of Ukraine. The reform replaced the feed-in tariff-based support for renewable development with an auctioning system, which privileged large capital investment instead of the decentralized expansion of renewables prioritized earlier. After the elimination of the feed-in tariff, former supporters of the 'quota system' mentioned earlier fell silent, revealing the instrumental character of Germany's policy instrument competition: to end a highly successful regime to fund renewable expansion that had accounted for 47% of CO_2 reduction in Germany, compared with a 10% drop related to emissions trading. [47]

Anti-wind power campaigns

Beyond fighting the RESA, the efforts of research institutes and think tanks to fight the energy transition have been accompanied by single-issue initiatives and right-wing groups alike, which can be illustrated by their dedicated campaigns against wind energy. Several organizations including Windwahn (wind delusion), Vernunftkraft (rationality power), and Wildtierstiftung (wildlife foundation) focused on trying to break the momentum of Germany's energy transition.[48]

Windwahn is an online platform that aims to organize civic initiatives (CI) under one roof and sees itself as a mouthpiece for these CI. Its website features a map listing more than 1,100 associations and initiatives and explicitly welcomes other initiatives 'that act according to the motto 'wind power yes, but . . .'. The website also includes factual reasons to oppose wind energy. However, headlines such as 'Myth of cheap green electricity', 'Energy turnaround as a danger for the whole of Germany', 'Dark lull approaches', and 'Wind power megalomania' predominate, illustrating the group's radical rejection of wind power expansion.[49]

Vernunftkraft calls itself a 'federal initiative for sensible energy policy' and is an umbrella organization for fourteen state and regional associations that oppose the RESA, wind energy, and other renewable energy projects. In contrast to Windwahn, Vernunftkraft argues mostly factually. It reinforces the local conservation concerns it raises by deploying professionals to lobby the government. Politically, Vernunftkraft has been supported by the AfD, EIKE, parts of the CDU, and parts of the FDP as well as within the Ministry of Economics.[50]

Finally, the Wildtierstiftung is committed to nature conservation and education and represents the moderate edge of the spectrum critical of wind energy. However, from 2012 to 2019, the foundation was headed by Fritz Vahrenholt, a prominent climate change denier, a former renewable energy industry manager at RWI Innogy, and a long-term member of the supervisory board of Aurubis AG. Vahrenholt and the foundation's head of communications, Michael Miersch, attacked the government's climate policy goals and used anti-wind and other campaigns to support fossil industry positions. After the foundation dismissed Vahrenholt in 2019, climate change denial no longer played a role in the organization's work, as a look to the Wildtier-Webinar, the Blog, or the list of publication show. Since his departure, Vahrenholt has engaged in a country-wide anti-climate policy campaign termed 'save our industry'.

Right-wing extremist mobilization

The forces of business-related climate policy delay and climate denialist groups have recently been joined by right-wing extremist organizations targeting climate protection as part of their platform of degrowth, a decentralized economy, population control, and an end to immigration. These groups have organized to violently obstruct the climate justice movement. For example, the regional organization Pro-Lausitzer Braunkohle e.V., which advocates for the continued use of coal, organized counter-demonstrations against the German climate justice group Ende Gelände in 2016 during its occupation of the coal mining company Leag in the Lausitz.

Local far-right groups participated in these counter-demonstrations, physically attacking activists. Various civic initiatives promoted on social networks (e.g. Fridays for Hubraum, or 'cubic capacity') and 'No ban on gasoline and diesel vehicles in Germany' served as channels for radicalization in which fantasies of harm and sometimes even murder are voiced against climate activists. The latter group is administered by the Automobilclub Mobil Germany, a competitor of the larger General German Automobile Club. Together, these groups push for Germany to embrace stereotypes of petro-masculinity.[51]

CLIMATE POLICY OBSTRUCTION DISCOURSES: THREE INSM CAMPAIGNS

Over the years, the INSM has increasingly engaged with German climate policy. In addition to its campaign against the RESA, mentioned above, the think tank continued to criticize German climate action and accompanied their arguments with high-profile media campaigns in 2012, 2017, and 2019. A review of the discourses displayed in these three campaigns offers clear examples of the types of narratives fossil interest groups use to intentionally obstruct climate action, which stand in contrast to their official support for it.

INSM's first campaign was launched with the slogan: 'Stop the RESA—do the energy transition', presenting the RESA as its opposite: an obstacle to climate protection. The INSM claimed that the RESA promotes inefficient technologies and thus makes the energy transition too expensive. By providing a counter-narrative based on an alternative Competitive Model for Renewable Energies, the INSM aimed to promote market-based instruments instead of government regulation and thus fight off the feed-in tariff, which was becoming increasingly unpopular amongst German industry due to the growing uncertainty created by obstructionist attacks from various quarters. Thus, the narratives of cost inefficiency and ineffectiveness included a more appealing narrative: market solutions that would purportedly result in better climate protection with fewer restrictions upon industry.

In the media campaign accompanying this discursive framing, an electrical outlet superimposed with symbolic images served as a visual motif for print ads and posters in public spaces. For example, under the question: 'How does German energy policy affect the price of electricity?' the INSM placed a picture of a time bomb over the power socket. In another commercial, the iconic image of Edvard Munch's painting *The Scream* appeared over the socket. Above, it said: 'Help! The energy transition is becoming unaffordable'.

The follow-up campaign in 2017 refrained from such dramatic imagery, but the organization continued to adhere to its cost criticism and the alleged ineffectiveness of the RESA. However, the INSM no longer contrasted the RESA with its own market-based model. Instead, the group extended its argument to other concerns. While the organization still strongly emphasized the alleged additional burden for electricity customers and especially industry, it now also stoked fears of a loss of industrial competitiveness. With reference to cheaper energy costs as a 'central location factor', the organization created the spectre of the relocation of industry and a concomitant loss of employment while ignoring the well-established negative effects of unchecked climate change on jobs and the economy.

In its 2019 campaign, INSM focused more on the federal government's climate policy in general. In addition to the narratives of energy poverty, inefficiency, and loss of competitiveness, the organization generated yet another image. While it described climate change as 'currently the greatest challenge facing humanity', it also referred to a 2°C target for limiting global warming in the Paris Agreement. Through this rhetorical figure, the organization lowered the bar for emissions reductions needed (it is 1.5°C in the Paris treaty). Moreover, INSM's campaign highlighted the need for international efforts to fight climate change while also sidestepping responsibility, stating that Germany's share of global CO_2 emissions is marginal, a staple argument of fossil interest groups in many countries whose historical emissions, like Germany's, are substantial.

In addition, the INSM relied on another aggressive media campaign to promote its positions on German climate policies. It began targeting the leader of the Green Party, Annalena Baerbock, during her election campaign of 2019. In INSM's parodic print campaign, Baerbock appeared dressed as the biblical figure Moses, holding up two stones engraved with the Ten Commandments. These commandments stated that 'you may not drive a combustion engine', 'you may not fly', and other such restrictions, ending with 'you may not even think that there is an end to prohibitions'. This image, titled 'Why we do not need a state religion', appeared in leading German newspapers such as *Frankfurter Allgemeine Zeitung*. Here, INSM's pictorial language converged with the narratives of the German denial organization EIKE, which has stated that 'not the climate is endangered, but our freedom'.[52]

As part of their latest campaign, in a 23 February 2023 article 'Five ways to a better energy policy',[53] the INSM reflected the fundamental redirection of German climate policy after Russia's invasion of Ukraine and the subsequent inflation crisis. In contrast to its previous positions, the group now promoted a faster expansion of renewable energy. But its messages still contained a toned-down criticism of cost inefficiencies and state subsidies. Now, the organization placed a stronger emphasis on technology options by promoting hydrogen development, the expansion of liquified natural gas (LNG) terminals, fracking, and carbon capture and storage (CCS).

CONCLUSION: INDUSTRY HEADWINDS AGAINST CLIMATE AMBITIONS

As this chapter has shown, existing climate policies and environmental protection in Germany have been strongly contested. Initial demands by environmentally conscious civil society groups, the rise of the Green Party, and the recent emergence of Fridays for Future and other climate movements have faced headwinds from powerful industry associations, neoliberal think tanks, employer lobby groups, and conservative civil society movements since the 1980s. Especially since the mid-2010s, a solid neoliberal opposition to the country's energy transition has developed that has proven more influential than the fringe climate denial position of a few actors. However, Germany displays a diverse range of opponents of renewable energy projects whose members have ties to factions of the major political parties including the Christian and Social Democrats and the smaller, right-leaning liberal FPD.

Positions beyond and between the left–right spectrum make orientation difficult. Not all conservatives are climate obstructionists. Some far-right groups conceive of climate protection as a matter of homeland security. The dogmatic character of certain 'citizen initiatives' against renewable energy projects suggests the involvement of organized obstructionists. There is a trend of 'covert' networks of anti-renewables lobbyists throughout Germany who—on behalf of companies—file lawsuits, advise CI, and act as experts. Similar to 'astroturf' organizations in the United States and other countries, some activist groups set up to oppose wind farms and solar panels in Germany that appear to be grassroots movements are actually sustained by (fossil) interest groups.[54]

Certainly, the strongest efforts have been orchestrated against Germany's RESA. Through academic opposition (e.g. from RWI), partisan think tanks (e.g. CEP), public media campaigns (e.g. INSM), and continuous lobby pressure from powerful companies and industry associations (represented by the industry-financed think tank IW), the once radical act to expand decentralized renewable energy production eventually morphed into a soft measure unlikely to help meet Germany's goal of becoming climate neutral by 2045.

Until recently, the 'grey' group of obstructionist actors has portrayed the continued use of fossil fuels as necessary to ensure reliable, affordable power and domestic energy security.[55] Following the Russian invasion of Ukraine, German energy policy has been contradictory, with a focus on both the diversification of fossil gas supplies and a reinvigoration of efforts to increase the use of renewable energy. The BMWK concluded long-term contracts with gas and oil suppliers and continues building LNG terminals on Germany's coasts. While regulatory hurdles against the expansion of wind power have been removed, the Liberal Party-led Ministry for Digital and Transport succeeded in erecting another barrier by blocking the European phase-out of combustion engines by 2035, a demand from the automotive sector and the eFuels Alliance. Similarly, the plan to phase out fossil gas heating ran into strong opposition and has since been both weakened and further delayed.

To better understand the ambiguities in the policy positions of the major industrial sectors and political parties in Germany, it will be necessary to study systematically the revolving door between political and business careers and the alliances between inner-party groupings and outside interest groups. For example, while car, steel, fossil energy, and chemical industry interests play a role in the SPD via its works council and union representatives, the links between industry and the Christian Democratic and Free Democratic parties run mainly through management circles. Future research is needed to better understand the structural dimensions of and strategic efforts in the transport and construction industries in addition to the energy sector. While the fight against the RESA shows the capacity of obstructionist forces to fight and win uphill battles, the *Energiewende* is still the policy arena with the best record of forwarding Germany's climate policy agenda. The focus of climate policymaking urgently needs to shift to transport, heating, and housing. Much more research is needed on the lobby groups in these areas, which have so far succeeded in blocking or delaying decarbonization.

NOTES

1. UBA (2022), *Umweltbewußtsein in Deutschland*. Umweltbundesamt (UBA): Dessau-Roßlau.
2. Amory Lovins (1977), *Soft Energy Paths*, 1st ed. New York: Harper & Row.

3. Florentin Krause, Hartmut Bossel, Karl-Friedrich Müller-Reißmann (1981), *Energie-Wende*, 3rd ed. Frankfurt am Main: S. Fischer.
4. Donella H. Meadows, ed. (1972), *The Limits to Growth*. London: Earth Island Limited.
5. Robert Jungk (1986), *Der Atom-Staat*, 71st ed. Reinbek bei Hamburg: Rowohlt.
6. Dieter Plehwe (2022), 'Reluctant Transformers or Reconsidering Opposition to Climate Change Mitigation? German Think Tanks between Environmentalism and Neoliberalism', https://doi.org/10.1080/14747731.2022.2038358.
7. Volkmar Lauber and Lutz Mez (2006), 'Renewable Electricity Policy in Germany, 1974 to 2005', *Bulletin of Science, Technology & Society*, 26: 105–120.
8. Greg Garrard, Goodbody, Axel, Handley, George B., et al. (2019), *Climate Change Scepticism*. London: Bloomsbury Academic.
9. UBA (2023), *UBA Forecast*. UBA: Berlin.
10. Daniel Delhaes (2023), 'Deutschland dürfte seine Klimaziele 2030 und 2045 verfehlen', *Handelsblatt*. https://www.handelsblatt.com/politik/deutschland/klimapolitik-deutschland-duerfte-seine-klimaziele-2030-und-2045-verfehlen/29273700.html
11. Deutschlandfunk, *Auf dem Weg zur Klimaneutralität* (2021, 24 June). https://www.deutschlandfunk.de/auf-dem-weg-zur-klimaneutralitaet-die-neuen-klimaziele-fuer-100.html
12. Kristoffer Ekberg, Bernhard Forchtner, Martin Hultman, et al. (2022), *Climate Obstruction: How Denial, Delay and Inaction are Heating the Planet*. New York: Taylor & Francis; and Achim Brunnengräber (2015), 'Klimaskeptiker in Deutschland und ihr Kampf gegen die Energiewende', https://refubium.fu-berlin.de/handle/fub188/20150.
13. William F. Lamb, Giulio Mattioli, Sebastian Levi, et al. (2020), 'Discourses of Climate Delay', *Global Sustainability*, 3: 1–5, doi:10.1017/SUS.2020.13.
14. Stella Schaller and Alexander Carius (2019), 'Convenient Truths. Mapping Climate Agendas of Right-Wing Populist Parties in Europe', https://adelphi.de/en/publications/convenient-truths. Accessed 8 November 2023.
15. Susanne Götze and Annika Joeres (2022), *Die Klimaschmutzlobby*. München: Piper; andMatthias Quent, et al. (2022), *Klimarassismus*. München: Piper.
16. goClimate.de, 'Emissionen nach Sektoren einfach erklärt'. Accessed 1 April 2023.
17. Weert Canzler and Dirk Wittowsky (2016), 'The Impact of Germany's Energiewende on the Transport Sector: Unsolved Problems and Conflicts', *Utilities Policy*, 41: 246–251.
18. Anja Baisch (2021), *Fossile Strategien*. Hamburg: Tredition GmbH.
19. Claudia Kempfert (2023), *Schockwellen*, 1st ed. Berlin: Campus Verlag.
20. Tobias Haas (2019), 'Struggles in European Union Energy Politics: A Gramscian Perspective on Power in Energy Transitions', *Energy Research & Social Science*, 48: 66–74.
21. Achim Brunnengräber, Albert Denk, Lucas Schwarz, Dörte Themann (2023), 'Monumentale Verdrängung', *Blätter für deutsche und internationale Politik*, 9–12. https://www.wiwo.de/unternehmen/banken/banken-gls-bank-verlaesst-klima-allianz-vorwuerfe-an-wall-street/28967554.html
22. Ansgar Graw (2019, 22 December), 'Deutschland Wildtier Stiftung Kollateralschaden eines Rauswurfs', *Welt*; and Dennis Fischer (2021, 29 November), 'Aurubis erstreitet sich kostenlose CO2-Zertifikate', *Energate*

22. *Messanger*. https://www.wiwo.de/unternehmen/banken/banken-gls-bank-verlaesst-klima-allianz-vorwuerfe-an-wall-street/28967554.html
23. Greenpeace, 'Die DWS - Deutschlands klimaschädlichster Vermögensverwalter?', https://www.greenpeace.de/klimaschutz/finanzwende/deutsche-bank-tochter-dws-greenwashing-betrug. Accessed 1 April 2023.
24. UNEP, 'Net-Zero Banking Alliance – Members', https://www.unepfi.org/net-zero-banking/members/. Accessed April 1, 2023.
25. WirtschaftsWoche (2023), *GLS Bank verlässt Klima-Allianz – Vorwürfe an Wall Street*.
26. Massimo Bognanni, 'Saubere Ziele, fragwürdige Anlagen', https://www.tagesschau.de/investigativ/wdr/atomfonds-101.html. Accessed 1 April 2023.
27. influenceMap (2021), *German Automakers Dominate the Fight to Weaken Climate Regulation*.
28. Lobbypedia (2023), 'Verband der Automobilindustrie', https://lobbypedia.de/wiki/Verband_der_Automobilindustrie. Accessed 8 November 2023..
29. Lobbycontrol (2023), 'Pipelines in Die Politik', https://www.lobbycontrol.de/pipelines-in-die-politik-die-macht-der-gaslobby-in-deutschland/. Accessed 8 November 2023.
30. Christina Deckwirth, 'Wie die Gaslobby das Heizungsgesetz entkernt hat', https://www.lobbycontrol.de/lobbyismus-und-klima/wie-die-gaslobby-das-heizungsgesetz-entkernt-hat-109931/. Accessed 8 November 2023.
31. eFuels Alliance (2023, 23 May), 'Porsche Joins the eFuel Alliance', https://www.efuel-alliance.eu/fileadmin/Downloads/Pressemitteilungen_2023/20230523_PM_Vorstandssitzung_EN_final.pdf. Accessed 8 November 2023.
32. Dieter Plehwe and Kardelen Günaydin (2022), 'Whither Energiewende Strategies to Manufacture Uncertainty and Unknowing to Redirect Germany's Renewable Energy Law', *IJPP*, 16: 270.
33. AfD, 'Themenposition: Energie', Umwelt, Klima, https://www.afd.de/energie-umwelt-klima/. Accessed 1 April 2023.
34. Núria Almiron, Jose A. Moreno, Justin Farrell (2022), 'Climate Change Contrarian Think Tanks in Europe: A Network Analysis', *Public Understanding of Science (Bristol, England)*, https://journals.sagepub.com/doi/10.1177/09636625 22113781; and Alexander Ruser (2022), In: Michael Hoelscher, Regina A. List, Alexander Ruser, Stefan Toepler (eds.), *Civil Society*, 1st ed., pp. 349–358. Cham: Springer International Publishing.
35. Wikipedia, 'Naomi Seibt', https://de.wikipedia.org/wiki/Naomi_Seibt. Accessed June 29, 2023.
36. Veith Selk and Jörg Kemmerzell (2022), 'Retrogradism in Context: Varieties of Right-Wing Populist Climate Politics', *Environmental Politics*, 31: 755–776.
37. Reinhard Schneider (2023), *Die Ablenkungsfalle*. München: oekom verlag.
38. Bernd Hirschl and Thomas Vogelpohl (2019), 'Energiewende in Deutschland und Europa'. In: Weert Canzler and Jörg Radtke (eds.), *Energiewende. Eine sozialwissenschaftliche Einführung*, pp. 69–95. New York: Springer.
39. Kai Roger Lobo (2011), *Die Elektrizitätspolitik und ihre Akteure von 1998 bis 2009*.
40. Rolf Gerhardt, Gesamtwirtschaftliche, sektorale und ökologische Auswirkungen des Erneuerbare Energien Gesetzes (EEG), https://docplayer.org/7207276-Gesamtwirtschaftliche-sektorale-und-oekologische-auswirkungen-des-erneuerbare-energien-gesetzes-eeg.html. Accessed 1 April 2023.
41. Lobbycontrol, 'Kampagne der INSM und des RWI gegen die Förderung des Ökostroms', Berlin: Freie Universität (Dissertation). https://lobbypedia.de/

wiki/Kampagne_der_INSM_und_des_RWI_gegen_die_F%C3%B6rderung_des_%C3%96kostroms. Accessed 1 April 2023.
42. Manuel Frondel and Christoph M. Schmidt (2008), 'CO2-Emissionshandel: Auswirkungen auf Strompreise und energieintensive Industrien', *RWI Positionen*. Essen: RWI Positionen, No. 26. https://www.econstor.eu/bitstream/10419/52546/1/666558469.pdf
43. Claudia Kemfert (2015), *Kampf um Strom*, 8th ed. Hamburg: Murmann.
44. Claudia Kemfert and Jochen Diekmann (2009), *Förderung erneuerbarer Energien und Emissionshandel: wir brauchen beides*. Berlin: Wochenbericht, Deutsches Institut für Wirtschaftspolitik.
45. EFI (2014), *Gutachten zu Forschung, Innovation und technologischer Leistungsfähigkeit Deutschlands 2014*, Gutachten. Berlin: Expertenkommission Forschung und Innovation (EFI).
46. Dieter Plehwe (2021), 'Der Streit um Umwelt und Klima. Rechte Think Tanks haben nicht nur politische Interessen', *WZB-Mitteilungen*, 172: 33–35.
47. Nathalie Grün, Tonja Iten, Felix Nipkow (2019), *Mythen und Fakten zu Deutschlands Energiewende*. Zürich: Schweizerische Energiestiftung; and Wolfgang Riedl and Hans-Josef Fell (2018), *Erneuerbare Energien – die tragende Säule für die CO2-Emissionsreduktion in Deutschland*. Berlin: Energy Watch Group.
48. Nicolai Kwasniewski (2021, 11 February). 'Die Anti-Windkraft-Bewegung', *Spiegel*. https://www.spiegel.de/wirtschaft/windenergie-so-verhindert-die-anti-windkraft-bewegung-neue-anlagen-a-46d88419-3b1d-427d-b6c0-cf696fec283c
49. Dörte Themann, Di Nucci, M. R., Krug, M. (2021), Gegenwind von Rechts? Windenergie im Spannungsfeld zwischen Klima-, Natur- und Heimatschutz. In: Heike Leitschuh, et al. (eds.), *Ökologie und Heimat*, pp. 113–126. Stuttgart: S. Hirzel Verlag.
50. Manfred Redfels (2021), 'Die Gegner der Energiewende', Greenpeace Recherche. https://www.greenpeace.de/publikationen/gegner_der_windkraft.pdf. Accessed 8 November 2023.
51. Kristoffer Ekberg, et al. (2023), *Climate Obstruction*. New York: Routledge.
52. EIKE, 15. 'Internationale EIKE Klima- und Energiekonferenz', https://eike-klima-energie.eu/15-internationale-klima-und-energiekonferenz. Accessed 1 April 2023.
53. INSM, 'Five Ways to a Better Energy Policy', https://www.insm.de/insm/kampagne/angebotspolitik/energie. Accessed 29 June 2023.
54. Manfred Redelfs, 'Greenpeace-Recherche: Lobbyisten führen Kreuzzug gegen Windkraft', https://www.greenpeace.de/ueber-uns/leitbild/investigative-recherche/netz-windkraftgegner; and Moritz Neujeffski and Max Goldenbaum, 'Klimawandelleugner*innen in Deutschland', https://www.nf-farn.de/klimawandelleugnerinnen-in-deutschland. Accessed 25 July 2023.
55. IEA (2023), *Energy Security*. Paris: International Energy Agency.

7

Climate Obstruction in the Netherlands

Strategic and Systemic Obstruction of Dutch Climate Policies (1980–Present)

MARTIJN DUINEVELD, GUUS DIX,
GERTJAN PLETS, AND VATAN HÜZEIR

INTRODUCTION: CLIMATE ACTION AND INACTION IN THE DUTCH POLDER

As Figure 7.1 shows, in 2017, two years after the 2015 Paris Agreement was adopted, the Netherlands emitted 191 million metric tonnes of carbon dioxide equivalents (MMT CO_2e) of greenhouse gases (GHGs). Compared with 1990 levels of 220 MMT CO_2e, that reduction amounted to 1.1 MMT CO_2e per year.[1] Dutch industry made the largest contribution to this relatively modest decline in emissions, which include a significant decrease in non-CO_2 emissions.[2] The years from 2019 to 2021 would see a more sudden drop in emissions.[3] This was partly a consequence of the COVID-19 pandemic lockdowns, however. In addition, the decline provides a distorted view of the impact of the Netherlands' CO_2 reduction efforts because GHG emissions connected to shipping, aviation, and other types of transportation are not included. This sector is especially relevant for a historically mercantile country like the Netherlands, with its large seaports in Rotterdam and Amsterdam and a major hub-oriented airport (Schiphol). Just as in other Western European nations, moreover, the Netherlands' emissions

Martijn Duineveld, Guus Dix, Gertjan Plets, and Vatan Hüzeir, *Climate Obstruction in the Netherlands* In: *Climate Obstruction across Europe*. Edited by: Robert J. Brulle, J. Timmons Roberts and Miranda C. Spencer, Oxford University Press. © Oxford University Press 2024. DOI: 10.1093/oso/9780197762042.003.0007

Netherlands Greenhouse Gas Emissions

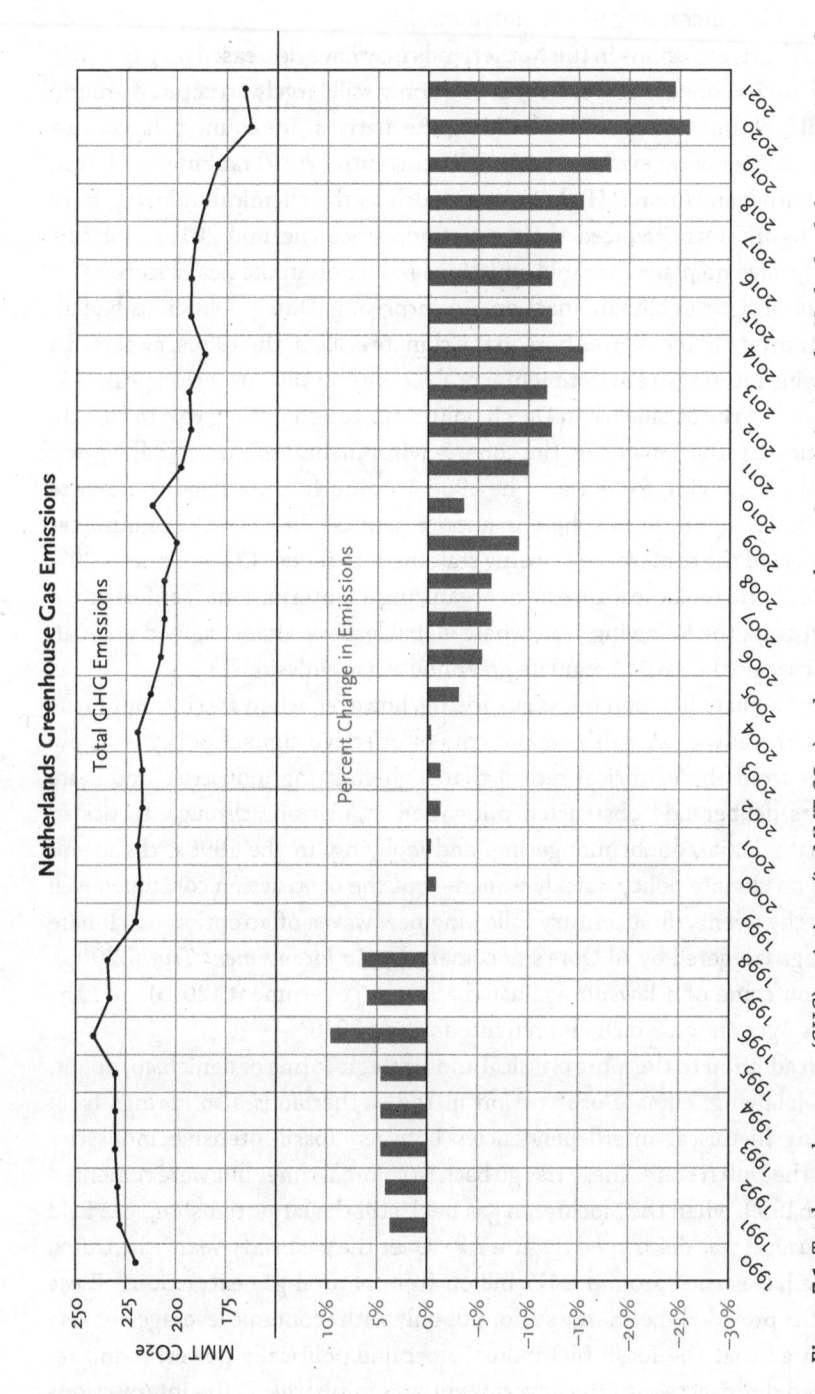

Figure 7.1 Total greenhouse gas (GHG) emissions (in MMT CO_2e) and percentage change in emissions in the Netherlands between 1990 and 2021, inclusive.

Source: Total GHG emissions based on data provided by Gütschow and Pflüger (2023) for Kyoto Six Greenhouse Gas Totals.

have been exported to the Global South over the past three decades as production has increasingly been outsourced.[4]

In short, emissions in the Netherlands may have decreased on paper, but the Dutch economy and society are not only still largely structured around fossil fuels but also behind in building alternatives. The country 'has been a slow adopter of renewable energy (RE), currently [2017] ranking 2nd last in the European Union'.[5] High emitters, such as the chemical industry, have not significantly reduced their emissions since the mid-2010s and currently have no plans for rapid emissions reduction in the near future.[6]

These signs of climate inaction are surprising. Dutch politicians had already begun to focus attention on the climate issue in the 1980s, as part of a growing interest in environmental problems generally. Attention peaked at the end of the decade when Dutch politicians took a leading role in climate politics internationally. At the time, 'environmental minister [Ed] Nijpels [was] . . . , trying to reorient the 1988 Toronto International conference on the Changing Atmosphere in a more political direction'.[7] The minister supported the conference's closing statement to reduce CO_2 emissions 20% by 2005. He took the lead, too, in organizing an international conference in Noordwijk the following year, where global leaders 'almost agreed upon an international treaty to regulate greenhouse gas emissions'.[8]

The climate inaction is less surprising, however, when we shift our focus from the advocates to the opponents of effective climate policy. It is evident from the historical record that high-emitting industries and state actors deliberately obstructed mitigation regulations through tactics of climate denial, doubt mongering, and lobbying. In the 1990s, the attention on climate policy quickly waned—but the obstruction continued well into the twenty-first century following new waves of attention on climate change triggered by Al Gore's documentary *An Inconvenient Truth* (2006), the outcome of a lawsuit against the Dutch government (2015), and the Paris Agreement, which entered into force in 2016.

In addition to the more classical and strategic forms of denialism, doubt, and lobbying, climate obstruction in the Netherlands also springs from strong historical interdependencies between fossil-intensive industries and the Dutch state. These ties go back to colonial times but were cemented after 1959, when the Slochteren gas field, still the largest onshore gas field in Europe, was discovered (Figure 7.2). Over the past sixty years, the Dutch state has earned around €417 billion from natural gas extraction.[9] These profits provided the energy sector not only with economic leverage but also ensured that the fossil fuel industry became politically powerful and received direct access to the government and ministries.[10] The intersections between industry, politics, and society at large, therefore, run deep, and

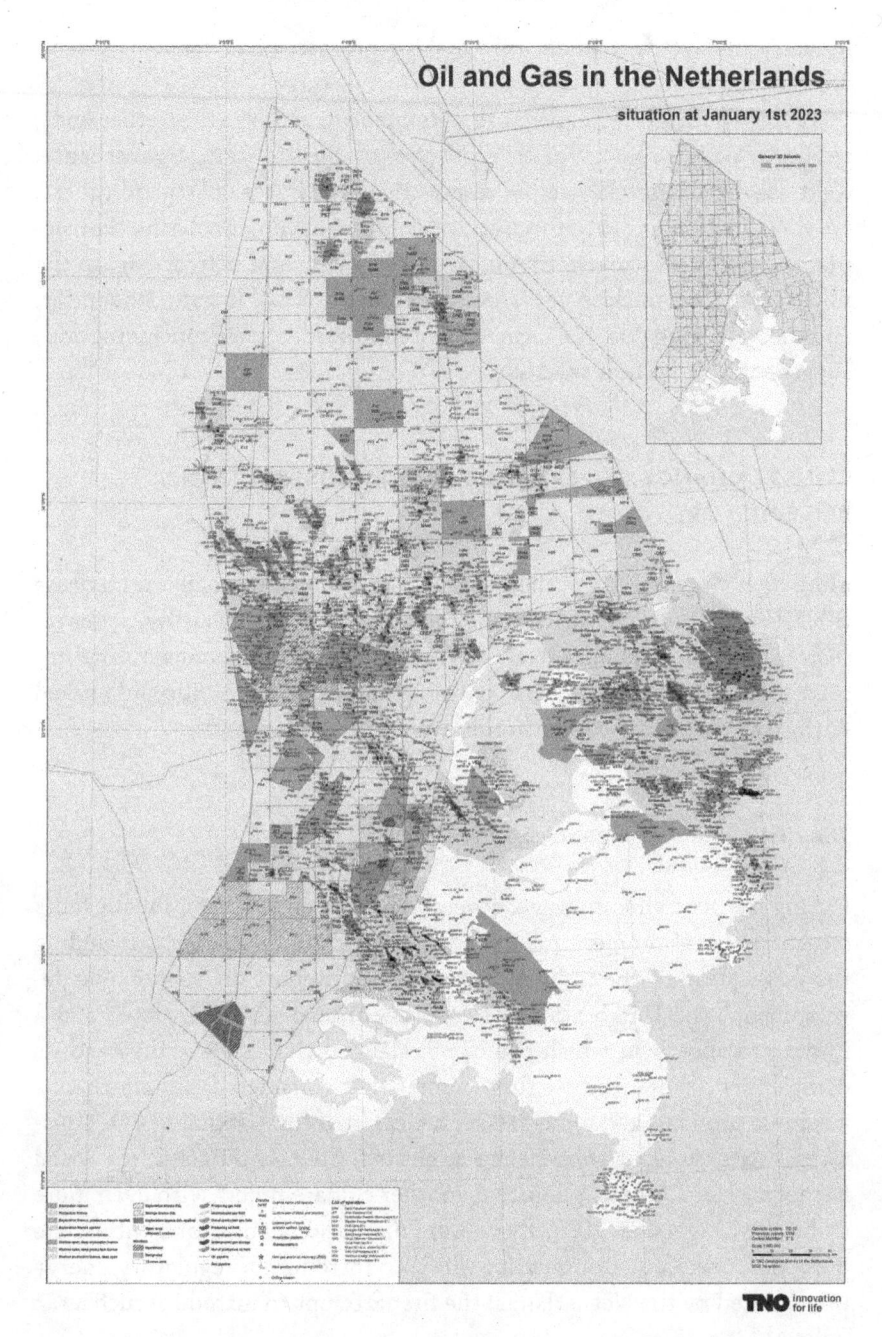

Figure 7.2 Petroleum (gas and oil) deposits in the Netherlands as of 2023. Although new gas fields are being discovered in the North Sea, the easternmost Slochteren field represents one of the largest land-based gas fields in Europe.
Source: https://www.nlog.nl/olie-en-gaskaarten-van-nederland.

industry involvement in decision-making processes has been completely normalized.

To cover both strategic and more systemic obstruction in the Netherlands, we begin with a history of three 'waves' of climate change governance. Next, we discuss the key actors responsible for climate obstruction there. We then analyse in depth three strategic forms of climate obstruction: denial and doubt, discursive framings, and lobbying and networking. In the final section, we analyse governance ideologies, fossil interdependencies, and the 'revolving door' as forms of systemic obstruction, concluding with suggestions for further research.

CLIMATE CHANGE GOVERNANCE AND ITS OBSTRUCTION: A BRIEF HISTORY

More than sixty years ago, the Royal Netherlands Meteorological Institute (KNMI) was already discussing the role of CO_2 in climate change,[11] yet research into climate change itself was limited and the phenomenon was not seen as an urgent problem. This pattern changed in the 1980s and peaked during the first climate wave in the latter part of the decade.

The first climate change wave (1987-1989)

During the first climate wave, several national and international events created societal momentum for addressing climate change,[12] including the publication of the influential Brundtland Report on 'sustainable development', the Dutch scientific report 'Concern for Tomorrow', and a Christmas speech in which the queen claimed that 'slowly, the earth is dying'. This resulted in the first cabinet that considered climate change a serious problem and aimed to set a clear goal for stabilizing CO_2 emissions.[13] After a new government was elected, the new minister, the Social Democrat Hans Alders, published another climate report with even more ambitious targets. In 1991, there were discussions and plans within the European Economic Community (EEC) to introduce a regulatory energy tax. Chaired by the Netherlands, the first attempt to introduce such a tax failed.[14]

The first wave of climate mitigation ambitions also gave rise to the climate obstructionist actors and their strategies and tactics. At the time of the (almost successful) multilateral Noordwijk climate conference, the 'godfather' of Dutch climate scepticism, chemistry professor

Frits Böttcher, began to receive funding from the fossil fuel industry and became a key 'merchant of doubt' in the Netherlands. In the early 1990s, climate sceptical arguments also made their appearance in both the House of Representatives (far right) and the Senate (Social Democrats).[15]

The first major obstruction of the proposed climate policies sprang from the conflict between the Ministry of the Environment and the Ministry of Economic Affairs. The latter, very much on the side of industry, feared that the former would become too powerful in 'determining energy policy via climate policy'.[16] The introduction of an energy tax was successfully resisted by the ministry, the Confederation of Netherlands Industry and Employers (then VNO), and the business community. Alluding to scepticism, Alexander Rinnooy Kan, VNO's chairman, argued that 'the greenhouse effect is certainly not uncontroversial'.[17] The Ministry of the Environment continued to plead for the energy tax but now faced the CEOs of major chemical and steel industries such as Akzo, DSM, Hoechst, Hoogovens, and Shell[18]—who lobbied Prime Minister Lubbers and other ministers not to implement the energy tax. A spokesman for the prime minister said afterward that 'no firm commitments' had been made but that 'the Netherlands will not be a guiding country' in Europe.[19]

The second major obstruction during the first climate wave was a surge of sceptic voices, including the right-wing, populist party leader Pim Fortuyn[20] (who got his inspiration from Frits Böttcher), and scientists/researchers Arthur Rörsch, Hans Labohm, and Salomon Kroonenberg. Partly, this surge was set against the backdrop of an ongoing rise of populism in the Netherlands in the early 2000s.[21] With climate change already ranking low on the political agenda, these voices 'made policymakers emphasize the importance of finding win-win solutions between the economy and the environment in climate policy' and push at the European level for a 'clean, clever, competitive' storyline of eco-efficiency during the Netherlands' 2004 EU Council Presidency.[22] Although it was not predominantly geared toward obstructing climate change policies, the populist rise can be seen as 'a sharp turning point in the framing and agenda setting of climate change in the Dutch public debate'.[23] Following the 2001 terrorist attacks in the United States and the assassination of Pim Fotuyn in 2002, the rise of Dutch populist parties saw a stronger polarization of society in which environmental issues in general and climate change in particular were portrayed as an 'elitist concern of the establishment'.[24] In the years thereafter, two consecutive right-wing cabinets cut back green ambitions and green budgets, symbolized by the replacement of a minister of the environment by a state secretary.

The second climate change wave (2006–2011)

In Al Gore's film *An Inconvenient Truth*, the Netherlands is pictured as half-flooded after one of the extreme climate scenarios discussed becomes reality. The film played in cinemas across the Netherlands' and triggered the second climate wave. It led Dutch Prime Minister Jan Peter Balkenende (a Christian Democrat) and the British Prime Minister Tony Blair (a Social Democrat) to call on their EU colleagues to address climate change.[25] The new Dutch cabinet again included a minister of environment, the Social Democrat Jacqueline Cramer, who presented an 'ambitious climate program aimed at 30% reduction in GHG emissions by 2020'.[26]

In line with the earlier emphasis on competition and eco-efficiency, this second climate wave was permeated with a 'green growth' ideology that took climate change as an opportunity for Dutch businesses. Besides the government and environmental nongovernmental organizations (NGOs), Dutch business leaders, too, now seemed to be on board in calling for change.[27] Their support, however, was reluctant at best. Leading up to the 2006 Dutch elections, the chairman of the VNO called on politicians to adhere less faithfully to the Kyoto Protocol, warning that 'soon we will be the only country that obediently sticks to Kyoto'.[28]

The ongoing obstruction by the VNO was aided by a sharp shift in the public framing of climate change in 2009. The controversy known today as 'Climategate', which centred on the hacked emails of climate scientists, led to a debate in which sceptical voices rang louder than before.[29] Climate scepticism now entered mainstream media, and a new climate sceptic website 'Climategate.nl' was established as a platform for discussing the emails.[30] In Dutch politics, political parties on the far right began to call for postponing decision-making on climate policy altogether. The far-right Party for Freedom (PVV) was the strongest denialist voice in parliament and gained real political power after the 2010 elections. A conservative minority coalition, authorized by the PVV, dissolved the Ministry of the Environment and stayed almost completely silent on climate change in the new coalition agreement.[31]

The third climate change wave (2015–2019)

As a result of a lawsuit filed by the Dutch NGO Urgenda, the court in The Hague ruled, in June 2015, that the state must do more to reduce GHG emissions in the Netherlands. Later that year, on 12 December, the

Paris Agreement was adopted,[32] 'requiring countries to come up with increasingly ambitious national climate plans . . . [to limit] the temperature increase to 1.5°C above preindustrial levels'.[33] Later, in 2017, a newly installed Dutch coalition government decided to develop a comprehensive and ambitious policy package to tackle climate change.[34] In line with new EU regulations, the overall goal was to reduce GHG emissions by 49% of 1990 levels by 2030.[35] To do so, conservative-liberal Minister of Economic Affairs and Climate Eric Wiebes appointed Nijpels, the 'first wave' minister of environment, to lead a Climate Assembly. The assembly consisted of a series of 'sector tables' on industry, electricity, construction, agriculture, and mobility at which civil servants had to co-design plans with major industrial stakeholders to decarbonize the Dutch economy.[36] The fossil fuel industry and the major high emitters were well-represented: Shell, RWE, BP, ExxonMobil, and Gasunie (a transboundary pipeline conglomerate) had a direct seat at the table. All the other high emitters were there, too, from Tata Steel and Yara (producer of fertilizers) to representatives of the ports of Rotterdam and Amsterdam.[37] After two years of dialogue, a National Climate Agreement was reached in 2019. By the time the Dutch government collapsed in the summer of 2023, however, it was still making mitigation plans with individual companies and industrial sectors.[38]

A seat at the table and close ties to the Ministry of Economic Affairs enabled the high emitters to lobby against and delay many regulatory policies that could curb emissions more quickly.[39] When the government has acted, it favoured 'positive' measures appreciated by industry, such as subsidies for more 'sustainable' oil refineries through technological solutions, or technofixes, such as carbon capture and storage (CCS).[40] Until recently, the government refused to abandon fossil subsidies estimated to be between €39.7 and €46.6 billion per year[41] and even sought to speed up the process for obtaining new drilling licenses for gas fields in the North Sea.[42] There is no indication that it is considering stricter regulations that could enforce a planned phase-out of fossil fuels or sectoral decline of polluting industries.[43]

The renewed emphasis on climate policymaking also relaunched climate denialism in the Netherlands. A new organization, CLINTEL, was established there in 2019, which operates on both the national and international levels (discussed later). The organization is affiliated with (former) politicians from the right-wing People's Party for Freedom and Democracy (VVD) but exerts influence on parties on the far right (PVV, Forum for Democracy).[44]

THE KEY DUTCH CLIMATE OBSTRUCTIONISTS

Over the course of climate governance history, several individuals and organizations became prominent players in directly opposing climate policy or in undermining such policies through misinformation or the promotion of fossil interests in the public sphere. In this section, we discuss the most important actors in Dutch climate obstruction.

The Dutch merchants of doubt

The Dutch merchants of doubt have been active since the first climate wave in the late 1980s.[45] In terms of size, number of publications and activities, and degree of financialization, they pale in comparison with their American counterparts.[46] As mentioned, the godfather of the Dutch sceptics was Frits Böttcher, a long-time advisor to Shell.[47] Böttcher was politically well integrated as a member of the Dutch conservative party (VVD) and government advisory councils.[48] In the 1990s, Böttcher received more than half a million euros from Shell and other Dutch multinationals[49] for a 'CO_2 project'. The project ended in 1998.[50] During that period and thereafter, he wrote climate sceptic reports, books, and opinion pieces and helped to establish a national[51] and international network of climate sceptics that included Fred Singer, the oil-funded denialist in the United States.

Böttcher's 'successor', Guus Berkhout, has a strikingly similar profile in the sense that both men 'are scientists, only started promoting climate scepticism after retirement, have a past at Shell, have been active members in the VVD and have never done climate science research'.[52] Together with journalist Marcel Crok and supported by Hans Labohm, 'Netherlands' most famous climate sceptic',[53] Berkhout founded the climate sceptic organization CLINTEL. Funded by two wealthy real estate owners, the organization campaigns against climate legislation. In doing so, they maintain close contacts with the Heartland Institute, the Canadian Friends of Science (an oil-industry-funded think tank), the European Climate Realist Network, and many known climate sceptics.[54]

The Ministry of Economic Affairs

The Ministry of Economic Affairs and Climate Policy, formerly known as the Ministry of Economic Affairs, is a powerful ministry in the Dutch political landscape. It describes itself as a ministry that 'promotes the Netherlands as a country of enterprise with a strong international competitive position and an eye for sustainability'.[55] In the history of climate governance, however, the ministry has proven to be a steady climate policy obstructor.[56] Where the former Ministry of the Environment favoured stricter regulations, the Ministry of Economic Affairs has always actively opposed an energy tax.[57] As a civil servant at the Ministry of the Environment recalled: '[Economic Affairs] blindly assumed what was put forward by Shell and the Confederation of Netherlands Industry and Employers. If industry didn't want it, the Ministry of Economic affairs didn't want it'.[58]

The Confederation of Netherlands Industry and Employers

The Confederation, now known as VNO-NCW, is the largest Dutch employers' organization and claims to represent 'the common interests of Dutch business, both at home and abroad'.[59] Representing the stakes of Dutch multinationals, including many industrial high emitters, 'successive cabinets have always taken the objections of VNO/NCW and the energy-intensive industry very seriously'.[60] Translated to the context of climate policy, this has meant that 'very few climate measures have been taken in the past twenty-five years to which this organisation raised major objections. Thanks in part to their influence, the hefty subsidies on fossil energy have also never been abolished'.[61] Since the first climate wave, and continuing to this day, VNO-NCW has been obstructing regulatory climate policies and measures through its privileged position in the policymaking process.[62]

Industry lobby groups

In addition to VNO-NCW as an official representative body, there are two important but largely invisible lobby groups in which companies join forces. The lobby group ABDUP—Akzo, Bataafse (Shell), DSM, Unilever, and Philips—is one of the oldest in the Netherlands, with long-standing access to key political players in The Hague. Since the 1980s, they have approached ministries or welcomed top officials to their own meetings and helped to shape 'the design of long-term visions and associated political agendas, and often provided the chairs of government advisory committees'.[63] Once a year, the 'President's Consultation' took place in luxury hotels, with the CEOs and prominent Dutch politicians such as Prime Minister Mark Rutte.[64]

The second lobby group, PHAUSD—a collaboration between the compagnies Philips, Hoogovens, Akzo, Unilever, Shell and DSM—was formed in 1978 with the explicit aim of monitoring developments in environmental legislation.[65] In that capacity, it regularly communicated with high-ranking civil servants in the Minister of Economic Affairs.[66] PHAUSD's lobbying practices can be characterized as policy 'sabotage', as it tended to mobilize to block new proposals for binding environmental legislation in favour of voluntary covenants between the government and industry.[67]

Shell

Shell presents itself as 'a global group of energy and petrochemical companies' that takes 'an innovative approach to help build a sustainable energy future'.[68] It is ranked number seven 'in the top 20 companies of carbon dioxide emitters since 1965'.[69] Formerly known as Royal Dutch Shell, it is not the only fossil company operating in the Netherlands. However, it has a special place in Dutch climate obstruction due to its strong historic links to politics and society. As explained earlier, the company has direct access to high-level politicians[70] and key ministries and works closely in public–private partnerships in hydrocarbon extraction.[71] Shell also held memberships in lobby groups that campaigned against climate action and undermined European renewable energy targets.[72]

In addition, Shell is also very visibly present in Dutch society. To protect its so-called licence to operate, Shell engages in advertisement campaigns that highlight its allegedly sustainable profile and sponsorship relations with cultural institutions,[73] forest agencies, and major newspapers.[74] In education, it provides teaching materials to schools, organizes energy festivals for children, serves on university boards, and is heavily involved in academic education and research.[75]

STRATEGIES, TACTICS, AND DISCURSIVE FRAMINGS

In the Netherlands, we can distinguish three main forms of strategic climate obstruction: the use of denial and doubt tactics, discursive framings that favour the interests of the fossil industry, and lobbying and networking campaigns.

Denial and doubt tactics

The Dutch merchants of doubt, introduced earlier, use several arguments and tactics in their campaigns to obstruct climate policies. To make their arguments, they draw predominantly on American sources.[76] For example, they argue that 'CO_2 is good for plants', question whether human activity influences global warming, and promote scientifically disproven alternative explanations for the phenomenon. In addition, they discredit climate scientists, dismissing them as guild-driven alarmists, and characterize the Intergovernmental Panel on Climate Change (IPCC) as a politically motivated body.[77] Since the 1990s, these Dutch merchants of doubt have spread disinformation via opinion pieces in newspapers and through contact with like-minded journalists and powerful political players.[78] The journal and conferences of the Netherlands' professional association of engineers, KIVI, played a supportive role for these Dutch climate sceptics.[79] Currently, denialist voices are still present in Dutch society, presented on self-created websites, a conspiracy-driven public broadcasting network (Ongehoord Nederland), and a large, right-wing newspaper (*De Telegraaf*).

The sceptic voices of CLINTEL are represented in the Dutch Parliament by the populist and right-wing political parties Forum for Democracy (FvD), PVV, and the VVD.[80] In the 1990s, the merchants of doubt were successful obstructors as their work led to 'a lack of political support for regulatory measures with regard to CO_2 reduction'.[81] According to Pier Vellinga, a now-retired professor of climate science, Fritz Böttcher was 'instrumental' in delaying climate policy in the Netherlands in the 1990s. 'His publications reached all the way up to the Department of Economic Affairs . . . , [they] never implemented any effective policy concerning CO_2 reduction'.[82] Although the influence of the Dutch merchants of doubt declined after 'Climategate' in 2009, they were still able to influence the VVD's campaign platform as recently as 2017.[83]

Discursive framings

With the growing public acceptance of climate change, especially since the third climate wave, many large companies have distanced themselves from climate sceptical discourse. Most now publicly acknowledge climate change and present themselves as part of the solution. The discursive framing tactics used in public debate and marketing campaigns have shifted from denial to delay.[84] Responsibility for climate action is placed on consumers, far-off technological solutions are promoted, and more structural solutions such as downscaling production are never discussed. The sustainability agenda of Royal Dutch Airlines (KLM), for instance, mirrors the discourses of climate delay almost perfectly by, for example, 'overstating the optimism of the technological projections, with reliance on unproven technological advancements'[85] and redirecting the obligation to mitigate carbon emissions 'to the consumers, the government, other airlines, and other industries'.[86]

The industries' discursive framings are particularly visible in what is popularly known as 'greenwashing' or, in industry-speak, as preserving their 'social licence to operate'.[87] For more than twenty years, for example, Shell has been hiring PR agency Edelman, known for its innovative tactics, to build public trust and keep restrictive legislation at bay.[88] Edelman developed the concept of a 'Generation Discover Festival' for Dutch children.[89] In this festival, Shell promoted a vision of the future of energy in which natural gas is a solution to climate change.[90] Discursive framings that lend legitimacy to Shell and its products also spring from their sponsorship of museums. One of the main Dutch science museums, Boerhaave, organized a Shell-sponsored temporary exhibition heralding the company as part of historical progress while downplaying its contemporary environmental impact.[91]

Scientific expertise is also enlisted to maintain public support for fossil fuels. After large-scale protests in 2012 in the north of the Netherlands due to heavy earthquakes caused by gas extraction, the fossil industry (Shell, ExxonMobil, and GDF Suez) partnered with the Dutch government and the Rotterdam School of Management in a two-year project that explicitly aimed for 'broader societal public support for gas as an energy carrier and a broadly supported "licence to operate" for the gas sector'.[92] The involvement of fossil industries in children's education, cultural exhibitions, and science enables these industries to frame their past, present, and future in a way that embodies an image of objectivity and positive values more convincingly than direct corporate statements.[93]

Lobbying and networking

The Dutch climate obstructors also seek to maintain their position in networks of government and universities to create informal opportunities to exchange information and protect their interests. The lobbying group PHAUSD, for example, had real 'lobby power' because of its direct relationships with high-level civil servants and ministers at the Ministry of Economic Affairs.[94] When binding environmental legislation instead of voluntary covenants was proposed, 'the reaction of the industry would be to bypass the policy process at the ministries. By using their connections, they would directly pressure the minister or representatives in parliament'.[95]

Companies also actively work to create and maintain networks with governments at different scales. Shell's Generation Discover children's festival, for instance, also created openings to cooperate with organizations and local governments, thus giving legitimacy to the company.[96] And, as some of the most important science funding bodies in the Netherlands, Shell and other corporations are able to maintain close ties with universities, research institutes, and the Dutch Research Council (NWO).[97] From the 1990s onward, these ties were further institutionalized by creating positions for industry on the management boards of Dutch universities and allowing sponsored professorships.[98]

SYSTEMIC OBSTRUCTION

In the history of Dutch climate governance, one thing is clear: the close ties between the Dutch state and business have been a major factor in obstructing many proposed climate policies.[99] We call this 'systemic obstruction'. Whereas the tactics of sowing doubt, using discursive framing, lobbying, and networking can be seen as active, intentional forms of obstruction, systemic obstruction is much more a tacitly understood way of thinking and acting that is engrained in individuals, institutions, and their relationships. Less visible, systemic obstruction is what makes active interventions so much easier—or sometimes even unnecessary when ideas and interests are aligned.

Three forms of systematic obstruction can be identified. The first is a distinctive governance ideology and practice that evolved in the Netherlands. In that so-called *polder model*, various stakeholders—employer organizations and unions, for example—are asked to engage in conversation and negotiations that are handled in extra-parliamentary settings.[100] The 'Climate Assembly' installed after the third climate wave is an excellent example. To reduce GHG emissions, the major emitters were invited to discuss sectoral reduction plans because they were expected to know best.

The second form of systemic obstruction is a historically grown interdependency between the Dutch state and particular companies. The history of KLM Royal Dutch Airlines is one example. Despite its environmental burdens, and against economic logic, the growth of aviation has been the main imperative, and taxpayers' money has been used to save this company from going bankrupt on multiple occasions.[101] The strong interdependencies between Shell and the Dutch state also stand out. Shell was founded in the Dutch East Indies (now Indonesia) when the Shell transporting company and the oil company Royal Dutch merged in 1907.[102] The collaboration between state and oil sector was strengthened in 1923, with the joint venture Dutch-Indian Oil Company (NIAM). This collaboration, in turn, served as a template for the establishment of the Dutch Oil Company (NAM), a joint venture of Shell and Exxon (then Standard Oil), in 1947.

As a recent investigation demonstrated, the interdependencies remain strong to this day: 'The [Dutch] Government was found to be tightly interwoven with the fossil fuel system, with ownership and financial relations found in all segments of the fossil fuel value chain, from production and exploration to use and R&D, and at the local, regional, as well as national levels of government'.[103] In 2022, a parliamentary investigation into gas extraction and earthquakes in the province of Groningen showed that the informal networks of the NAM and the Ministry of Economic Affairs led them to prefer economic yields and efficient extraction over citizen safety.[104]

The third form of systemic obstruction is the 'revolving door'. Again, Shell stands out: 'The "revolving doors" relationship between Shell and the Dutch government began in the early twentieth century and has been "flipping" ever since'.[105] Before he served as prime minister between 1933 and 1938, Hendrikus Colijn fought in the Dutch colonial wars to protect and expand petroleum concessions in Sumatra and was CEO of the Bataafse Petroleum Maatschappij, the Indonesian subsidiary of Shell, between 1914 and 1922.[106] A prominent postwar politician, Frits Bolkenstein worked for Shell from 1960 to 1976 before he became a minister and chair of the VVD and, as we have seen, a climate sceptic. To this list we can add many others.[107] The fact that there used to be a formal secondment for civil servants in which staff was exchanged between the Dutch Ministry of Foreign Affairs and Shell speaks for itself.[108]

CONCLUSION

The Netherlands seemed on its way to develop into a climate leader during its first wave of political attention to climate change. Two more waves followed, but each, unfortunately, were met with episodes of climate obstruction. The Dutch history of climate governance is therefore one of initial ambitions hampered by active doubt- and denial-generating tactics by the Dutch merchants of doubt, the networking and lobbying efforts of industries and lobby groups, and the narratives build on discursive framings by which companies and the political establishment, sometimes in tandem, have cultivated public support for fossil fuels.

Strategic obstruction has been made easier—or sometimes even unnecessary—by systemic climate obstruction, which aligns the Dutch state and the fossil fuel industry. The historic ties between the Ministry of Economic Affairs and the fossil lobby groups, for example, have enabled industries to gain easy access to the government to obstruct regulatory policies that would curb their emissions. Consequently, the Dutch government has focussed mainly on passing measures preferred by the fossil fuel industry, such as incentives and subsidies for techno-fixes supposedly designed to help it become more sustainable.

This chapter marks the beginning of a belated academic research enterprise focused on climate obstruction in the Netherlands. Apart from a few scientific articles, most of the available research in this area has been conducted by investigative journalists and NGOs. Social scientists can play a distinct yet complementary role in analysing climate obstruction by integrating existing investigations, deepening the existing body of theoretical work, and empirically studying new cases. A climate obstruction research agenda for the Netherlands should focus on both strategic and systemic obstruction as well as the ways in which these obstruction efforts have increasingly been resisted.

First, the field needs an ongoing mapping of the tactics that industries—from the chemical and fossil industries to aviation and 'Big Agro'—use to protect the status quo and curb stricter government regulation. Second, we need a more thorough analysis of the ways in which other societal actors seek to counteract obstruction tactics. For example, the protests in the northeast of the Netherlands after the gas-related earthquakes eventually led the government to stop gas extraction there. Similarly, many citizen initiatives are challenging the taken-for-granted ties between the fossil industry and their political, cultural, or scientific institutions. Although strategic and systemic obstruction of climate polices will not be gone overnight, it is now met by equally strategic attempts to obstruct obstruction.

NOTES

1. Or an average annual reduction rate of 0.49% over a twenty-seven-year period.
2. There were significantly less methane emissions in waste facilities and fluorinated gases in the 1990s because of a government ban and less nitrous oxide emissions in the production of nitric acid. See Centraal Bureau voor de Statistiek (2022, March), 'Uitstoot broeikasgassen 2,1 procent hoger in 2021',

https://www.cbs.nl/nl-nl/nieuws/2022/11/uitstoot-broeikasgassen-2-1-procent-hoger-in-2021.)

3. To 163 MMT (2020) and 166 MMT (2021)
4. P. Lucas, T. Maas, and M. Kok (2020), *Insights from Global Environmental Assessments: Lessons for the Netherlands*. The Hauge: PBL Netherlands Environmental Assessment Agency.
5. According to a report by Eurostad from 2017, cited in Sem Oxenaar and Rick Bosman (2020), 'Managing the Decline of Fossil Fuels in a Fossil Fuel Intensive Economy: The Case of the Netherlands'. In: Wood, Geoffrey, and Keith Baker (eds.), *The Palgrave Handbook of Managing Fossil Fuels and Energy Transitions*, p. 140. Basingstoke, UK: Palgrave Macmillan
6. Luuk Sengers and Evert de Vos (2023, April), 'De danse macabre van industrie en overheid', *De Groene Amsterdammer*, https://www.groene.nl/artikel/de-danse-macabre-van-industrie-en-overhei.
7. Art Dewulf, Daan Boezeman, and Martinus Vink (2017), 'Climate Change Communication in the Netherlands'. In: Art Dewulf, Daan Boezeman, and Martinus Vink (eds.), *Oxford Research Encyclopedia of Climate Science*, p. 7. New York: Oxford University Press, doi:10.1093/acrefore/9780190228620.013.455., p7
8. Bas van Beek, et al. (2023, September), 'For Nine Years, Multinationals Like Shell and Bayer Funded a Prominent Climate Denier', Follow the Money – Platform voor onderzoeksjournalistiek, https://www.ftm.nl/dutch-multinationals-funded-climate-sceptic.
9. CBS (2019, May), 'Natural Gas Revenues Almost 417 Billion Euros', Statistics Netherlands, https://www.cbs.nl/en-gb/news/2019/22/natural-gas-revenues-almost-417-billion-euros
10. For a more general picture on the Netherlands and the close ties between actors, see also Willeke Slingerland (2018), *Network Corruption: When Social Capital Becomes Corrupted: Its Meaning and Significance in Corruption and Network Theory and the Consequences for (EU) Policy and Law*. PhD-Thesis Vrije Universiteit Amsterdam.
11. Jeroen P. van der Sluijs, Rinie van Est, and Monique Riphagen (2010), *Room for Climate Debate: Perspectives on the Interaction between Climate Politics, Science and the Media*. The Hague, Netherlands: Rathenau Institute.
12. Dewulf, Boezeman, and Vink, 'Climate Change Communication in the Netherlands', p. 8.
13. Ibid., p. 7.
14. Because Britain and the southern member states disagreed. Wijnand Duyvendak (2011), *Het groene optimisme: het drama van 25 jaar klimaatpolitiek* Amsterdam: Bakker, p. 93.
15. van der Sluijs, van Est, and Riphagen, 'Room for Climate Debate', p. 20. 'The environmental minister responded that scientific uncertainties were an argument to engage in action, following the precautionary principle. However, action would prove tough in the episodes to come' (Dewulf, Boezeman, and Vink, 'Climate Change Communication in the Netherlands', p. 9). Such action, however, would face strong resistance in the years thereafter.
16. Duyvendak, *Het groene optimisme*, p. 83.
17. Ibid., p. 93.

18. Shell, in this chapter, refers to the international fossil fuel company formerly known as Royal Dutch Shell PLC, the holding company of the international Shell group, which is currently named Shell PLC.
19. Duyvendak, *Het groene optimisme*, p. 95.
20. Marcel Ham (2009, July), '"Ik ben jullie praatjes zat" — Milieudefensie', Milieudefensie.nl, https://web.archive.org/web/20090726065135/http://www.milieudefensie.nl/publicaties/magazine/2001/november/fortuyn.htm.
21. Duyvendak, *Het groene optimisme*, pp. 162, 168–169.
22. Dewulf, Boezeman, and Vink, 'Climate Change Communication in the Netherlands', p. 12.
23. Ibid., p. 11.
24. Ibid.
25. Duyvendak, *Het groene optimisme*, p. 194.
26. '[(Base year: 1990), a 20% share of renewable energy by the same year, and an annual energy saving rate of 2%'. See Dewulf, Boezeman, and Vink, 'Climate Change Communication in the Netherlands', p. 14.
27. Duyvendak, *Het groene optimisme*, p. 198.
28. Ibid., p. 206. See also Geraadpleegd (2023, 8 September), 'Klimaatverandering vraagt om nieuwe industriële revolutie – YouTube', geraadpleegd, https://www.youtube.com/watch?v=pamff3J3b64.; Heeft de EC het juiste beleid voor aanpak klimaatprobleem?, 2007, https://www.youtube.com/watch?v=4APsCCkJaZI.
29. Maarten A. Hajer (2012), 'A Media Storm in the World Risk Society: Enacting Scientific Authority in the IPCC Controversy (2009–10)', *Critical Policy Studies* 6, 4: 452–464.
30. Duyvendak, *Het groene optimisme*, p. 235.
31. Hajer, 'A Media Storm'.
32. At the United Nations Climate Change Conference in Paris (COP 21).
33. Dewulf, Boezeman, and Vink, 'Climate Change Communication in the Netherlands', p. 16.
34. Yara van Heugten and Jan Daalder (2023), 'De afspraken uit het Klimaatakkoord dienen vooral grote bedrijven, blijkt uit nieuw rapport', Follow the Money – Platform voor onderzoeksjournalistiek, maart, https://www.ftm.nl/artikelen/tno-klimaatakkoord-niet-geschikt-voor-fundamentele-transitie.
35. Douwe Truijens, et al. (2023, March), 'Wie schreef het klimaatakkoord?' TNO, https://energy.nl/publications/wie-schreef-het-klimaatakkoord/.
36. The minister also invited several environmental NGOs, but their negotiation power was limited by design. In contrast to the industrial actors, their signature on the final climate agreement was a nice bonus—but not necessary. See Douwe Truijens, et al., 'Wie schreef'.
37. Heugten and Daalder, 'De afspraken uit het Klimaatakkoord dienen vooral grote bedrijven, blijkt uit nieuw rapport'.
38. Ibid.
39. Platform Authentieke Journalistiek (2019, September), 'Lobby van multinationals blijkt kind aan huis bij ministeries', Follow the Money – Platform voor onderzoeksjournalistiek,, https://www.ftm.nl/artikelen/abdup-lobby-verweven-met-ministeries.
40. Reyes, Oscar, and Balanyá Belén (2016), 'Carbon Welfare: How Big Polluters Plan to Profit from EU Emissions Trading Reform', *Corporate Europe Observatory*

Report, pp. 22–23; Chris Hensen (2023, August), 'Geen enkel land dat inzet op CO2-opslag moet zich rijk rekenen', *NRC*, https://www.nrc.nl/nieuws/2023/08/18/geen-enkel-land-dat-inzet-op-co2-opslag-moet-zich-rijk-rekenen-a4172300.
41. https://www.volkskrant.nl/nieuws-achtergrond/meeste-fossiele-subsidies-naar-plastic-en-grootverbruikers-elektriciteit-slechts-klein-deel-wordt-afgebouwd~b5662648/ //// https://nltimes.nl/2023/09/04/dutch-govt-spends-eu375-billion-per-year-fossil-fuel-subsidies.
42. Alman Metten, 'Belastingvoordelen voor fossiele brandstoffen nóg veel groter', Mejudice, https://www.mejudice.nl/artikelen/detail/belastingvoordelen-voor-fossiele-brandstoffen-nog-veel-groter. Accessed 24 July 2023.
43. Cf. Holly Jean Buck (2021), *Ending Fossil Fuels: Why Net Zero Is Not Enough*. New York: Verso Books.
44. Ministerie van Economische Zaken en Klimaat (2023, July), 'Antwoorden op Kamervragen over fouten in laatste IPCC-rapport – Kamerstuk – Rijksoverheid.nl', kamerstuk (Ministerie van Algemene Zaken), https://www.rijksoverheid.nl/documenten/kamerstukken/2023/07/03/beantwoording-kamervragen-over-ernstige-fouten-in-het-laatste-ipcc-rapport. In the juridical domain, CINTEL tried to be accepted as a third party in court cases of Friends of the Earth against Shell. Although that attempt failed, a foundation of 'Concerned Energy Users', with similar climate denialist credentials, did manage to get accepted. (Paul Luttikhuis (2023, April), 'Stichting van "bezorgde energiegebruikers" toegelaten tot rechtszaak Milieudefensie tegen Shell', *NRC*, https://www.nrc.nl/nieuws/2023/04/25/stichting-van-bezorgde-energiegebruikers-toegelaten-tot-rechtszaak-milieudefensie-tegen-shell-a4163077.)
45. van der Sluijs, van Est, and Riphagen, 'Room for Climate Debate'.
46. Jan Paul van Soest (2014), *De twijfelbrigade: Waarom de klimaatwetenschap wordt afgewezen en de wereldthermostaat 4 graden hoger gaat*. Amsterdam: Uitgeverij Mauritsgroen.; Platform Authentieke Journalistiek (2020 March), 'De boekhouding van "klimaatscepticus" Böttcher', Follow the Money – Platform voor onderzoeksjournalistiek, https://www.ftm.nl/artikelen/geldstromen-boekhouding-bottcher.
47. Platform Authentieke Journalistiek (2020, March), 'De boekhouding van "klimaatscepticus" Böttcher', Follow the Money – Platform voor onderzoeksjournalistiek, https://www.ftm.nl/artikelen/geldstromen-boekhouding-bottcher. See also Alexander Beunder (2020, February), 'Hoe Frits Böttcher met steun van tientallen bedrijven de basis legde voor de klimaatscepsis in Nederland', *de Volkskrant*, , sec. Topverhalen vandaag, https://www.volkskrant.nl/nieuws-achtergrond/hoe-frits-bottcher-met-steun-van-tientallen-bedrijven-de-basis-legde-voor-de-klimaatscepsis-in-nederland~b1accbaf/.
48. He was known to a wide audience as 'the co-founder and former chair of the Dutch branch of the Club of Rome'(see Platform Authentieke Journalistiek, 'De boekhouding van "klimaatscepticus" Böttcher'.)
49. Like AkzoNobel, Hoogovens, NAM, Gasunie, Texaco, and Schiphol.
50. Beunder, 'Hoe Frits Böttcher met steun van tientallen bedrijven de basis legde voor de klimaatscepsis in Nederland'.
51. Which included journalist Simon Rozendaal, economist Hans Labohm, and VVD politician Hans Wiegel (Pointer (2020, February), 'Klimaattwijfel zaaien met hulp van oliegeld en populistisch rechts netwerk', KRO-NCRV, https://pointer.kro-ncrv.nl/klimaattwijfel-zaaien-met-hulp-van-oliegeld-en-populistisch-rechts-netwerk; van Soest, *De twijfelbrigade*, p. 49.

52. Pointer, 'Klimaattwijfel zaaien met hulp van oliegeld en populistisch rechts netwerk'.
53. According to his Wikipedia page: 'Hans Labohm', Wikipedia, September 2023, https://nl.wikipedia.org/w/index.php?title=Hans_Labohm&oldid=65901604.
54. Pointer, 'Klimaattwijfel zaaien met hulp van oliegeld en populistisch rechts netwerk'; Mayte Moreno-Soldevila (2022), 'Androcentrism and Conservatism within Climate Obstructionism. The Case of the Think Tank CLINTEL in The Netherlands', *Ámbitos. Revista Internacional de Comunicación*, 55: 41–57, doi:10.12795/Ambitos.2022.i55.03. See, for example, Frits Bolkestein (2017, January), 'About Intellectuals and Climate Alarmism • Watts Up With That?', https://wattsupwiththat.com/2017/01/20/about-intellectuals-and-climate-alarmism/.
55. Landbouw en Innovatie Ministerie van Economische Zaken (2023, August), 'Ministry of Economic Affairs and Climate Policy', organisatie, Government.nl (Ministerie van Algemene Zaken), https://www.government.nl/ministries/ministry-of-economic-affairs-and-climate-policy.
56. Duyvendak, *Het groene optimisme*.
57. Ibid., pp. 92–96. According to Minister of the Environment, from 1994 to 1998, Margreeth de Boer, 'compulsory measures such as a CO_2 tax were on the table every now and then', but there was never enough support for them. . . . It was mainly the Department of Economic Affairs that thwarted such measures' (Platform Authentieke Journalistiek, 'De boekhouding van "klimaatscepticus" Böttcher').
58. Duyvendak, *Het groene optimisme*, p. 83.
59. VNO-NCW (2015, October), 'VNO-NCW in brief', https://www.vno-ncw.nl/over-vno-ncw/english.
60. Duyvendak, *Het groene optimisme*, p. 289.
61. Ibid.
62. Ibid., pp. 226, 331.
63. Platform Authentieke Journalistiek (2019, September), 'Lobby van multinationals blijkt kind aan huis bij ministeries', Follow the Money – Platform voor onderzoeksjournalistiek, https://www.ftm.nl/artikelen/abdup-lobby-verweven-met-ministeries.
64. And among others Minister of Economic Affairs Maxime Verhagen, Minister of Finance Jeroen Dijsselbloem and European Commissioner on climate Frans Timmermans (Platform Authentieke Journalistiek.)
65. Platform Authentieke Journalistiek.
66. Duyvendak, *Het groene optimisme*, p. 74.
67. Ibid.
68. Shell (2023, 30 August), 'About Us | Shell Global', geraadpleegd, https://www.shell.com/about-us.html.
69. Rhodante Ahlers and Ilona Hartlief (2021), *Still Playing the Shell Game: Four Ways Shell Impedes the Just Transition*. Amsterdam: SOMO; Matthew Taylor and Jonathan Watts (2019, October), 'Revealed: The 20 Firms behind a Third of All Carbon Emissions', *The Guardian*, sec. Environment, https://www.theguardian.com/environment/2019/oct/09/revealed-20-firms-third-carbon-emissions.
70. Platform Authentieke Journalistiek (2019, September), 'Lobby van multinationals blijkt kind aan huis bij ministeries', Follow the Money – Platform voor onderzoeksjournalistiek, https://www.ftm.nl/artikelen/abdup-lobby-verweven-met-ministeries.

71. Joost Jonker and Jan Luiten van Zanden, (2007) 'Van nieuwkomer tot marktleider, 1890–1939: Geschiedenis van Koninklijke Shell, deel 1', Amsterdam: Boom; The House of Representatives of The Netherlands, et al. (2023, February), 'Groningers before Gas: Parliamentary Committee of Inquiry into Natural Gas Extraction in Groningen' Den Haag: The House of Representatives of The Netherlands.
72. Jelmer Mommers and Damian Carrington (2017, February), 'If Shell Knew Climate Change Was Dire 25 Years Ago, Why Still Business as Usual Today?', The Correspondent, https://thecorrespondent.com/6286/if-shell-knew-climate-change-was-dire-25-years-ago-why-still-business-as-usual-today/692773774-4d15b476. From the mid-1970s, Shell was aware of climate change and the contribution of their fossil products; the company 'shaped a series of influential industry-backed publications that downplayed or omitted key risks; emphasized scientific uncertainties; and pushed for more fossil fuels, particularly coal'. Desmog (2023, 30 August),'Lost Decade: How Shell Downplayed Early Warnings Over Climate Change – DeSmog', https://www.desmog.com/2023/03/31/lost-decade-how-shell-downplayed-early-warnings-over-climate-change/.)
73. In this way, Shell has also taken over some responsibilities of the neoliberalizing Dutch state since the 1990s. Gertjan Plets en Marin Kuijt (2022, March), 'Gas, Oil and Heritage: Well-Oiled Histories and Corporate Sponsorship in Dutch Museums (1990–2021)', *BMGN – Low Countries Historical Review*, 137, 1: 50–77, doi:10.51769/bmgn-lchr.7028.
74. Femke Sleegers (2021), 'The Dirty Truth about Shell's Children's Marketing', *Future Beyond Shell* (blog), https://futurebeyondshell.org/greenwashing-the-dirty-truth/. See also René Didde (1989), *Als het tij keert: Shell en Nederland, macht & verbeelding*. Amsterdam: Ravijn; Rhodante Ahlers and Ilona Hartlief, 'Still Playing the Shell Game'; Mommers and Carrington, 'If Shell Knew Climate Change Was Dire'; George Monbiot (2019, June), 'Shell Is Not a Green Saviour: It's a Planetary Death Machine', *The Guardian*, sec. Opinion, https://www.theguardian.com/commentisfree/2019/jun/26/shell-not-green-saviour-death-machine-greenwash-oil-gas.
75. Vatan Hüzeir and Germain Fraser (2017), *A Pipeline of Ideas: How the Rotterdam School of Management Facilitates Climate Change by Collaborating with the Fossil Fuel Industry*. Rotterdam: Changerism; NOS (2023, February), 'Bedrijven betalen salaris 200 hoogleraren, transparantie ontbreekt vaak', https://nos.nl/artikel/2464193-bedrijven-betalen-salaris-200-hoogleraren-transparantie-ontbreekt-vaak.
76. van Soest, *De twijfelbrigade*; Dewulf, Boezeman, and Vink, 'Climate Change Communication in the Netherlands'; Böttcher, 'De boekhouding van "klimaatscepticus" Böttcher' – Follow the Money – Platform voor onderzoeksjournalistiek'.
77. van Soest, *De twijfelbrigade*; Dewulf, Boezeman, and Vink, 'Climate Change Communication in the Netherlands'.
78. Böttcher was an often-seen visitor to the Energy Department at the Minister Economic Affairs. We are not aware if his successors were still welcome there. Böttcher, 'De boekhouding van "klimaatscepticus" Böttcher – Follow the Money – Platform voor onderzoeksjournalistiek'.
79. Pointer, 'Klimaattwijfel zaaien met hulp van oliegeld en populistisch rechts netwerk'.
80. Ibid.

81. Böttcher, 'De boekhouding van "klimaatscepticus" Böttcher – Follow the Money – Platform voor onderzoeksjournalistiek'.
82. Ibid.
83. 'The sentence that climate change leads to "rising sea levels and heavy rainfall" was removed from the text at the insistence of Frisian VVD member Gert-Jaap van Ulzen, also a board member of the climate sceptic think thank Groene rekenkamer'. See Belia Heilbron, Thomas Muntz, and Frank Straver (2017, March), 'VVD schrapt zin over klimaatverandering na kritiek', Trouw, https://www.trouw.nl/nieuws/vvd-schrapt-zin-over-klimaatverandering-na-kritiek~b401a357/.)
84. William F. Lamb, et al. (2020), 'Discourses of Climate Delay', *Global Sustainability*, 3: 1–5, doi:10.1017/SUS.2020.13.
85. Jimena Natalia Diamint (2023), *In the Clouds or on Solid Ground?: Mapping KLM's Dominant and Counter-Discourses on Aviation and Climate Change*, p. 71. Dissertation Wageningen University, WorldCat.org, https://edepot.wur.nl/631194.
86. Ibid., p. 76.
87. Which is one of Shell's top three strategic ambitions stated in its 2019 business strategy report. See Shell (2023, 30 March), 'Our Business Strategy', https://reports.shell.com/sustainability-report/2019/introduction/our-business-strategy.html. See also Mike Gaworecki (2016, July), 'Inside Shell's PR Strategy To Position Itself As A "Net-Zero Emissions" Leader', *DeSmog* (blog), https://www.desmog.com/2016/07/10/inside-shell-s-pr-strategy-position-itself-net-zero-emissions-leader/.
88. Tobacco Tactics (2023, 30 March), 'Edelman – TobaccoTactics', https://tobaccotactics.org/wiki/edelman/.
89. According to the organization, 35,000 people visited the 2020 Rotterdam edition of the festival, 'which included a record number of over 300 school classes with about 13,000 pupils'. See Sleegers, 'The Dirty Truth'.
90. Nationaleonderwijsgids (2023, 27 March), 'Fossielvrij Onderwijs tegen misleidende reclame Shell-kinderfestival in beroep', https://www.nationaleonderwijsgids.nl/cursussen/nieuws/50891-fossielvrij-onderwijs-tegen-misleidende-reclame-shell-kinderfestival-in-beroep.html. Shell did not promote gas in the last edition because they lost a court case initiated by the Dutch advertising watchdog. See Yamilla van Dijk (2019), 'Shell afgestraft na misleidende reclame: "GTL draagt bij aan beter milieu"', AD.nl, https://www.ad.nl/den-haag/shell-afgestraft-na-misleidende-reclame-gtl-draagt-bij-aan-beter-milieu~a423acbd/.)
91. Plets and Kuijt, 'Gas, Oil and Heritage'.
92. Vatan Hüzeir and Germain Fraser, *A Pipeline of Ideas*, pp. 17–18.
93. Mark W. Rectanus (2002), *Culture Incorporated: Museums, Artists, and Corporate Sponsorships*. Minneapolis: University of Minnesota Press.
94. Platform Authentieke Journalistiek, 'Lobby van multinationals blijkt kind aan huis bij ministeries'.
95. Gijs A. Diercks (2012), 'Explaining Dutch Failure and German Success in Renewable Energy Policymaking: An Agency/Structure Perspective', Utrecht University, p. 49.
96. 'Employees explaining, they are collaborating with many other parties to organise the festival together. And every year, the festival was opened by a leading executive or public official. Politicians, journalists, scientists, mayors and

ministers were invited and gave speeches, like the Dutch Prime Minister and the mayor of Rotterdam' (Sleegers, 'The Dirty Truth').

97. A well-studied example is the Rotterdam School of Management (RSM). In 1966, a select group of multinational companies, closely linked to each other through a federation of employers, donated between €300,000 and €1,000,000 each to establish a new graduate school of management. Royal Dutch Shell was one of the founding companies. Many fossil fuel companies, such as Petrobras, Saudi Aramco, Vale, and Shell, are clients of the institute and two of them—Shell and Petrobras—are executive members of its communication network. See Hüzeir and Fraser, *A Pipeline of Ideas*.

98. Jeanne Westerberg (1997 May), 'Universiteit wordt kennisleverancier', *de Volkskrant*, sec. Topverhalen vandaag, https://www.volkskrant.nl/nieuws-achtergrond/universiteit-wordt-kennisleverancier~b3f0a529/. Resident-director Jan Slechte of Shell Nederland, for instance, became chairman of the supervisory board of the Technical University of Delft; and Cor Herkstroter, President-Director of Royal Dutch Shell, is appointed chairman of the supervisory board of Erasmus University Rotterdam Nederland and chairman of the supervisory board of the Technical University of Delft. (Ardi Vleugels and Jasper Been (2023, 31 July), 'Voor het eerst in kaart gebracht: wie betalen onze hoogleraren?', FD.nl, https://fd.nl/samenleving/1459676/voor-het-eerst-in-kaart-gebracht-wie-betalen-onze-hoogleraren.) For an overview in progress of ties between Dutch Universities and the fossil industry, see Solid Sustainability Research (2023, March), 'Fossiele industrie en de academie: een onderzoek naar banden tussen universiteiten en olie- en gasbedrijven – Solid Sustainability Research', https://www.solid-sustainability.org/nl/banden-fossiel-unversiteiten. For the ties between Wageningen University and the Fossil industry, see Brigitte W. (2022, June), 'Fossil Fuelled WUR', *The Jester* (blog), https://www.thejesterwageningen.nl/2022/06/10/fossil-fuelled-wur/.

99. Duyvendak, *Het groene optimisme*; Dewulf, Boezeman, and Vink, 'Climate Change Communication in the Netherlands'.

100. Martijn Duineveld and Guus Dix (2022, July), 'Voor een leefbare planeet moet het poldermodel de prullenbak in', *Beleid en Maatschappij* 49, 3: 246–249, doi:10.5553/BenM/138900692022049003006. See also Florian Kern and Michael Howlett (2009, November), 'Implementing Transition Management as Policy Reforms: A Case Study of the Dutch Energy Sector', *Policy Sciences*, 42, 4: 391–408, doi:10.1007/s11077-009-9099-x.

101. Ties Joosten (2022), *De blauwe fabel. Waarom we de KLM al een eeuw lang tegen elke prijs in de lucht houden*. Amsterdam: Follow the Money.

102. Sam Gerrits (2021), *De aarde en het gas een geschiedenis van fossiele brandstoffen, geld en macht*. Groningen: Uitgeverij Passage.

103. Oxenaar and Bosman, 'Managing the Decline of Fossil Fuels', p. 158.

104. J. A. Vijlbrief (2023, July), 'Brief van de staatssecretaris Economische Zaken en Klimaat. Tweede Kamer, vergaderjaar 2022–2023', www.tweedekamer.nl%2Fdownloads%2Fdocument%3Fid%3D2023D31720&usg=AOvVaw1B73qZgDykDXfQzZhj8GoE&opi=89978449. The House of Representatives of The Netherlands, et al., 'Groningers before Gas. Parliamentary Committee of Inquiry into Natural Gas Extraction in Groningen'.

105. Ahlers and Hartlief, *Still Playing the Shell Game*.

106. Ibid., p. 44.

107. Such as Wouter Bos (Labor Party), the former Minister of Finance and Vice Minister President; Eric Wiebes (VVD), former Minister of Economic Affairs and Climate; Sigrid Kaag (social liberals), Minister for Foreign Trade and Development Cooperation; and Wobke Hoekstra (Christian democrat), Minister of Foreign Affairs and candidate European Commissioner responsible for the European Green Deal.
108. Ahlers and Hartlief, *Still Playing the Shell Game*, p. 44.

8
Climate Obstruction in Poland

A Governmental–Industrial Complex

KACPER SZULECKI, TOMAS MALTBY, AND JULIA SZULECKA

INTRODUCTION: ADDICTED TO COAL?

Despite the recent rapid deployment of renewable energy sources, primarily solar, Poland remains Europe's most coal-dependent economy.[1] For more than two decades, governments treated this admittedly challenging departure point as an argument for the 'unique treatment' of Poland in European and global climate protection efforts. Since the nation's accession into the European Union in 2004, consecutive Polish governments have been veto players on more ambitious climate policy initiatives and decarbonization targets.

The International Energy Agency (IEA) argues that if the world is to follow a pathway to limit global warming to 1.5°C, all members of the Organisation for Economic Cooperation and Development (OECD) need to have phased out coal by 2030.[2] Meanwhile, at the 2018 United Nations climate summit (COP 24) hosted by Poland, President Andrzej Duda stated that 'there is no plan today to fully give up on coal' and that Polish supplies would last 200 years.[3]

After Poland vetoed the European Union's 2050 net zero emissions target in 2019, European Union managed to adopt it later, with a caveat: the European Council noted that 'one Member State [Poland], at this

---

Kacper Szulecki, Tomas Maltby, and Julia Szulecka, *Climate Obstruction in Poland* In: *Climate Obstruction across Europe*. Edited by: Robert J. Brulle, J. Timmons Roberts and Miranda C. Spencer, Oxford University Press.
© Oxford University Press 2024. DOI: 10.1093/oso/9780197762042.003.0008

stage, cannot commit to implement this objective'.[4] Poland's 2040 energy strategy, finalized in 2021, envisages electricity generation from coal in 2030 at a level higher than the European Commission's assessment of the European Union's total coal budget,[5] with 11% of its energy still provided by coal in 2040,[6] and coal mining phased out only by 2049.[7] Poland is one of only two EU member states with no coal power phase-out target, while twenty-one of twenty-seven other member states have committed to phasing out coal by 2030 as part of the Powering Past Coal Alliance, a voluntary grouping of states, regions, and cities aiming to accelerate coal phaseout.[8] The 'dirtiest' coal plant in Europe in terms of emissions is Bełchatów, in central Poland,[9] and Poland was also the only EU member state that added new coal capacity in 2021.[10]

Within Polish society, there are clear signals of a change in societal attitudes toward the climate crisis, especially since 2018, when Poland hosted COP 24 in Katowice. This event coincided with the emergence of new climate protest initiatives, linked to the global Fridays for Future and 'School Strike for Climate' movement as well as Extinction Rebellion's protests. Droughts, heat waves, and Europe's poorest air quality—thirty-six of Europe's fifty most air polluted cities are located in Poland[11]—have also contributed to raising awareness of human activity's environmental impacts. The divergence between government policy and the expectations of ambitious climate action among a growing part of society is becoming increasingly apparent.

As of 2023, Poland remains the sole EU country not committed to the net zero 2050 objective, citing 'the difficult starting point of the Polish transition and its social and economic aspects'.[12] In this chapter, we argue that Poland's insufficient climate protection efforts cannot be justified by a difficult point of departure.[13] They are instead the result of different forms of climate obstruction, some of them strategic and intentional.

Historical emissions

The collapse of Poland's centrally planned communist economy between 1988 and 1990 resulted in many energy- and carbon-intensive industrial facilities closing, contributing to a sharp reduction in national emissions between 1987 and 1990 and a further drop between 1996 and 2002 before stabilizing for the next two decades (Figure 8.1).[14] This historic change meant, however, that Poland was able to meet both European and international (i.e. the Kyoto Protocol's) reduction targets without additional effort or a conscious climate policy. When the European Union set a collective

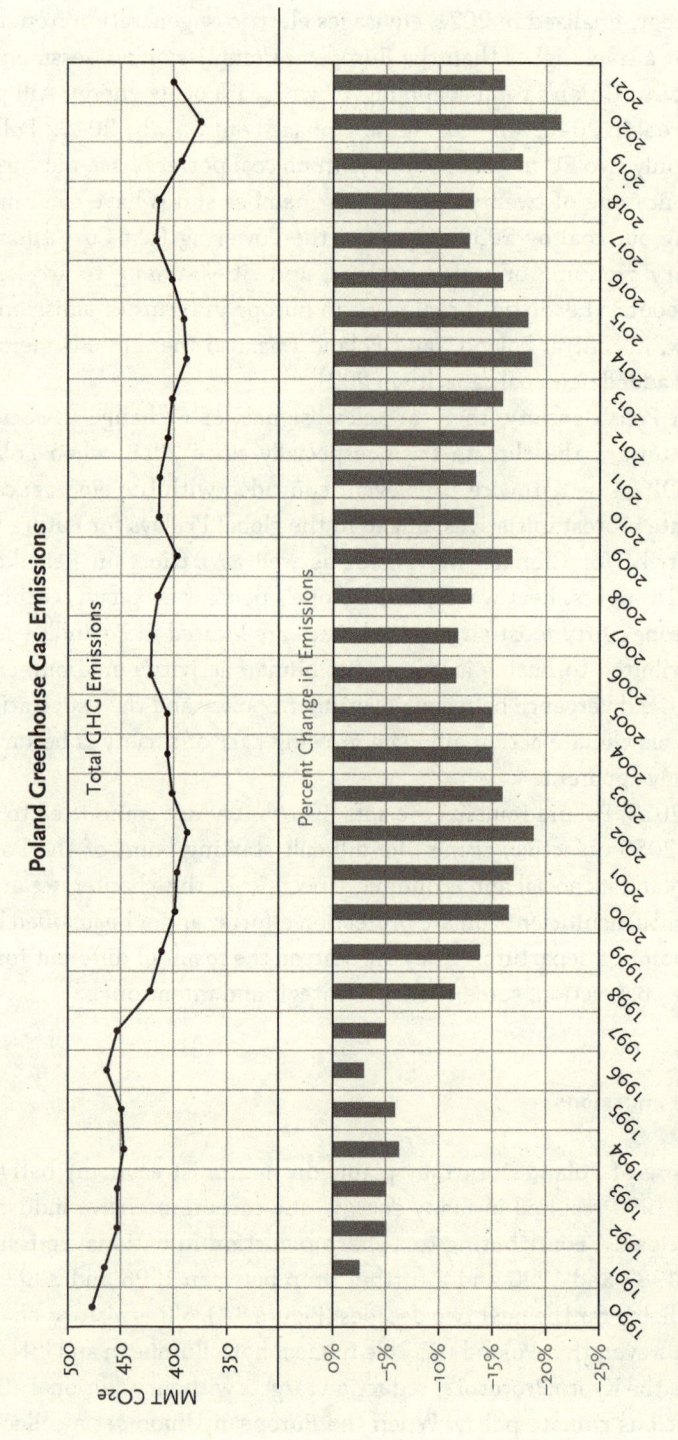

Figure 8.1 Total greenhouse gas (GHG) emissions (in MMT CO_2e) and percentage change in emissions in Poland between 1990 and 2021, inclusive.
Source: Total GHG emissions based on data provided by Gütschow and Pflüger (2023) for Kyoto Six Greenhouse Gas Totals.

2020 net emissions reduction target of 20% based on a 1990 baseline and successfully achieved this goal with a 32% reduction,[15] Poland decreased its emissions by 20.1%.[16] It also achieved its 2020 renewable energy target of 16% (reaching 16.1%) but fell short of its energy efficiency target.[17]

Poland's per capita greenhouse gas (GHG) emissions peaked in 1980 at more than 500 million metric tonnes of carbon dioxide equivalents (MMT CO_2e) and were 477 MMT CO_2e in 1990.[18] The country's share of global responsibility is 0.96%.[19] On a pathway to limit global warming to 2°C, Poland's fair-share emissions for 2030 are a reduction from 1990 levels of 45% (to 260 MMT CO_2e), 55% for a 1.5°C scenario, or 67% (to 155 MMT CO_2e) for a 1.5°C LED scenario.[20] However, the European Union's collective proposed reduction of 55% by 2030[21] is considered insufficient by Climate Action Tracker.[22] Poland's goal is far less ambitious at 'approximately 30%', and far from the global or EU fair share.[23]

HISTORICAL BACKGROUND: HOW POLAND BECAME COAL LAND

When Poland regained its independence after World War I, a significant part of Upper Silesia—which had become an industrial powerhouse of the German Empire thanks to large coal reserves—was awarded to the Poles, together with its mines and industrial facilities. This gave Poland indigenous coal resources, large enough to make the country energy independent, and serving as a base for industrial expansion and a profitable export industry.

Then, during the reconstruction efforts after World War II, Poland conducted a massive electrification program based mainly on hard-coal power plants. Post-war territorial changes increased indigenous coal resources considerably and by 1980, the country had become the second-largest producer of coal in Europe, after the Soviet Union. Coal mining and heavy industry were promoted by the communist authorities as the foundation of post-war prosperity, but also as a source of national and working-class pride.[24] As such, the notion of coal as 'black gold', Poland's invaluable treasure, and the figure of the selfless miner sacrificing himself for the benefit of society became deeply engraved in the Polish collective imagination of technological progress.[25] Coal is inextricably linked by politicians with Poland's role in the European Union and the world—Prime Minister Beata Szydło stated that coal was 'a synonym of development and modernity'.[26] This notion prevails, and the symbolic importance of coal is acknowledged even by Green Party MPs: 'We are dependent on coal. Not only in the energy sector and in the economic sense, but also in our national mentality . . . associated with some idea of patriotism, independence, and energy security'.[27] Attacks on coal are cast as 'an attack on sovereignty',[28] particularly by the far-right political forces.[29]

In the 1970s, the communist authorities constructed several large-scale hydropower plants and explored developing nuclear energy capacity to limit coal dependency. However, protest campaigns in the second half of the 1980s and the transition from communism to capitalism halted these plans in 1990.[30] Meanwhile, lignite (brown coal) was gaining prominence, and the Bełchatów plant, completed in 1988, became Europe's largest coal power plant and one of the largest in the world. As many as 388,000 people—roughly 1% of the entire population—were employed in the coal sector in 1990, when close to 100% of electricity was generated from coal.[31]

Following the political transition and in response to ecological catastrophe, to which the coal sector greatly contributed, the period 1990–1991 saw an eruption of environmental legislation in all domains, including energy.[32] In 1990, an energy policy until 2010 stated that 'environmental protection should be the main factor influencing the choice of energy sources' and indicated renewables as the preferred solution.[33]

By 2021, the share of electricity generated from coal was down to 72%[34] and the number of people directly employed in hard and brown coal mining had fallen to under 75,000,[35] though this still represented approximately half the coal mining jobs in the European Union.[36] Nineteen million people continued to use coal for winter heating, and 80% of private homes in the European Union using coal are located in Poland.[37] Eighty-seven percent of all coal consumed in EU homes in 2019 was in Poland, using 10 million tonnes, half mined in the country.[38] In 2021, Poland had thirty-four coal mines,[39] seventeen coal-powered plants, and more than twenty combined heat and power facilities using coal.[40] The majority of coal power plants were built between 1960 and 1980 and are nearing retirement. It is estimated that, by 2030, Poland may lose 41,000 jobs in the sector.[41] Employment in, loyalty toward, and support for the sector are heightened by a social multiplier effect,[42] and miners are highly respected in society, more so than teachers, doctors, and professors.[43]

As a result, the dominant perception of what is in Poland's national interest is a *just transition*.[44] This is within a context in which coal miners wield significant political power (mining jobs are concentrated in the southern region of Upper Silesia—which elects 12% and 13% of seats in the lower (Sejm) and the upper (Senate) houses of parliament, respectively). The 2015–2023 government was close to the 'Solidarity' trade union,[45] and consequently, a just transition was framed as a gradual one taking place over decades. As a civil servant in the Ministry of Climate noted, 'We lived in a kind of illusion that our policies are generally beneficial until now. That is not true . . . [the energy transition] will take years or decades of gradual evolution'.[46] From a Polish perspective, as another senior Climate Ministry expert stated, transition is conditioned on equity: '[the] direction set in the climate and energy policy is quite clear—this transition should take place only if it doesn't leave anyone behind'.[47]

In June 2021, the Łódź region issued a 'territorial just transition plan' to end lignite mining and shut down the Bełchatów coal power plant by 2036, with support from the European Union's Just Transition Fund. However, in 2021, 80% of the local population expressed fear of mass unemployment as a result of the closure.[48]

CLIMATE OBSTRUCTION: FROM DENIALISM TO SKEPTICISM

There is a dispute on this [climate change] because there is a question mark as to the human cause of these changes—I am on the side of those who think there is a big question mark on this. . . . The claim that coal-fired power plants make the temperature warmer . . . in my opinion, there is no sufficient basis for this.[49]
—Governing coalition MP, 2022

Poland has a history of climate obstruction in the European Union: along with other Central and Eastern European members, in 2009, Poland threatened to block the European Union's 2020 Strategy for reducing GHG emissions and expanding renewables and energy efficiency.[50] In 2011, Poland vetoed the European Commission's roadmap to (mostly) decarbonize the economy by 2050 on the grounds that the economic costs of an energy transition were too high.[51] A year later, the Polish presidency of the Visegrad Group (with Czechia, Hungary, and Slovakia) authored a 'Concept Paper on the Climate and Energy 2030 Vision' reiterating this opposition and emphasizing the importance of analysing the 'costs of ambitious headline targets for 2030'.[52] In the 2015 electoral campaign, the right-populist Law and Justice (PiS) politicians proposed an 'opt-out' from the European Union's climate policy and the renegotiation of the 2020 Strategy. While in power, the PiS government attempted to block the European Union's 2030 and 2050 decarbonization plans, though ultimately unsuccessfully.[53]

In sum, Polish authorities have opposed most calls for more ambitious climate action, including the Emissions Trading System (ETS) and binding renewable energy targets.[54] Although these EU policies were ultimately implemented, they were weakened as a result of Polish-led obstruction, as they were in the case of the European Union's 2030 Climate and Energy Framework, negotiated in 2014, in what might be termed a 'Polonization' of EU policy in place of the 'Europeanization' of Polish ambitions.[55]

Among Poland's right-wing politicians, EU climate policy is continuously criticized: as an elite 'fashion' characterized by 'hypocrisy of people who usually belong to the elite and don't give a damn about the interests of those who have to pay for it' and causing 'chaos'[56]; and as 'highly ideological', a product of 'political postmodernism'.[57] The European Union's emphasis on leading by example in climate policy[58] is portrayed as 'kamikaze politics'[59] and a 'threat' to Polish national interests, as 'it has nothing to do with climate protection, but is an element of the economic policy of countries such as Germany. . . . The very model of EU transformation is unfair and pathological. And it is now bankrupt'.[60]

Opposition politicians have often sided with government parties in contesting European energy and climate policy. This suggests the existence of a broad cross-party coalition that is likely to persist after changes in government.[61] This consensus has been undermined by PiS's anti-EU stance and increasing political polarization, which has led to a partisan framing of climate policy: 'the opposition is against the government . . . [so] the government is not so willing to adopt climate policies'.[62]

The attitude of political elites may have been both a reflection and a cause of broader societal disinterest in climate action, at least before 2018. The initially climate-sceptic and openly denialist attitudes that dominated in the 1990s and 2000s were epitomized by the fact that the Polish Academy of Sciences was one of the last national science institutions in the world to issue an official statement, in 2007, acknowledging the anthropogenic character of climate change.[63] Research by McCright and colleagues found that the salience of climate change is lower in Central and Eastern Europe than the rest of Europe, with less concern among citizens and politicians.[64] According to a 2015 Eurobarometer poll, 69% of EU citizens considered climate change to be 'a very serious problem', compared with 56% in Poland.[65] This finding supports that of Kvaløy and colleagues that, of forty-seven countries sampled, respondents in Poland were the least concerned about climate change,[66] although, by 2023, the gap between the European Union and Poland on concern about climate change had narrowed to 8%.[67] A challenge remains in that the majority of the public agree with the government's cautious approach to the pace of energy transitions: in 2023, 55% of those polled agreed that Poland should choose its own pace to achieve climate neutrality, even if that means after 2050.[68]

Although earlier research emphasized 'denialism' and 'contrarianism' as a defining feature of Poland's climate debate,[69] this characterization may be misleading, and such claims are based on a selective reading of the most extreme views and ideas present in the public debate. To do so is tempting, as many high-profile political actors in Poland have over the past decade produced many astonishing denialist statements.[70] For example, the annual 'climate nonsense' prize, awarded by the Climate Education Foundation, in 2014 went to Zbigniew Ziobro, the minister of justice, for saying that carbon dioxide cannot be harmful because we consume it in carbonated beverages; in 2016, the honour was given to Janusz Korwin-Mikke, an MEP who stated that most scholars say that global warming, if it exists at all, has nothing to do with human activity.

However, our own research on Polish media found that *primary obstruction*, or open climate denialism (including *trend scepticism*, questioning the existence of global warming, and *attribution scepticism*, doubting human responsibility for climate change) plays a relatively minor though not insignificant or unimportant role in the Polish debate. As an opposition MP stated, 'Opinions that deny the scientific facts of climate change ... are intended solely to cause controversy and unnecessary discussion on obvious phenomena'.[71] Meanwhile, *secondary obstruction*, or what may also be termed *response scepticism* or *delayism*, is mainstream. Our research finds that, in the 2014–2016 period, 17% of the sampled discourse in the press and TV included representation of views that deny climate change is occurring. Furthermore, 25% of the discourse included views that accept that it is happening but deny the role of humans. When considering the COPs hosted by Poland, an analysis shows that 14% of media discourse on climate change around the event included the views that denied climate change was occurring in 2013, but this proportion had decreased significantly, to 6%, by 2018.[72] Public-opinion polling echoes this change: in 2009, 65% of Polish citizens thought that climate change was primarily caused by human involvement (26% disagreed), but this figure gradually increased over the next decade, to 75% in 2018 (with 18% attributing it to natural causes).[73]

The dominant feature of Polish climate debate is less clearly denialism but rather response scepticism. In August 2018, an MP and state secretary responsible for energy, Piotr Naimski, argued that 'any binding stance that would be accepted at the conference in Paris will be harmful to Poland, so a failure of the [COP] summit is in Poland's interest'. The Polish President, Andrzej Duda, at the same time stated that 'Decarbonisation is completely not in our interest'. Similarly, 40% of media discourse during the Warsaw conference in 2013 was related to acknowledging the problem but considering no policy response necessary, though significantly this proportion had fallen by half, to 21% by 2018.[74]

A note of optimism relates to a change in perception among politicians and civil servants in recent years, particularly after 2018: 'For 4 years between 2015 and 2019'—an opposition MP says—'the PiS government did not want to acknowledge that the coal age is over'.[75] However, as two civil servants from the Ministry of Climate (established in 2019, to coordinate Polish policy domestically and in the European Union and UN) noted, 'in recent years we can witness the effects of it every summer'[76] and 'people see it [climate change] and demand action from local authorities. . . . These matters became the subject of political debate. It has significantly changed. Our ministry has a top and priority status'.[77]

Apart from the visible effects of climate change, such as droughts and heat waves, the change can be attributed to air pollution: an MP argued that 'in terms of climate policy . . . the attitude in Poland to the fact that the coal age is history happened not because of the European Court of Justice, European politics or the Youth Climate Strikes—but because of smog'.[78] As a result, despite the rhetorical prominence of climate obstruction, 'today there is no one in the government who would question the energy policy . . . we all know that we step away from coal'.[79] In this context, the remaining opposition to any sort of climate policy is visible on the far right, among the ultraconservative MPs from Konfederacja and, more importantly in Solidarna Polska, a junior coalition partner in the 2015–2023 government whose position on the energy transition, according to a former government minister 'resembles the behaviour of textile workers protesting machines in the 19th century'.[80]

However, there is a form of secondary climate obstruction not often discussed in the literature on climate scepticism and denial—acknowledging both the scale of the issue and the requirement to respond but locating the time to respond at some unspecified near future moment when technology permits; a future vision of 'clean coal,' for example—'an attempt to make an effective argument out of something that cannot be properly argued'.[81] We observed a very significant increase in the media discourse representing these views, from 23% to 44%, after the 2015 election, in which the populist Law and Justice party won. Similarly, media analysis around the two COPs in 2013 and 2018 indicated that 50% of all discussion was focused on action to be taken not now, but in the future.[82]

KEY ACTORS IN CLIMATE OBSTRUCTION

Poland's climate policy landscape is characterized by a large degree of segmentation—meaning that actor coalitions that dominate it are divided and lack common discursive framings of climate policy or visions of energy transition. Among these actor coalitions, which should be understood as dynamic and related but still clearly discernible—one is dominant in terms of its agenda-setting and political power. This coalition is concentrated in governmental institutions, particularly the ministries responsible for climate and energy policy (Table 8.1), agencies, and state-owned energy companies and utilities. Together, they constitute a unique 'governmental–industrial complex' (GIC).[83] We conceptualize the GIC as a powerful *discourse coalition*, a group of actors that share common storylines, problem definitions, and preferences for certain solutions. Actors in discourse coalitions 'try to impose their views of reality on others, sometimes through debate and persuasion'—as pro-governmental think tanks and GIC-linked media do in this case—'but also through manipulation and the exercise of power,'[84] which is particularly important for a discourse coalition built around core state institutions and the fossil industry.

Since 2020, responsibilities for decarbonization, drafting climate policy, and steering the energy system have been divided between the Ministry of Climate and the Environment and the Ministry of State Assets. Together with the Chancery of the Prime Minister and the Government Plenipotentiary for Strategic Energy Infrastructure (a post created by the Law and Justice government), they constitute the core public actors influencing climate action. Other important public actors include the national regulator, the Office for Energy Regulation (URE), and the state-owned transmission system operator PSE. However, important voices in the debate on climate policy come from other ministries as well as the two chambers of parliament—the Sejm and the Senate.

Poland's major energy companies are state controlled (Table 8.2); the state owns a majority of their shares or legally controls them through the Ministry of State Assets. The oil company Orlen has, since 2016, become

Table 8.1 THE POLISH MINISTRIES RESPONSIBLE FOR CLIMATE, ENERGY, AND THE ENVIRONMENT BETWEEN 1999 AND 2024, INCLUSIVE

Environment	Climate	Energy
Ministry of Environment (1999–2019)		Ministry of the Economy (2003–2015)
	Ministry of Energy (2015–2019)	
Ministry of Climate (2019–2020)		Ministry of State Assets (2019—present)
Ministry of Climate and Environment (2020—present)		

Table 8.2 OWNERSHIP OF ENERGY AND MINING COMPANIES IN POLAND

Company	State treasury shares
Petroleum and gas companies	
PKN Orlen	49.90%
PGNiG	Acquisition by Orlen
Lotos Group	Acquisition by Orlen
Electricity companies	
PGE Polska Grupa Energetyczna	60.86%
Enea	52.29%
Tauron	30.06%
Energa	0% (Orlen 90.92%)
Mining companies	
Polska Grupa Górnicza (coal)	100%
KGHM (metals)	31.80 %

Source: Authors own elaboration based on publicly available information on companies' websites in 2023.

a 'national champion' that accumulates stocks in other energy companies, expands to other European countries, and wields significant political power. State companies control more than 75% of the power market, which leads even the most moderate mainstream energy analysts to call this setup an oligopoly.[85] The state also owns 100% of the shares in the largest coal mining company, PGG.

Institutional links are only one level; the other is personal, a circulation of elites through a 'revolving door'. An example is Deputy Minister of Agriculture Janusz Kowalski, an outspoken critic of climate policy. As a civil servant explained in 2021, Kowalski's flamboyant rhetoric is not merely aimed at attracting attention: 'Kowalski's circle has a significant influence on what goes on behind the scenes. It's not just him, but it is the whole camp with some informal relationships. . . . Kowalski worked in many places, at PGNiG . . . and his connections are still there'.[86]

The status of these companies is contested. Formally corporate entities listed on the stock exchange, they are under government control and often must operate according to logic contrary to shareholders' interests and economic efficiency. Since 2016, the PiS government has changed the charters of the four major (and partly state-owned) energy companies, introducing a clause saying that they constitute an 'instrument of national energy security'.[87] This change implied that they would no longer be subjected primarily to economic market logic but might be forced to act in the 'national energy security' interest, left undefined. The insistence on state ownership and prioritizing 'national energy security' is consistent with an 'energy sovereignty' paradigm that emphasizes the national character of energy systems, energy resources, and the identities of energy companies. Renewable sources are then often framed as foreign, as one civil servant attests: 'Several times I have participated in discussions in the Sejm or in the Senate, and I can see the attitude of deputies or senators when it comes to renewable energy. The main question is—who produces these turbines? These panels? Well, most of it is German or Scandinavian. Exactly—"so what benefit do we have from this, apart from the fact that we will have green energy?"'[88]

This resource nationalism and 'energy xenophobia' is amplified by the right-populist government's foreign policy rhetoric, in which energy plays an important role. A member of parliament noted that:

> The political imagination of Poles is appropriated by sheer and biased propaganda ... the enemy is the European Union, which imposes the climate package on us, the enemy is Germany, which pursues its interests. We, the only righteous ones, are surrounded by enemies. . . . Our coal should stay while the EU scandalously tries to destroy it, along with undermining our sovereignty.[89]

State ownership is also the cause of opaque boundaries between public administration, politics, and the energy sector. During the transition from communism, Poland experienced only a gradual development toward a professional civil service. Most public institutions remain politicized, and each election brings significant staff changes on all levels. Calls for technocratic 'governments of experts' are often popular, so ministers do not have to have a parliamentary mandate (they are often not elected politicians) although it is seen as positive if they have experience in the policy arena. Hence, experience in the energy sector is seen as valuable in the ministries dealing with energy. The 2019–2021 climate minister Kurtyka was described as not having a 'political base', which resulted in a situation where 'various energy companies may have a greater influence than they should have on the entire course of activities related to our energy policy'.[90] Among civil servants, a gradual energy transition is seen as necessary: 'We, as an administration, must always counterweigh and maintain balance. Now we are at the starting point. The greater the dynamics of such a process, the greater the costs. We, therefore, need to adapt our pace of transformation to the resources we have'.[91]

With state companies under the control of the government, politicization extends beyond the civil service. First, while ministers and their deputies cannot sit on state company supervisory boards, directors of ministerial departments and other civil servants can. Their formal role is to safeguard the interests of the State Treasury. However, board membership is also lucrative and is used as a premium for loyalty within ministries. Furthermore, sitting on boards is also often an anchor for future positions in the company itself.[92]

Regarding climate obstruction, the most important outcome of the GIC's existence is energy sector incumbents' shaping of governmental policy through regulatory capture. The Climate Ministry should be the most climate-ambitious part of government on decarbonization but is 'strongly influenced by transmission and distribution network operators. They believe that renewables are a challenge, a problem'.[93] The result of this influence is an energy policy orientation that seeks to sustain the political economic status quo of the energy sector. Energy transition is framed as possible only to the extent that it can be achieved by the state-owned energy companies. In practice, that means delaying a coal phase-out for as long as possible and replacing coal generation with energy sources that can be controlled only by large players such as nuclear power plants (in partnership with private business) and offshore wind farms. The role of gas plants, also state owned, is as a transition fuel, with distributed renewables playing only an auxiliary role. As a deputy minister stated in 2022: 'What is the basis for energy production? In my opinion, it should be coal, not natural gas. Coal should be at the centre of the energy transformation in Poland until the construction of a nuclear power plant'.[94]

There is a challenge to these positions, particularly from environmental nongovernmental organizations (NGOs) as well as think tanks that provide counterevidence and independent data and pressure the government for transparency about the data it uses and the assumptions behind its policy projects, allowing for alternative interpretations and often divergent policy conclusions regarding the viability of techno-fixes. However, to date, the GIC core sets the tone of the debate. Independent or semi-independent think tanks and organizations also exist but rely on state financing, putting them in the orbit of the GIC, and present more or less explicit climate obstructionist arguments. Following the degree of dependence from greater to lesser, they include:

- The GIG Institute and the Institute for Fuel Technology and Energy, which are overseen by the Ministry of State Assets;
- the Polish Economic Institute and the National Economic Chamber (KIG), which rely on direct funding from the state budget;
- the Centre for Climate and Energy Analyses, which is part of the governmental institution set up for reporting emissions (KOBIZE);

- the Polish Electric Energy Committee, financed by the energy sector companies, mostly state owned;
- the conservative Jagiellonian Club and Jagiellonian Institute, which, although often openly critical of the government, organize events in partnership with state energy companies;
- the independent liberal or libertarian institutes such as the Warsaw Enterprise Institute and the Forum for Civic Development (FOR), which represent business interests outside the state sector.

Although there are no Polish institutions listed among the Global Warming Sceptic Organizations, the Atlas Network, or in the DeSmog Climate Disinformation Database, these think tanks generate expertise and arguments that can be classified as secondary obstruction, discussed in the following sections.

Discursive framings: Preaching delay

I don't think there are many who genuinely believe in the climate crisis. [and] I think there are also many people who think we can't afford [an energy transition]. And that our influence—of Poland and the Polish industry—is so petty in the world that our actions will not help.[95]

—Civil servant, State Forests, 2020

In the Polish climate discourse, delayism is visible in attempts to either shift responsibility onto others or, more often, to extend the timeframe of necessary action and political intervention well beyond 2050. Although Poland does not have a net zero target year and does not foresee a coal generation phase-out, much of climate obstructionist discourse is focused on the apparently unsustainable pace of the energy transition and EU climate policy. While Poland is a signatory of the Paris Agreement, such calls for a more 'realistic' and 'considerate' transition pathway continue to be visible. Meanwhile, state-controlled energy companies are torn between market realities and the EU regulatory environment in which they operate—clearly set for a decarbonized future on one hand and the unpredictable Polish regulatory environment and political pressure from the government on the other. As a result, most of them have some kind of climate strategy or sustainability policy. Tellingly, in the gas company Lotos, 'climate risks' that were defined by a special task group 'are transformation-related, not physical in character'[96] (i.e. it is fossil fuel phase-out that is the climate-associated risk, not climate change itself).

There are also visible discrepancies between what the companies, operating as corporate organizations, do and what their politically nominated managers say. For example, the largest power company, PGE, which plans to achieve carbon neutrality by 2050, was headed by Wojciech Dąbrowski who, on occasion, openly denied the impact of CO_2 on global warming and suggested that climate policy is a form of external pressure against Polish sovereignty.[97]

When it was sued by Greenpeace as a climate polluter, PGE convened an expert body of scholars who issued a formal statement questioning the scientific consensus behind anthropogenic climate change.[98] Such open denialism is, as noted already, increasingly rare. All the major energy companies have refocused their investment strategies on low-carbon sources or negative emissions, even as their representatives continued to call for a more cautious or delayed decarbonization strategy. As the Orlen spokesperson, Adam Czyżewski, put it:

> The energy transition is not a race, but a crossing that has to be done in a coordinated way. . . . [D]ecarbonization is also security, but the security of the future . . . the goal is to move towards climate neutrality, but to achieve this you need fossil and renewable energy sources. The transition cannot be rushed, because it is not about one country achieving its goals, but about it happening for all.[99]

The Orlen CEO, Daniel Obajtek, echoed this sentiment in a 2021 speech where he said: 'Let us not expect change to happen year-on-year or in ten years. This cannot be achieved by a rapid revolution, but perhaps a more rapid evolution'.[100] Czyżewski has also used arguments that push responsibility onto others, be it non-EU states or consumers. He emphasized the need for non-European countries to follow the European Union's example, with further ambition being conditional on this joint effort. He also underlined the fact that an energy transition is 'primarily' a shift in consumption habits.

Other energy sector actors present similar arguments, including trade unions. A leader of All-Poland Alliance of Trade Unions (OPZZ) noted that EU climate policy 'makes no sense in a situation where most industrialized countries like China, Russia or the US are not reducing emissions',[101] while, in 2019, the leader of the largest union, Solidarity, asked the Polish delegation to the European Council to negotiate 'a change in the time horizon' of the net zero target.[102] While very different in their political position, some leading think tanks echo similar themes in their public statements. The conservative but independent Jagiellonian Club suggested in 2022 that 'Poland should tame the ambitions of the Union. We should tone down the goals of the European Green Deal'.[103] The National Economic Chamber, which before 2013 voiced openly climate-skeptical views, has since switched positions, proposing to increase Poland's ambitions well beyond the government's energy strategy. And yet, in November 2022, the Chamber organized an event entitled 'With coal—safely towards green transition' at which experts argued for the need to revise the speed of EU climate policy and called for reflection on its direction.[104]

A further justification for delay amongst GIC actors is the effect of decarbonizing on and society's responsibility for its citizens and workers. This emphasis on a 'just transition', which is primarily an argument for delay and not for a more socially transformative process, is visible among trade unionists, politicians, and energy company experts alike: 'Environmentalism—yes; climate—yes; but jobs and livelihoods are most important', said an OPZZ official during the 2013 COP 19 in Warsaw.[105]

Much more subtle forms of climate obstruction can be observed in the think tank sector, where most experts question neither the overall climate protection goals nor EU climate policy but contribute to the daunting picture of the impossibility of transforming Poland's energy sector in time. In February 2020, the Warsaw Enterprise Institute organized two events criticizing the government for blocking the expansion of onshore wind, which it deemed the cheapest and quickest tool of decarbonization, but still suggested that phasing out coal 'is not possible before 2040, and perhaps even 2050 even if nuclear enters the mix'.[106] A 2020 Polish Economic Institute study concluded that Poland could become climate neutral by 2056, according to an optimistic scenario, while it might take until 2067 under other assumptions.[107] However, the consulting firm McKinsey has presented a cost-effective pathway for Poland to achieve climate neutrality by 2050.[108]

In the expert community, this sort of soft 'impossibilism'—portraying ambitious climate action as beyond reach and futile—is often presented as realism, contrasted with 'irrational' and ideologically or faith-driven environmentalism. That kind of dichotomy is particularly clear in the GIC and its expert network efforts to promote nuclear energy as central to Poland's decarbonization, in contrast to renewable energy, which is seen as disruptive for the market and the sector and ultimately unreliable. In a recent vote on the revised EU Renewable Energy Directive, Poland was one of the two members voting against, stating that renewables 'jeopardise both the stability of the grid and overall energy security'.[109]

STRATEGIES AND TACTICS OF OBSTRUCTION: TECHNO-FIXER AT HOME, IMPOSTER ABROAD

In Poland, calls for delay can be discursively grounded in references to national security,[110] but may even be disguised as ambitious climate policies that focus on a specific and narrow area of climate action or a promising future technology (a 'silver bullet'). They may also be seen in attempts at attracting attention in global climate negotiations and presenting the country as a global climate action leader, all while drawing attention away from the more demanding problem of economy-wide decarbonization.

Techno-fixes as tools of obstruction

Future visions can play the role of a tool of social control through the process of rational planning (which presents the way to desirable and expected outcomes) but also by exporting the problems beyond the 'here and now' reality: we will be able to solve the problem of the tension between our coal-based economy and climate change in the near future, thanks to technology development—'clean coal technology'.[111] Related research has argued that 'believing that science will solve environmental problems tends to be associated with lower environmental concern', so such 'future vision skepticism' is significantly correlated with lower climate ambition.[112]

Poland's decarbonization strategy, as presented in the most recent official documents, will require a technology mix including elements that are not yet commercialized or are even still merely hypothetical. A 2021 roadmap presented by the Centre for Climate and Energy Analyses included large-scale implementation of carbon capture, utilization, and storage (CCUS) combined with bioenergy, industrial electrification, hydrogen use, electric vehicles, and structural changes in agriculture as prerequisites for Poland to achieve climate neutrality by 2050,[113] in addition to a vast expansion of renewables, batteries, grids, and possibly also nuclear energy.

For many years, however, key actors in Polish climate politics pointed to specific 'silver bullet' technologies as answers to the challenge of climate mitigation or, more often, referred to them as preferred solutions for the future, justifying limited or no action in the present. As noted earlier, the first of such technologies have been 'clean coal' technologies, to be developed by domestic R&D and allowing Poland to maintain a coal sector in an energy transition. In 2013, Poland's GIG Research Institute opened a Centre for Clean Coal Technologies, which received large grants from the state budget and EU funds. What these technologies might be in practice remained unclear, but they are still referenced by some GIC actors as an argument against EU climate policy and coal phase-out. In 2021, a Solidarity trade union leader from the mining company PGG complained that 'nobody thinks how beneficial it would be for Poland to invest in low and "zero emission" sources of energy from coal'.[114] The head of the Jagiellonian Institute summarized such remarks as 'incantations, meant to work only on the domestic forum' as 'today there is not and will never be a debate on clean coal technologies, dismantling the ETS', and underlined that Poland 'lost lots of time on such empty gestures'.[115]

Due to its carbon-intensive energy generation, Poland was one of the early leaders of CCUS development, and the first pilot projects were already planned under the Civic Platform government of 2007–2011. Little progress has been made since, though state-controlled energy companies (Lotos, Enea, and Tauron) mention CCUS projects in their strategies. Many modelled pathways toward net zero rely on large-scale CCUS deployment, which remains hypothetical—a problem not unique to Poland.

The Polish government's contribution to the project of carbon removal is establishing carbon forestry as a flagship climate policy. In 2016, State Forests, a state-owned enterprise governing all publicly owned forests, presented the elaborate Forest Carbon Farms project with the intention of turning industrial timber plantations into carbon sinks. That same year, at COP 22 in Marrakesh, the Polish delegation promoted the idea together with the concept of a 'waste-free coal power industry'.[116] The environment minister, Jan Szyszko, is often dubbed the project's 'godfather'.[117] Forest Carbon Farms were officially established by ordinance in 2017,[118] and state energy companies were encouraged to buy carbon credits from the Carbon Farms program. This policy contradicted domestic academic expertise suggesting that 'climate change mitigation through carbon sequestration in forests may adversely affect available water resources in Poland, due to high evapotranspiration of forests'.[119] Nevertheless, government-controlled and pro-government media vigorously promoted the idea and portrayed it as Poland's unique contribution to global climate action and as a solution that would allow the country to maintain its coal sector while also meeting international obligations. However, after Szyszko's death in 2019, the idea of carbon forestry quickly lost prominence, and, in 2023, only the energy company Tauron mentioned it as part of its climate policy.

Lately, the most prominent 'silver bullet' technology promoted in Poland is nuclear energy. According to the government as well as many mainstream GIC-linked experts, large nuclear power plants are the only realistic way to decarbonize the power sector. The 'Polish Energy Policy by 2040' strategy envisages as much as 9,000 megawatts (MW) of nuclear capacity by the mid-2040s, making the Polish national megaproject one of the most ambitious globally. However, it will only meet 20–25% of the energy demand, and, if it replaces coal as late as the 2040s, the cumulative emissions from the Polish power sector will have far exceeded the country's carbon budget.

Nuclear's role as an element of obstruction lies in the fact that while it is in theory compatible with renewables, it has gained a larger share of media and industry attention, whereas insufficient and slow deployment of renewables has not been addressed. The director of the Polish Economic Institute laid out the self-contradictory official line, saying that nuclear 'supported by RES', or renewable energy sources, is necessary to maintain energy security, even though most models, including the government's own, see the roles of renewables and nuclear in reverse order. The Law and Justice government has underestimated the growth of distributed renewables and undermined their further expansion—particularly by introducing a very restrictive siting policy for onshore wind that has effectively banned it, limiting it to 0.28% of Poland's land area,[120] and by failing to initiate grid expansion enabling more renewables deployment. Although all energy companies are making plans to expand their renewables capacity, in the case of several of them this has meant buying up existing wind farms from smaller players or investing in offshore projects to be completed in the 2030s. In this case, the state power companies, although all are strongly in favour of onshore wind, were not able to influence lawmakers; political opposition against wind farms, particularly from the far-right junior party Solidarna Polska, eventually weakened a 2023 amendment of the siting law.

Meanwhile, virtually all Polish energy companies now boast nuclear ambitions. PGE is charged with building the first Polish nuclear power plant and supervising the national, large-scale nuclear projects, which remain the only ones that have secured state financing, although its conditions are not yet set. Orlen, in 2023, announced plans to construct as many as seventy-six small modular reactors (SMRs) in twenty-six localities, adding up to 22,000 megawatts (more than double the government's plan for the mid-2040s), with the first envisaged to come online in 2028. So far, however, not a single reactor made by Orlen's American partner has been built. Enea and Tauron are also planning SMR construction. A nuclear sector analyst counted 126 announced nuclear reactors of different types and sizes planned for Poland, all of which should be operational by 2043 (if plans are realized).[121] Independent experts, even those favouring nuclear energy, have been highly sceptical, as is one of the ruling party MPs: 'It cannot be said in 2021 that we will indicate the location by 2022 or 2023, and then within 10 years, we will build a nuclear power plant. This is unreal. We see it after Flamanville [France] or Olkiluoto [Finland]—it is unreal'.[122]

Climate imposter: Poland's international climate diplomacy

Despite its clear denialist tendencies in the public climate debate, Poland has been surprisingly active in the United Nations Framework Convention on Climate Change (UNFCC) process. There are two reasons for this. The first was economic: the post-communist economic crisis and restructuring led to significant declines in emissions, and Poland was able to decouple emissions from economic growth; between 1988 and 2016, its gross domestic product (GDP) more than doubled, while GHG emissions had fallen by more than 30% by 2002.[123] The Kyoto Protocol 'was perceived [by politicians] as a good opportunity for Poland to capitalize' on this record.[124] The second reason was agenda-setting power: governments in Warsaw have been broadly engaged in UNFCCC negotiations. Jan Szyszko, the late three-time environment minister, was elected president of the COP 5 in Bonn (1999), leading intersessional talks for a year,[125] while Poland hosted three summits: in Poznan (2008), Warsaw (2013), and Katowice (2018) using regional-rotation rules to shape the character of negotiations.[126] National energy companies were sponsors of these summits, and Polish governments framed their proposals as a rational alternative to 'ideologically driven' climate action.[127] At the 2018 COP, Poland proposed the Silesia Declaration on a 'just transition', a call for a cautious pace to mitigate social and economic costs,[128] and Just Transition is the first of three pillars of the 2040 Polish energy strategy.[129] This approach led one government representative to argue that Polish views were 'always a part of the mainstream in the matters of global policies . . . in international UNFCCC negotiations our approach is much more similar to others'.[130]

CONCLUSION: PHASING OUT, PHASING DOWN, OR CLINGING ON?

The challenge Poland faces in phasing out coal, on which it is highly reliant, is real. However, it is also a situation that some other European countries have faced previously. This point of departure, often used to justify delaying climate action, is not a sufficient excuse. It also does not explain the character of Poland's climate policy debate. The apparent degree of climate obstruction has instead been created and perpetuated by a dominant coalition of governmental institutions, agencies, and state-owned energy companies and utilities that constitute a GIC, surrounded by think tank experts and journalists who are dependent on state and energy company financing and thus promote the GIC arguments. While there has been a shift in discourse and policy emanating from the GIC, it continues to promote obstructionist strategies and offers 'silver bullets' in the form of promising future technologies such as 'clean' coal and unrealistic plans for new nuclear power plants. This pattern is underpinned by Poland's stated commitment to a just (and gradual) energy transition, which, rather than illustrating concern for the vulnerable citizens left behind, is one more of the GICs 'climate imposter' tactics.

Further research is needed to examine whether societal changes, including the effects of COVID since 2020 and Russia's full-scale invasion of Ukraine in 2022, are leading to a more substantive engagement for a more progressive and ambitious Polish climate policy. The 2023 parliamentary elections, which resulted in a change in government, and a planned revision of the 2040 energy strategy will provide some evidence here, but the political economy of the energy sector, with the close and often obscured connections between governmental bodies and energy companies, is likely to remain in place.

NOTES

1. Eurostat, Production of electricity and derived heat by type of fuel, 2023, https://ec.europa.eu/eurostat/databrowser/view/NRG_BAL_PEH__custom_7652146/default/table?lang=en
2. International Energy Agency, 'Net Zero by 2050', https://iea.blob.core.windows.net/assets/7ebafc81-74ed-412b-9c60-5cc32c8396e4/NetZeroby2050-ARoadmapfortheGlobalEnergySector-SummaryforPolicyMakers_CORR.pdf. Accessed 26 July 2023.
3. Christian Davies (2019, 26 September), 'Fight the Power: Why Climate Activists Are Suing Europe's Biggest Coal Plant', *The Guardian*, sec. Environment, https://www.theguardian.com/environment/2019/sep/26/fight-power-climate-activists-europe-biggest-coal-poland-bechatow.
4. European Council (2019, 12 December), 'European Council Meeting Conclusions', p. 2, https://www.consilium.europa.eu/media/41768/12-euco-final-conclusions-en.pdf.
5. Małgorzata Kasprzak (2021, 15 March), 'Disappointing lack of ambition in PEP 2040', *Ember*, https://ember-climate.org/insights/commentary/disappointing-lack-of-ambition-in-polands-energy-policy-until-2040/
6. DW (2021, 28 April), 'Poland Clinches Deal to Phase out Coal by 2049', https://www.dw.com/en/poland-clinches-historic-deal-to-phase-out-coal-by-2049/a-57367983.
7. Climate Action Tracker, 'Policies & Action', https://climateactiontracker.org/countries/eu/policies-action/. Accessed 26 July 2023.
8. Powering Past Coal Alliance, 'Our Members', https://poweringpastcoal.org/members/. Accessed 26 July 2023.

9. Felix Reitz (2020, 24, June), 'European Coal in Structural Decline', Europe Beyond Coal, https://beyond-coal.eu/2020/06/24/european-coal-in-structural-decline/.
10. Climate Action Tracker, 'Policies & Action', https://climateactiontracker.org/countries/eu/policies-action/. Accessed 26 July 2023.
11. World Bank (2019), 'Air Quality in Poland: What Are the Issues and What Can be Done?', https://documents.worldbank.org/pt/publication/documents-reports/documentdetail/426051575639438457/air-quality-in-poland-what-are-the-issues-and-what-can-be-done.
12. Polish Government Ministry of Climate and Environment (2021), 'Energy Policy of Poland until 2040', https://www.gov.pl/attachment/a1e42067-c749-4dbe-82bf-211d46821d9d p. 4.
13. We draw on the following sources and types of data: (1) existing scholarly and grey literature on Polish climate policy; (2) a focused document and media analysis, informed by input received from an email survey of several climate and environmental NGOs; (3) unpublished data gathered within the project 'Anatomy of Disbelief', which explored Polish climate scepticism. These include the results of a quantitative media analysis and broad content analysis of mainstream media around four politically important periods: the time around the 2013 and 2018 COP summits organized in Poland, as well as the pre- and post-2015 national election period; and (4) fifteen semi-structured interviews with Polish politicians and civil servants, conducted between 2019 and 2022, within the aforementioned project by Jakub Bodziony.
14. Hannah Ritchie and Max Roser (2020, 11 May), 'Poland: CO_2 and Greenhouse Gas Emissions', Our World in Data, https://ourworldindata.org/co2/country/poland.
15. European Environment Agency (2022), 'Trends and Projections in Europe 2022', EEA Report No 10/2022, https://www.eea.europa.eu/publications/trends-and-projections-in-europe-2022/at_download/file
16. Eurostat, 'Net Greenhouse Gas Emissions, https://ec.europa.eu/eurostat/databrowser/view/sdg_13_10/default/table?lang=en Accessed 26 July 2023.
17. European Environment Agency, 'Trends and Projections in Europe 2022'.
18. Instytut Ochrony Środowiska (2019), 'Krótka Historia Działań i Wyzwania Na Rzecz Ochrony Klimatu w Polsce 1988–2018–2050'. Warsaw: Instytut Ochrony Środowiska—Państwowy Instytut Badawczy, https://ios.edu.pl/wp-content/uploads/2019/03/Kr%C3%B3tka-historia-dzia%C5%82a%C5%84-i-wyzwania-na-rzecz-ochrony-klimatu-w-Polsce_PL.pdf.
19. Climate Equity, 'Climate Equity Reference Calculator', https://calculator.climateequityreference.org/. Accessed 26 July 2023.
20. Ibid.
21. European Parliament and Council (2021, 30 June), 'Regulations', https://eur-lex.europa.eu/legal-content/EN/TXT/PDF/?uri=CELEX:32021R1119&from=EN
22. Climate Action Tracker, 'Policies & Action', https://climateactiontracker.org/countries/eu/policies-action/. Accessed 26 July 2023.
23. Polish Government Ministry of Climate and Environment, 'Energy Policy of Poland until 2040'.
24. Magdalena Kuchler and Gavin Bridge (2018), 'Down the Black Hole: Sustaining National Socio-Technical Imaginaries of Coal in Poland', *Energy Research & Social Science*, 41: 136–147, https://doi.org/10.1016/j.erss.2018.04.014.

25. Ibid.
26. Ibid.
27. Interview 4, MP, 20 June 2020. All interview transcripts on file with the authors. Please see note 13 for an overview of data sources.
28. Interview 8, MP, 21 June 2021.
29. Piotr Zuk and Kacper Szulecki (2020, August), 'Unpacking the Right-Populist Threat to Climate Action: Poland's pro-Governmental Media on Energy Transition and Climate Change', *Energy Research & Social Science*, 66: 101485, https://doi.org/10.1016/j.erss.2020.101485.
30. Kacper Szulecki, Janusz Waluszko, and Tomasz Borewicz (2022), *The Chernobyl Effect: Antinuclear Protests and the Molding of Polish Democracy, 1986–1990*, Protest, Culture & Society, vol. 32. New York: Berghahn Books.
31. Hanna Brauers and Pao-Yu Oei (2020), 'The Political Economy of Coal in Poland: Drivers and Barriers for a Shift Away from Fossil Fuels', *Energy Policy* 144: 111621, https://doi.org/10.1016/j.enpol.2020.111621.
32. Julia Szulecka and Kacper Szulecki (2022), 'Between Domestic Politics and Ecological Crises: (De)Legitimization of Polish Environmentalism', *Environmental Politics* 31, 7: 1214–1243, https://doi.org/10.1080/09644 016.2019.1674541; Julia Szulecka and Kacper Szulecki (2017), 'Polish Environmental Movement 1980–2017: (De)Legitimization, Politics & Ecological Crises', ESPRi Working Paper, http://dx.doi.org/10.2139/ssrn.3075126.
33. Monitor Polski, 'Uchwała Sejmu Rzeczypospolitej Polskiej z dnia 9 listopada 1990 r. w sprawie założeń polityki energetycznej Polski do 2010 r.', https://isap.sejm.gov.pl/isap.nsf/download.xsp/WMP19900430332/O/M19900332.pdf. Accessed 26 July 2023.
34. US Government (2022, 22 July), 'Poland—Country Commercial Guide: Energy Sector', https://www.trade.gov/country-commercial-guides/poland-energy-sector.
35. Agencja Rozwoju Przemysłu S.A. Oddział Katowice, 'Wyniki techniczno-ekonomiczne działalności oraz inwestycje w górnictwie węgla kamiennego w Polsce w okresie styczeń–wrzesień 2022 r.' https://www.wug.gov.pl/bhp/nadz orowane_zaklady.
36. European Commission (2018), 'EU Coal Regions: Opportunities and Challenges Ahead', p. 21, https://publications.jrc.ec.europa.eu/repository/handle/JRC112593.
37. Maciek Nabrdalik and Marc Santora (2018, 22 April), 'Smothered by Smog, Polish Cities Rank Among Europe's Dirtiest', *The New York Times*, sec. World, https://www.nytimes.com/2018/04/22/world/europe/poland-pollution.html.
38. Jo Harper (2022, 27 April), 'Warsaw and Budapest Split over Russian Energy Ties', *Deutsche Welle*, https://www.dw.com/en/warsaw-and-budapest-split-over-russian-energy-ties/a-61595947.
39. Agencja Rozwoju Przemysłu S.A. Oddział Katowice, 'Wyniki techniczno-ekonomiczne'.
40. Paweł Madejski, 'Elektrownie w Polsce', http://galaxy.agh.edu.pl/~madejski/elek trownie-w-polsce/. Accessd 26 July 2023.
41. European Commission,' EU Coal Regions'.
42. Dana R. Fisher (2006), 'Bringing the Material Back in: Understanding the U.S. Position on Climate Change', *Sociological Forum*, 21, 3: 467–494, https://www.jstor.org/stable/4540952.

43. Sucha24 (2016, 24 November), 'Ranking of the Most Respected Professions: First Place for a Firefighter, Surprising Profession at the End of the Ranking', https://sucha24.pl/wydarzenia/item/5684-ranking-najbardziej-szanowanych-zawodow-pierwsze-miejsce-dla-strazaka-dziwi-zawod-plasujacy-sie-w-koncowce-rankingu
44. Serwis Rzeczypospolitej Polskiej, 'Uchwała w sprawie „Polityki energetycznej Polski do 2040 r."', https://www.gov.pl/web/polski-atom/uchwala-w-sprawie-polityki-energetycznej-polski-do-2040-.
45. Interview 13, MP. 7 July 2022.
46. Interview 3, Civil Servant, 3 June 2020.
47. Interview 6, Civil Servant 16 January 2021.
48. Money.pl (2021, 15 December), 'W 2036 r. chcą zamknąć Elektrownię Bełchatów. Mieszkańcy już dziś mówią o wzroście bezrobocia w regionie', www.money.pl, 15 December 2021, https://www.money.pl/gospodarka/w-2036-r-chca-zamknac-elektrownie-belchatow-mieszkancy-juz-dzis-mowia-o-wzroscie-bezrobocia-w-regionie
49. Interview 15, MP, 19 August 2022.
50. Pierre Bocquillon and Tomas Maltby (2017), 'The More the Merrier? Assessing the Impact of Enlargement on EU Performance in Energy and Climate Change Policies', *East European Politics*, 33, 1: 88–105, https://doi.org/10.1080/21599165.2017.1279605.
51. Arthur Neslen and Frédéric Simon (2012, 8 March), 'Poland Defies Europe over 2050 Low-Carbon Roadmap', www.euractiv.com, https://www.euractiv.com/section/development-policy/news/poland-defies-europe-over-2050-low-carbon-roadmap/.
52. Henryka Mościcka-Dendys, et al. (2013), *Report of the Polish Presidency of the Visegrad Group. July 2012–June 2013*. Warsaw: Polish Ministry of Foreign Affairs, p. 44.
53. Andrzej Ancygier and Kacper Szulecki (2015, 30 October), 'New Polish Government's Energy Policy: More State Control', *Energy Post*, https://energypost.eu/new-polish-governments-energy-policy-expect-state-less-market/.
54. Jon Birger Skjærseth (2018), 'Implementing EU Climate and Energy Policies in Poland: Policy Feedback and Reform', *Environmental Politics*, 27, 3: 498–518, https://doi.org/10.1080/09644016.2018.1429046; Tomas Maltby and Pierre Bocquillon (2020), 'EU Energy Policy Integration as Embedded Intergovernmentalism: The Case of Energy Union Governance', *Journal of European Integration*, 42, 1: 39–57.
55. Jon Birger Skjærseth (2014), 'Implementing EU Climate and Energy Policies in Poland: From Europeanization to Polonization?' Fridtjof Nansen's Institut, https://www.fni.no/publications/implementing-eu-climate-and-energy-policies-in-poland-from-europeanization-to-polonization.
56. Interview 12, MP, 17 May 2022.
57. Interview 13, MP, 7 July 2022.
58. Tim Rayner, Kacper Szulecki, Andrew J. Jordan, and Sebastian Oberthür (2023), 'The Global Importance of EU Climate Policy: An Introduction'. In: Tim Rayner, Kacper Szulecki, Andrew J. Jordan, and Sebastian Oberthür (eds.), *Handbook on European Union Climate Change Policy and Politics*, pp. 1–21. Cheltenham, UK: Edward Elgar, https://doi.org/10.4337/9781789906981.00011
59. Interview 14, MP, 12 August 2022.

60. Interview 15, MP, 19 August 2022.
61. Kamil Marcinkiewicz and Jale Tosun (2015), 'Contesting Climate Change: Mapping the Political Debate in Poland', *East European Politics*, 31, 2: 187–207, https://doi.org/10.1080/21599165.2015.1022648; Aron Buzogány and Stefan Ćetković (2021), 'Fractionalized but Ambitious? Voting on Energy and Climate Policy in the European Parliament', *Journal of European Public Policy*, 28, 7: 1038–1056, https://doi.org/10.1080/13501763.2021.1918220.
62. Interview 2, Civil Servant, 15 July 2020.
63. Zbigniew W. Kundzewicz, James Painter, and Witold J. Kundzewicz (2019), 'Climate Change in the Media: Poland's Exceptionalism', *Environmental Communication*, 13: 366–380, https://www.tandfonline.com/doi/abs/10.1080/17524032.2017.1394890.
64. Aaron M. McCright, Riley E. Dunlap, and Sandra T. Marquart-Pyatt (2016), 'Political Ideology and Views about Climate Change in the European Union', *Environmental Politics*, 25, 2: 338–358, https://doi.org/10.1080/09644 016.2015.1090371.
65. European Commission (2015, November), 'Special Eurobarometer 435, Climate Change', https://europa.eu/eurobarometer/surveys/detail/2060.
66. Berit Kvaløy, Henning Finseraas, and Ola Listhaug (2012), 'The Publics' Concern for Global Warming: A Cross-National Study of 47 Countries', *Journal of Peace Research*, 49, 1: 11–22, https://doi.org/10.1177/0022343311425841.
67. European Commission (2023, July), 'Special Eurobarometer 538, Climate Change', https://europa.eu/eurobarometer/surveys/detail/2954.
68. CBOS (2023, March), 'Attitudes Towards Green Transformation', https://www.cbos.pl/SPISKOM.POL/2023/K_030_23.PDF
69. Andrzej Ceglarz, Rasmus E. Benestad, and Zbigniew W. Kundzewicz (2018), 'Inconvenience versus Rationality: Reflections on Different Faces of Climate Contrarianism in Poland and Norway', *Weather, Climate, and Society*, 10, 4: 821–836, https://doi.org/10.1175/WCAS-D-17-0120.1.
70. Piotr Zuk and Kacper Szulecki (2020), 'Unpacking the Right-Populist Threat to Climate Action: Poland's pro-Governmental Media on Energy Transition and Climate Change', *Energy Research & Social Science*, 66: 101485, https://doi.org/10.1016/j.erss.2020.101485.
71. Interview 1, MP, 20 December 2019.
72. Research conducted as part of the project Anatomy of Disbelief, with data gathering conducted by Wit Hubert and Grzegorz Bryda and overseen by Aleksandra Wagner. We analysed press and TV media coverage on climate change issues. The corpus included the fifteen most influential press titles in Poland and seven main TV stations. Dailies: *Rzeczpospolita*, *Gazeta Wyborcza*, *Polska the Times*, *Gazeta Polska*, *Nasz Dziennik*; tabloids: *Fakt*, *Super Express*; weeklies: *Neweek*, *Polityka*, *Gość Niedzielny*, *Do Rzeczy*, *W Sieci*, *Wprost*, *Uważam Rze*; television: TV-TVP1, TVP2, TVP info, TVN, TVN24, TV Trwam, Polsat. Articles and programmes were chosen on the basis of the following keywords: 'greenhouse effect', 'global warming', 'climate change', 'climate change policy'; in chosen periods (COP 2013, COP 2018) these keywords figured at least once in 1,254 items (analytical record), 532 of them were press articles and 772 were TV programmes. Statements were categorised into (A) Climate Delay, (B) Climate Action, (C) Passive Climate Denial, (D) Overt and Active Climate Denial.

73. CBOS (2018, November), 'Polacy wobec zmian klimatu', https://www.cbos.pl/SPISKOM.POL/2018/K_158_18.PDF, p. 6
74. Anatomy of Disbelief project data.
75. Interview 8, MP, 21 June 2021.
76. Interview 11, Civil Servant, 21 December 2021.
77. Interview 6, Civil Servant, 16 February 2021.
78. Interview 10, MP, 24 June 2021.
79. Ibid.
80. Ibid.
81. Interview 2, Civil Servant, 15 May 2020.
82. Anatomy of Disbelief project data.
83. Kacper Szulecki (2018), 'The Revolving Door between Politics and Dirty Energy in Poland: A Governmental-Industrial Complex'. In: Pam B. Quintanilla and Patrick Cummins-Tripodi (eds.), *Revolving Doors and the Fossil Fuel Industry: Time to Tackle Conflicts of Interest in Climate Policy-Making*, pp. 98–106. Brussels: Greens/EFA.
84. Maarten Hajer (1993), 'Discourse Coalitions and the Institutionalization of Practice: The Case of Acid Rain in Great Britai'. In: Frank Fischer and John Forester (eds.), *The Argumentative Turn in Policy Analysis and Planning*, pp. 43–76, at 45. Durham, NC: Duke University Press.
85. WNP (2017, 5 October), 'Krajowy rynek energii zamieniony w oligopol?', http://energetyka.wnp.pl/krajowy-rynek-energii-zamieniony-w-oligopol,307737_1_0_0.html.
86. Interview 7, Civil Servant, 16 February 2021.
87. Kacper Szulecki (2020), 'Securitization and State Encroachment on the Energy Sector: Politics of Exception in Poland's Energy Governance', *Energy Policy*, 136: 111066, https://doi.org/10.1016/j.enpol.2019.111066.
88. Interview 7, Civil Servant, 16 February 2021.
89. Interview 8, MP, 21 June 2021.
90. Interview 7, Civil Servant, 16 February 2021.
91. Interview 6, Civil Servant, 1 February 2021.
92. Szulecki, 'The Revolving Door', pp. 98–106.
93. Interview 7, Civil Servant, 16 February 2021.
94. Interview 15, MP, 19 August 2022.
95. Interview 2, Civil Servant, 15 May 2020.
96. Lotos, 'Sokoły z Gdańska: Ograniczający swój wpływ', https://odpowiedzialny.lotos.pl/268/sokoly_z_gdanska/ograniczajacy_swoj_wplyw. Accessed 26 July 2023.
97. Interia Zielona (2021, 6 October), 'Kongres 590: "Interesy stojące za unijną polityką ws. klimatu wymierzone w polską suwerenność"', https://zielona.interia.pl/polityka-klimatyczna/news-kongres-590-interesy-stojace-za-unijna-polityka-ws-klimatu-w,nId,5565251.
98. Nauka O Klimacie (2020, 28 December), 'Górniczy eksperci PGE GiEK negują zmianę klimatu', https://naukaoklimacie.pl/aktualnosci/gorniczy-eksperci-pge-giek-neguja-zmiane-klimatu-448/.
99. CIRE:PL (2022, 29 September), 'Dr. Adam Czyżewski, główny ekonomista w PKN ORLEN S.A. w panelu dyskusyjnym pn. Finansowanie inwestycji w zakresie bezpieczeństwa energetycznego oraz transformacji energetycznej', https://www.cire.pl/artykuly/serwis-informacyjny-cire-24/dr-adam-czyzewski-glowny-eko

nomista-w-pkn-orlen-sa-w-panelu-dyskusyjnym-pn-finansowanie-inwestycji-w-zakresie-bezpieczenstwa-energetycznego-oraz-transformacji-energetycznej.
100. Daniel Obajtek (2021, 7 October), 'Obajtek: Nie uciekniemy przed transformacją energetyczną, ale czas na hamulec refleksji', https://biznesalert.pl/obajtek-nie-uciekniemy-przed-transformacja-energetyczna-ale-czas-na-hamulec-refleksji/.
101. wGospodarce.pl (2014, 15 September), 'Związki zawodowe odrzucają politykę klimatyczną UE', https://wgospodarce.pl/informacje/16113-zwiazki-zawodowe-odrzucaja-polityke-klimatyczna-ue.
102. Solidarnosc Katowice (2019, 13 November), 'Polska nie może zgodzić się na tzw. neutralność klimatyczną', https://solidarnosckatowice.pl/polska-nie-moze-zgodzic-sie-na-tzw-neutralnosc-klimatyczna/.
103. Gazeta.PL (2022, 27 June), Grzegorz Sroczyński, 'Zielona transformacja to będzie dla nas koszmar. Wydatki, wydatki, wydatki', https://next.gazeta.pl/next/7,151003,28625406,zielona-transformacja-to-bedzie-dla-nas-koszmar-wydatki-wydatki.html.
104. Krajowa Izba Gospodarcza (2022, 7 November), 'Relacja z XI Meetingu Gospodarczego KIG „Z węglem – bezpiecznie, w zieloną transformację"', https://kig.pl/uslugi/konferencje-kongresy-szkolenia/wydarzenia-cykliczne/europejski-meeteing-gospodarczy-2022/
105. Forsal.PL (2013, 22 November), 'Guz: Porozumienie klimatyczne—tak. Ale miejsca pracy są ważniejsze', https://forsal.pl/artykuly/747864,opzz-solidarnosc-zwiazki-guz-porozumienie-klimatyczne-opinia.html.
106. WEI (2020, 16 March), 'Polska energetyka u progu transformacji', https://wei.org.pl/2020/blogi/gospodarka/admin/polska-energetyka-u-progu-transformacji/.
107. Pie.net (2020), 'Time for Decarbonisation', Polish Economic Institute, Warsaw, https://pie.net.pl/wp-content/uploads/2020/08/PIE-Time-for-decarbonisation.pdf.
108. Hauke Engel et al. (2020), *Carbon-Neutral Poland 2050: Turning a Challenge into an Opportunity*. Warsaw: McKinsey & Company.
109. Council of the European Union (2023), 'Draft Directive of the European Parliament and of the Council Amending Directive (EU) 2018/2001, Regulation (EU) 2018/1999 and Directive 98/70/EC as regards the promotion of energy from renewable sources, and repealing Council Directive (EU) 2015/652 – Statements, p. 4, https://data.consilium.europa.eu/doc/document/ST-13188-2023-ADD-1-REV-3/en/pdf.
110. Szulecki, 'Securitization and State Encroachment'.
111. Franco Ruzzenenti and Aleksandra Wagner (2018), 'Efficiency and the Rebound Effect in the Hegemonic Discourse on Energy', *Nature and Culture*, 13, 3: 356–377, https://doi.org/10.3167/nc.2018.130303.
112. Bruce Tranter and Kate Booth (2015), 'Skepticism in a Changing Climate: A Cross-National Study', *Global Environmental Change*, 33: 154–164, https://doi.org/10.1016/j.gloenvcha.2015.05.003.
113. CAKE Centrum Analiz Klimatyczno-Energetycznych (blog) (2021, 8 July), 'Nowa Analiza: Polska net-zero 2050', virtualmedia.pl, https://climatecake.ios.edu.pl/aktualnosci/news-cake/analysis-poland-net-zero-2050-the-roadmap-toward-achievement-of-the-eu-climate-policy-goals-in-poland-by-2050-summary/?lang=en.

114. Solidarnosc (2021, 5 August), 'Hutek: "Fit for 55" to krok w przepaść', http://knurow.solidarnoscgornicza.org.pl/2021/08/05/hutek-fit-for-55-to-krok-w-przepasc/.
115. Marceli Sommer (2021, 7 September), 'Roszkowski: Nawet polexit nie uratuje węgla [wywiad], DGB, https://serwisy.gazetaprawna.pl/ekologia/artykuly/8238583,marcin-roszkowski-zmiany-klimatyczne-transformacja-unia-europejska-wywiad.html
116. COP 22 (2016, 11 July) 'Leśne gospodarstwa węglowe' i 'bezodpadowa energetyka węglowa', Gram w Zielone, https://www.gramwzielone.pl/trendy/24032/cop22-lesne-gospodarstwa-weglowe-i-bezodpadowa-energetyka-weglowa.
117. Interview 1, MP, 20 December 2019.
118. Centrum Koordynacji Projektów Środowiskowych (2016, 6 May), 'Nowy projekt w CKPŚ – Leśne Gospodarstwa Weglowe', https://www.ckps.lasy.gov.pl/aktualnosci/-/asset_publisher/HTXX9aadlRBB/content/nowy-projekt-w-ckps-lesne-gospodarstwa-weglowe.
119. Zbigniew W. Kundzewicz and Piotr Matczak (2012), 'Climate Change Regional Review: Poland', *Wiley Interdisciplinary Reviews: Climate Change* 3, 4: 297–311, https://doi.org/10.1002/wcc.175.
120. Aleksandra Fedorska (2023, 12 February), 'Poland: Wind Power Runs Out of Steam', *DW*, https://www.dw.com/en/poland-wind-power-runs-out-of-steam/a-64666398.
121. Daniel RadomskiTwitter (18 March 2023), https://twitter.com/LombatSsc/status/1637002465458438148/photo/1
122. Interview 9, MP, 22 June 2021.
123. Instytut Ochrony Środowiska, 'Krótka Historia Działań i Wyzwania Na Rzecz Ochrony Klimatu w Polsce 1988–2018–2050'.
124. Kamil Marcinkiewicz and Jale Tosun (2015), 'Contesting Climate Change: Mapping the Political Debate in Poland', *East European Politics*, 31, 2: 195, https://doi.org/10.1080/21599165.2015.1022648.
125. UNFCCC (1999, 25 October), 'United Nations Conference on Climate Change Opens in Bonn, Germany', COP 5 Press Release No. 1, http://unfccc.int/cop5/media/pressre1e.html.
126. Aleksandra Arcipowska and Cynthia Elliott (2018, 5 December), 'Here's Poland's Recent History on Climate—and How They Can Steer the Future at COP 24', World Resources Institute, https://www.wri.org/insights/heres-polands-recent-history-climate-and-how-they-can-steer-future-cop24
127. Kacper Szulecki et al. (2016), 'Shaping the 'Energy Union': Between National Positions and Governance Innovation in EU Energy and Climate Policy', *Climate Policy*, 16, 5: 548–567, https://doi.org/10.1080/14693062.2015.1135100; Pierre Bocquillon and Tomas Maltby, 'The More the Merrier?
128. Polish Government (2018), 'Solidarity and Just Transition Silesia Declaration', https://www.documentcloud.org/documents/4936426-Solidarity-and-Just-Transition-Silesia. Accessed 26 July 2023.
129. Polish Government Ministry of Climate and Environment, 'Energy Policy of Poland until 2040', p. 8.
130. Interview 3, Civil Servant, 15 May 2020.

9
Climate Obstruction in Russia

Surviving a Resource-Dependent Economy, an Authoritarian Regime, and a Disappearing Civil Society

MARIANNA POBEREZHSKAYA AND ELLIE MARTUS

INTRODUCTION: THE FOUNDATIONS OF CLIMATE OBSTRUCTION IN RUSSIA

Russia is one of the world's largest producers and exporters of fossil fuels, including coal, oil, and gas, and the fourth-largest global emitter of greenhouse gases (GHGs).[1] Russia is also a recognized laggard in global climate politics. In 1990, the country emitted 3,170 million metric tonnes of carbon dioxide equivalents (MMT CO_2e) (Figure 9.1). However, due to the subsequent major economic and social crises following the fall of the Soviet Union, by 1992, emissions had involuntarily dropped to 2,530 MMT CO_2e. In 1998, they reached their lowest level yet, at 1,870 MMT CO_2e. Hence, to comply with its international commitments under the Kyoto Protocol not to exceed 1990 emissions levels, Russia did not need to do anything, yet could still access potential climate-related investments.[2] Later, as part of its nationally determined contributions (NDC) under the Paris Agreement, Russia committed to a 70% reduction in GHG emissions by 2030, again relative to 1990 levels, and has also proposed a target of net zero emissions by 2060. However, the NDC commitment has been rated 'critically

Marianna Poberezhskaya and Ellie Martus, *Climate Obstruction in Russia* In: *Climate Obstruction across Europe*. Edited by: Robert J. Brulle, J. Timmons Roberts and Miranda C. Spencer, Oxford University Press.
© Oxford University Press 2024. DOI: 10.1093/oso/9780197762042.003.0009

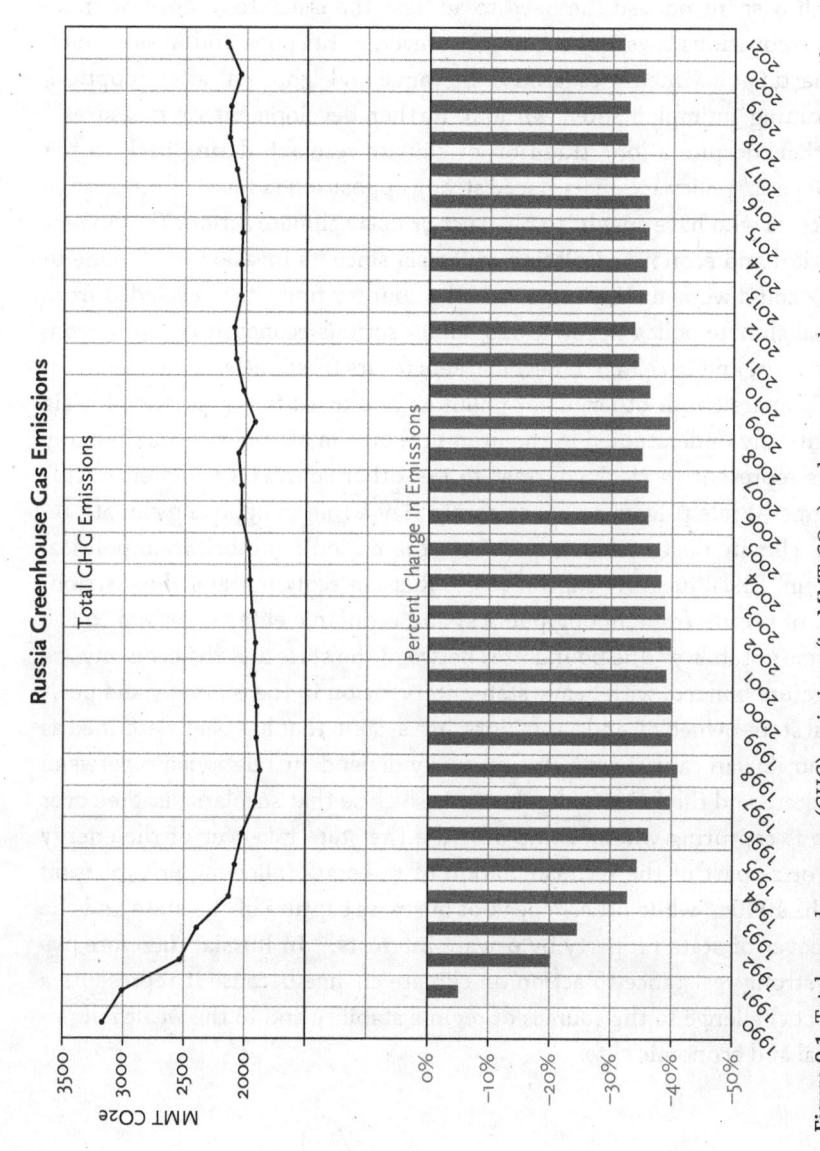

Figure 9.1 Total greenhouse gas (GHG) emissions (in MMT CO_2e) and percentage change in emissions in Russia between 1990 and 2021, inclusive.

Source: Total GHG emissions based on data provided by Gütschow and Pflüger (2023) for Kyoto Six Greenhouse Gas Totals.

insufficient' by the Climate Action Tracker because it requires little effort to achieve[3] and still leaves Russia one of the major global emitters.

Russia's domestic policy commitments also fall short.[4] It has produced several policy documents addressing climate, from the 2009 Climate Doctrine, which first introduced the need to address the issue, to a range of emissions reductions laws and decrees. However, as Korppoo and Alisson note, domestic policy measures 'tend to be vague and "ghosted" after adoption, remaining unimplemented without further development or measures'.[5] Further, despite a long tradition of climate research dating back to the Soviet era,[6] policy action has faced strong opposition in Russia from a range of actors who have sought to obstruct or delay climate action. The current political and economic isolation of Russia since its invasion of Ukraine in 2022 could worsen the situation as the country finds itself excluded from global climate policy negotiations, under serious economic pressure from sanctions, and in search of new markets for its fossil fuels.

Compared with other major polluters and fossil fuel exporters, Russia is critically understudied in the literature on climate politics and in many ways represents a stark contrast to the other countries explored in this volume. While private actors certainly play a role in opposing climate action, climate obstruction is built into the nation's authoritarian political system. Fossil fuels are central to the Russian economy, and the distribution of profits from among political and economic elites is central to the regime's stability.[7] The boundaries between the state and the economy are therefore blurred, with heavy state intervention in the economy and powerful state-owned oil and gas majors, in a system that has been described as authoritarian capitalism.[8] This mutually dependent relationship between the state and the fossil fuel industry is so close that scholars disagree over who is capturing whom: some describe the state takeover of the energy sector as part of the reconsolidation of the state following privatization in the 1990s,[9] while others speak of business capture of the state and the takeover of state property by private interests.[10] In Russia, therefore, we see strong resistance to action on climate change because it represents a direct challenge to the sources of regime stability and to the wealth of political and economic elites.

HISTORICAL EXAMPLES OF CLIMATE OBSTRUCTION

In 1997, the Kyoto Protocol, which set binding emission reduction targets for thirty-seven industrialized countries and economies in transition as well as the European Union, was announced at COP 3. Its conditions required ratification by countries that were collectively responsible for at least 55% of global GHG emissions. In 2001, the United States withdrew from the treaty, leading the international community to turn to Russia as one of the world's highest-emitting countries.[11] Russia ratified Kyoto in 2004, after years of deliberations and open climate obstruction at the national level.[12] Indeed, there were vocal opponents of Russia's involvement despite the Kyoto Protocol's very favourable conditions, stipulating that Russia needed to keep its emissions below the levels of the 1990 baseline year, a goal that, as noted earlier, it had already achieved.

Among the opponents of the Kyoto Protocol, two stood out: Yuri Izrael, a world-leading physicist who made a substantial impact in global climate science, and the economist Andrei Illarionov, a presidential economic adviser between 2000 and 2005. Izrael insisted that the Kyoto Protocol lacked 'a scientific base', and was just a 'political step'[13] which could undermine Russia's economic development. Illarionov doubted the anthropogenic nature of climate change and, on various occasions, called the Protocol 'an assault on economic growth, the environment, public safety, science, and human civilization'[14], an 'undeclared war against Russia',[15] and 'an international Auschwitz'.[16] According to him, Russia would exceed its GHG emissions quota and, therefore, would be forced to slow down or compensate for the overshoot.[17] It is believed that Illarionov played a key role in delaying the Protocol ratification by two years.[18]

Anti-Kyoto sentiment was also shared by some of the largest companies, including mining giant Norilsk Nickel, oil and gas major Yukos, and a few important governmental institutions including the Ministry of Energy (which became the Ministry of Industry and Energy in 2004), though other key bodies such as the Ministry of Natural Resources and to some extent the Ministry of Economic Development and Trade were more hesitant in their position.[19] Notably, after the Protocol's ratification, the opposing voices at the national level dissipated.

The convoluted Kyoto negotiations corresponded with an overall trend in Russia's environmental policies of the late 1990s–early 2000s, when the environment was 'frequently sacrificed to . . . resource exploitation, the chance to earn foreign revenue, and demand for cheap energy . . . [which was] further exacerbated by financial shortages, administrative inefficiencies, and public indifference'.[20] As discussed later, in the twenty years that followed, the situation has barely changed.

THE ROLE OF THE MAJOR ACTORS AND INSTITUTIONS INVOLVED IN CLIMATE OBSTRUCTION

We now turn to the four major actors and institutions involved in climate obstruction in Russia, examining the role of science, the media, government, and industry.

Science obstruction

The majority of the Russian scientific community has been clear in their support for the theory of anthropogenic climate change (ACC)[21] and actively contributed to the Intergovernmental Panel on Climate Change (IPCC) reports, while the sceptical voices that do exist remain a minority. However, we argue that these minor sceptical views among scientists have been disproportionately used by other interested actors, including industry and the government, as a tool of obstruction.

As noted, both Soviet and Russian climatologists have made a substantial impact in advancing climate science.[22] Against this backdrop, climate scepticism[23] within the scientific community presents an interesting case. For example, atmospheric physicist Kirill Kondratyev questioned the methodological value of climate modelling, which then allowed him to challenge ACC.[24] Prominent astrophysicist Khabibullo Abdusamatov insisted that the planet was not warming but would soon enter another ice age.[25] And, as mentioned, even one of the most established, world-famous Soviet and then Russian climatologists, Yuri Izrael, while not denying climate change per se, doubted how much humans had to do with it.[26]

The more recent appearance of climate denialism in the public sphere comes from the so-called science popularizers.[27] For example, Aleksandr Gorodnitskiy, a world-renowned geologist/oceanographer with a limited background in climatology, called ACC a 'myth' started by Al Gore, arguing that both Kyoto and the Paris Agreement are merely political manipulations.[28] As Wilson Rowe highlights, scepticism among Russian scientists was noticeable during the Kyoto deliberations[29] but became less vocal during Dmitry Medvedev's presidential term (2008–2012). During this period there was political acceptance of ACC, as demonstrated by, for example, the publication of the Climate Doctrine. While not holding the legal power of a law or presidential decree, this was an important step in setting guidance for future climate policy. During this time, we also saw the emergence of an economy-oriented narrative of climate mitigation policy co-benefits.

The Soviet climatologist Mikhail Budyko was one of the first scientists who, while making an undeniable contribution to the understanding of ACC, also 'determined that if science suggested ice removal was feasible, having limited consequences for broader natural systems, then potential socio-economic benefits were in the offing'.[30] The alleged benefits that climate change could bring Russia include the expansion of arable land in the North (as a warmer climate would make larger territories suitable for agriculture and prolong the harvesting season) and a shortening of the heating season, thus cutting energy expenses.[31]

One of the most powerful arguments in Russia was that ACC would provide easier access to natural resources in the north and the development of the Northern Sea Route (NSR). The point has 'travelled' to official political discourse, presenting climate change effects in the Arctic as an economic opportunity rather than a threat.[32] For example, at an international forum in 2017, President Vladimir Putin commented on Arctic warming, stating that 'climate change provides more favourable conditions for economic activity in this region' and cited the anticipated growth of shipping along the NSR from 1.4 million tonnes of goods in 2017 to 30 million tonnes by 2035.[33] It is also embedded within policy, with the 'Energy Strategy to 2035', for example, noting the significant potential benefits of the development of the NSR for the oil and gas sectors, giving year-round access to growing markets in the Asia Pacific.[34]

In addition to Yuri Izrael, other scientists also highlighted the potentially damaging nature of the Kyoto Protocol for the Russian economy, its supposed meaninglessness after the United States' withdrawal, and its 'unfairness' in calculating Russia's contribution to global emissions (e.g. disregarding its forests' carbon-absorbing capacity).[35] Within this approach of undermining the policy rather than ACC,[36] we also see the reappearance of the climate geoengineering debate. It initially entered scientific discourse in the early 1960s, while in the 2000s, the proponents of geoengineering restarted the discussion, exploring the benefits of spreading sulphate aerosol in the lower stratosphere.[37] This, according to Izrael, would create 'a kind of umbrella from the sun' dealing with climate change regardless of its origins and would be 'safe for health and many times cheaper than "Kyoto developments"'.[38]

The existence of scepticism and denialism among scientists and their contribution to climate obstruction can be explained by several factors, starting with the practical limitations of Soviet climatology, which was affected by 'the relative backwardness of Soviet computing technology'[39] and different approaches to environmental science[40] as well as the negative impacts of ideologies. Historically, in the Soviet Union 'climatology was largely shaped by Cold War conditions and Soviet science policy, which prioritized military research and neglected many other fields, including climatology'.[41] In turn, the fall of the Soviet Union resulted in an economic and political collapse provoking a 'brain drain' as well as the slowdown or cancellation of various research projects due to financial limitations. This said, on average and similar to global trends,[42] Russian scientists promoting sceptical or denialist narratives belong to older generations (those who were not able to switch to newer methods in climate research) and/or have a different disciplinary background, with limited climate-related experience.[43]

While some of these problems slowly became less relevant (e.g. there is a new generation of highly skilled Russian climatologists), others are re-emerging. For example, a new wave of 'brain drain' was triggered by the Russian invasion of Ukraine.[44] At the same time, researchers who have remained in Russia find themselves cut off from the international scientific community, external funding, and access to the most advanced equipment.[45] This situation will once again make climate scientific discourse more vulnerable to scepticism and denialism and/or desynchronization with global progress in climate knowledge.

Media obstruction

As Russia's political regime has become progressively more authoritarian over the past two decades, national media have been undergoing a corresponding process of showing less diversity, slowly becoming more in sync with the state agenda.[46] The country has several laws restricting the media industry, including Federal Law N-31, 'the fake news law'; Federal Law N-139, 'the internet blacklist law'; and Federal Law N-121, 'the foreign agent law'. The tightening of control has intensified since February 2022, forcing remaining independent media and individual journalists to flee the country or face a series of fines and restrictions, while Russia's regulatory bodies cut off access to major international media websites (including all Meta social media platforms).[47]

In Russia, there is no national media outlet with a clear sceptic position or consistent involvement in climate obstruction. However, the nature of Russia's political regime means that climate coverage is highly susceptible to variations in the state's attitude toward the problem. A study of climate coverage by the national newspaper *Izvestiya* from 1992 to 2012, for example, showed that, in the 1990s, there were very few mentions of climate change, but what did get published confirmed its anthropogenic nature and the problem's urgency. But between 2009 and 2012, 30% of all publications questioned ACC.[48]

Furthermore, Russia's media system represents an interesting case whereby the government-owned gas giant Gazprom controls a substantial number of media outlets, including forty-one TV channels, nine radio stations, six digital platforms, and eight print and online media outlets.[49] This portfolio included (until March 2022) one of the most independent radio stations, *Ekho Moskvy* (Echo of Moscow), which, despite being an example of high-quality liberal journalism, also transmitted climate denialism popularized by Yulia Latynina, a prominent journalist with a strong anti-regime stance and famous for her climate denialism.[50]

The sceptical narratives in both new and traditional Russian media go hand in hand with the conspiratorial thinking and/or antagonistic narrative of Cold War sentiment.[51] As demonstrated in the Kyoto example, this narrative has always been present in Russian public discourse but became more noticeable in the 2010s. For example, a major heat wave in central Russia in 2010 was explained as being a result of a 'climate weapon' deployed by the West, while international appeals to move away from fossil fuel dependency have been met with the argument that ACC is a myth created by profit-driven Western businesses (e.g. renewable energy companies).

A paucity of climate-related publications is another persistent trend in the Russian media. As Boussalis et al. illustrated in their 2016 study,[52] between 2000 and 2014, *The New York Times* alone published more articles on climate change than twenty-three major Russian national newspapers combined. Such modest media attention can also be seen as climate obstruction. With climate change consistently de-emphasized, the climate sceptic lobby has no need to be proactive to influence media coverage[53]; instead, it can simply continue to reenforce a 'climate "spiral of silence" that leads people who do not hear about the topic in daily life to avoid discussing it themselves'.[54]

Among other factors affecting climate coverage in Russia are the country's geographical characteristics (a diverse range of climatic zones from east to west and north to south), regional politics, and sociodemographic variations throughout the country[55] whereby climate change discourse can be affected by people's economic instability, the presence and position of the local climatologic community, and interference from federal and international stakeholders.

Like that of other authoritarian states,[56] Russian media climate discourse is substantially affected by the main 'newsmaker' in the country; hence, when President Putin casts doubts over ACC, the media reproduce this message without challenge. For example, in 2019, during the end of the year press conference, Putin, while highlighting Russia's commitments to GHG emissions reduction, stated that 'no one knows the reasons behind global warming', after which he alluded to natural processes that could be responsible.[57] Conversely, when Putin confirms ACC and claims 'we need to do everything we can to minimise our input', the media do not mention Russia's lukewarm climate policy but instead repeat the president's message that, in the past three decades, due 'to a radical restructuring of industry and energy, it was possible to reduce greenhouse gas emissions more than in other countries'.[58]

The sentiment fits into the narrative of Russia being 'a great ecological power',[59] which peaked during 2020–2021, when climate change for the first time became prominent on the state's agenda. As Head of the State Duma Committee on Ecology and Environmental Protection Vladimir Burmatov stated, 'Today the Russian Federation is one of the leaders on the climate agenda and it has something to show the world'.[60] Indeed, from 2020 until the invasion in February 2022, official rhetoric became much more in sync with a stronger climate policy, with Putin consistently reconfirming Russia's intentions to reach carbon neutrality by 2060.[61]

However, as a 2016 longitudinal study of the news media found,[62] challenging economic conditions have negatively affected climate coverage in Russia, with journalists either paying even less attention to the topic than usual or covering it only within the context of the international negotiations. In 2022, Russia managed to avoid the worst-case economic scenario despite the imposition of sanctions and the demands of the military campaign, yet the economic decline has been felt throughout the country, with the situation expected to deteriorate further.[63] For this and other reasons, the state public relations establishment has been focussed on justifying the invasion, monopolizing the media agenda, with climate coverage marginalized once again.[64]

Government obstruction

Tynkkynen and Tynkkynen, in their 2018 study of climate denial under Putin, highlight 'the specific interests of the energy sector in maintaining the status quo in domestic energy policy and in the general interests of Putin's regime in reducing the likelihood of criticism by the Russian people toward the hydrocarbon-based political and economic system'.[65] Indeed, as noted, the state depends on the fossil fuel industry for regime stability, and climate obstruction is therefore woven into the activities of government elites. We see this manifest in three core ways: through the position and public statements of Putin; via the restrictions imposed on nongovernmental organizations (NGOs), which limit meaningful participation of civil society in policy debates; and in the role of the powerful Ministry of Energy (MinEnergo).

The authoritarian nature of the political system means that Putin plays a leading role in setting broad strategic goals and framing the policy agenda. As noted, the president has made controversial statements in the past that have demonstrably slowed the development of Russia's climate policy. His remark, for example, that 'we shall save on fur coats and other warm things' sent mixed signals during an international climate change conference in Moscow in 2003, given that his overall message was about Russia's commitment to 'addressing climate change'.[66] Tynkkynen and Tynkkynen[67] argue that, around the start of Putin's third presidential term, we see a re-emergence of climate denialism at the highest level, with climate policies regarded as another potential threat to regime stability. As noted, Putin's discourse has also emphasized Russia's role as an 'ecological donor' or 'great ecological power' due to its existing contributions to global efforts to address climate change. This discourse has become part of the country's climate obstruction efforts because it is used to justify Russia's limited climate policy commitments and express doubts about international policy responses.

NGOs are generally regarded as playing a minimal role in shaping Russian climate politics.[68] However, prominent groups such as World Wildlife Fund (WWF) Russia, Greenpeace Russia, and Ecodefence have maintained climate programmes aimed at providing information to the public (e.g. Greenpeace's Green Deal of Russia, a proposed emissions reductions programme), participated in drafting policy (e.g. WWF was involved in the National Climate Adaptation Plan, aimed at mitigating the impact of climate change for Russia), and in some cases, held protests against coal mining (e.g. Ecodefence picketed in Novokuznetsk[69] and Germany to protest the importing of Russian coal[70]). Ecodefence is also among the plaintiffs in Russia's first climate lawsuit, which demands Russia reduce its GHG emissions.[71]

Thus, in recent decades, NGOs have increasingly been viewed as a threat to the regime, and efforts to restrict their operation have become a key element in the government's climate obstruction enterprise. This has manifested in a series of repressive laws, increased state scrutiny of civil society activities, and heavy administrative burdens for groups.[72] In 2012, the 'foreign agent law' was introduced, targeting groups receiving international funding and engaged in 'political activities' broadly defined; it carries the negative connotation that such people are spies or traitors.

New laws since 2012 have tightened the space for NGO activity even further.

While not specifically focused on the issue of climate change, but rather targeting the NGO community more broadly, these changes have ensnared individuals and groups campaigning on climate action. For example, Ecodefence was listed as a foreign agent in 2014, with individual members forced to leave the country.[73] Similarly, the Indigenous Peoples' Centre was put on the register of foreign agents in 2015 for 'organising discussions on climate change, its impact on indigenous peoples',[74] and prominent youth climate activist Arshak Makichyan was stripped of his Russian citizenship.[75] Finally, WWF Russia, one of the most prominent NGOs, was listed as a foreign agent in March 2023,[76] and in June its parent organization, WWF, was declared an 'undesirable organisation'. That designation meant it could no longer operate in Russia, forcing WWF Russia to disassociate itself from the global network.[77] This trend can be considered as part of climate obstruction in Russia because it limits civil society participation in public life and the ability of NGOs to provide input into policy decisions, including those individuals and groups actively campaigning for Russia to adopt a more ambitious climate agenda. This challenge is exacerbated by the framing used by Putin and others, who describe the IPCC and international climate cooperation as a form of 'Western dominance': something foreign and hostile to Russia's interests.

MinEnergo, the key bureaucratic stakeholder, is responsible for high-level energy strategy as well as policy development and implementation for specific power sectors including coal, electricity, oil, gas, renewables, and nuclear. This mandate has brought the ministry into conflict with the climate policy ambitions of other agencies within the government, including the Ministry of Economic Development (MED), which has been a central actor in driving domestic climate policy and shaping Russia's participation in the international climate discussions. However, MinEnergo has tempered some of the more ambitious forecasts and production plans put forward by the coal sector, for example.[78]

In terms of shaping policy debates, MinEnergo largely acts as an advocate for sectoral interests, particularly those of the fossil fuel industry. This is apparent from the key energy strategies and policy documents the ministry has produced, which emphasize the need to support Russia's fossil fuel industries and discuss climate change primarily as a national economic threat. As Romanova[79] notes, both the 'Energy Strategy' and the 'Energy Security Doctrine' (ESD) signal a recognition of the need to diversify export markets toward Asia to limit the impact of the European Union's 'political motivations' in shifting away from Russian oil and gas exports, while the '2019 ESD and most officials treat renewables and clean energy as (unfair) competition and, in some cases, as external political challenges'.[80] Previous documents such as the 'Economic Security Strategy to 2030' also recognize the economic threat of green technology and energy efficiency[81] but without calling for corresponding policy development around cutting Russia's own emissions.

MinEnergo has also successfully limited the climate policy ambitions of other ministries within government. In one prominent example, a draft law aimed at limiting GHG emissions from industry, developed by the MED, was met with strong opposition from industry, led by the Russian Union of Industrialists and Entrepreneurs (RUIE), a powerful industry association discussed later in this chapter. MinEnergo sided with these actors and was able to limit the obligations proposed in early drafts of the law, which included emissions quotas for industry, penalties for exceeding them, and the introduction of a market for trading carbon.[82] After sustained lobbying from MinEnergo and the RUIE, all these elements were removed from the final bill. In 2021, the Law on GHG Emissions was passed by the State Duma, the lower chamber of the Russian parliament, including only the mandatory disclosure of emissions by the largest companies and making all targets voluntary, without penalties for exceeding them.[83] In short, the combination of bureaucratic and elite-driven obstruction in Russia limits the space available for other actors to promote a pro-climate policy agenda. The situation is made even more challenging by the mutually dependent relationship between the state and the fossil fuel industry, discussed next.

Industry obstruction

Some of the strongest opposition to action on climate change comes from Russia's powerful industry actors, including private and state-owned companies and business associations. Russia is one of the world's largest oil- and gas-producing and export countries, with the oil and gas sectors dominated by large companies, including state-owned gas giant Gazprom and oil company Rosneft, as well as privately owned oil company Lukoil and gas producer Novatek. In addition to the government's involvement in state-owned energy companies, there are close connections between Putin's inner circle and independent (on paper) gas produces such as Novatek.[84]

Russia is also the world's largest exporter of coal, although, unlike the oil and gas sectors, the industry is mostly privately owned.[85] It is concentrated in a number of major coal-mining regions, including Kemerovo Oblast (Siberia), where it is an important source of employment and electricity.

While coal-fired power stations supply approximately 15% of Russia's domestic electricity overall,[86] in the major coal regions the percentage of coal in the electricity balance increases dramatically. For example, in 2021, in the city of Kemerovo, coal supplied 80% of the region's electricity.[87] The significant contribution to GHG emissions from burning coal means that the coal industry is seriously threatened by climate action, even in comparison with oil and gas, and has thus actively lobbied the government and sought to frame coal as essential for the Russian and global energy future.[88] Similar practices are found in other major coal-producing countries, which have been among the slowest to implement comprehensive climate policies.[89] Strategies and tactics adopted by industry take two primary forms: lobbying and other forms of interference in the policy process, and the use of discursive framings.

Lobbying

As we might expect, there is strong resistance from the fossil fuel sector to any suggestion of strengthening the climate policy agenda. Their resistance has been largely effective, as demonstrated in the example of the Law on GHG Emissions. These actors are motivated by a desire to promote the interests of their industry, including its expansion; to acquire increased government financial support; and to resist any form of regulation they perceive as burdensome. These priorities mean fossil fuel companies often come into conflict with government attempts to introduce stronger climate policies, such as curbing industry emissions.[90] Lobbying also extends beyond domestic politics, with Russia sending the second-largest number of 'fossil-fuel linked delegates' to COP 27.[91]

Another notable example of industry's climate obstruction concerns the RUIE, a powerful business association representing some of Russia's largest companies and regarded as the designated intermediary between business and government in Russia.[92] While it represents a range of companies,[93] many are connected to fossil fuels, and, as a result, it has been an active defender of fuel and energy interests. The RUIE has not always held a united or consistent position on climate change[94]; however, its executive has generally been sceptical of proposed government measures that might create additional regulations for business. Furthermore, the RUIE has direct access to policymakers, with representation on, for example, the high-level Interdepartmental Working Group on Climate Change.[95] This group was established in 2012 to coordinate policy implementation and provides a formal channel for industry to voice concerns over the direction of climate policy. The group had previously lobbied against Russia's participation in the Paris Agreement, though dropped its opposition when it became clear that Russia's commitments would be very limited and the economic consequences of failing to ratify Paris would be more severe.[96]

Importantly, industry lobbying aimed at climate obstruction involves not only attempts to curb the obligations imposed on industry through climate policy per se but also to increase government support in the form of subsidies, financial support, and access to preferable transport options. When it comes to coal, for example, serious rail bottlenecks create problems for exports,[97] and the industry has sought government help to find a solution. Companies have also lobbied for the expansion of coal production forecasts in policy documents, such as the 'Strategy for the Development of the Coal Industry (2020).' Finally, in addition to its involvement in policymaking, the fossil fuel industry has successfully resisted the implementation of laws on several occasions. Work by Korppoo[98] on gas flaring, for example, points to a case whereby policy was undermined by oil company noncompliance with associated petroleum gas regulations and weak oversight by government bodies. Furthermore, in the months after the invasion of Ukraine began, companies have lobbied to have climate regulations, including the new Law on GHG Emissions and other environmental laws and forms of reporting, delayed or reduced in hopes of limiting the effects of international sanctions.[99]

Discursive framings

Beyond lobbying, industry actors engage in other forms of climate obstruction through their use of discursive framing strategies. The most prominent example comes from the coal industry, which has questioned the economic rationality of climate policy and emphasizes the importance of coal as a source of employment and heating in major coal mining and export regions of the country.[100] This narrative is part of the broader discursive framing of climate as a 'second-order' problem, discussed further on.

Communication with the public and shareholders through their online presence and corporate reporting is illustrative of this approach. As Martus and Fortescue[101] discuss, they make no blanket denial of climate change but rather attempt to shift the narrative around coal and its future in the context of climate change, with an emphasis on the social and economic importance of coal at a regional level. As with the creation of astroturf organizations in other contexts,[102] in the past, Russian companies have provided financial support to 'grassroots' groups whose campaigns advocated the continued operation of the coal industry, such as the 'Right 4 Coal' campaign run by the Siberian Generating Company (a coal-fired power generating company).[103] In the context of war, we would expect these narratives around employment and energy security of the regions to intensify. As we discuss later, we are also seeing a new emphasis on anti-Western rhetoric, including by key fossil fuel actors cheering the end of the Western-led 'green agenda'.

OVERARCHING DISCURSIVE FRAMINGS EMPOWERING CLIMATE OBSTRUCTION

As demonstrated earlier, while tools and approaches might differ depending on the stakeholder, their contribution to climate obstruction in Russia is underpinned by four overarching narratives: (1) Russia as 'a great ecological power', (2) 'climate policy as a Western tool of dominance', (3) 'climate change as an opportunity', and (4) 'climate change as a second-order problem'. The way each of the actors contributes to these discourses is summarized in Table 9.1 and explored in more detail further on.

Russia as a great ecological power

The narrative of Russia as a 'great ecological power' first became evident during the Kyoto negotiations, where Russia's key role in bringing the agreement into force was framed as saving global climate governance.[104] The country's unintentional GHG emissions drop in the early 1990s provided a foundation for state leaders for the next three decades to continue referring to Russia as an environmental leader or donor.[105] Another issue driving this narrative is Russia's vast boreal forests, with stakeholders presenting the country as a giant carbon sink that has already done enough for the world. This narrative has fuelled debates within the scientific community over the method for calculating forests' GHG absorption. While significant uncertainty remains over the accuracy of the data,[106] the scientific debate has been leveraged politically using the most ambitious estimates.[107] Both the Paris Agreement NDC and updated targets within the 'Strategy for the Socio-Economic Development of Russia with a Low Level of Greenhouse Gases to 2050' link emissions reductions to the 'maximum possible absorption capacity of forests and other ecosystems'.[108]

The narrative of Russia being an 'ecological donor' is used by the government and industry actors as an excuse not to act and has become an

Table 9.1 EXAMPLES OF HOW VARIOUS ACTORS CONTRIBUTE TO THE IDENTIFIED DISCURSIVE FRAMINGS

Discursive framing	Actor			
	Science	Media	Government	Industry
Russia as a great ecological power	Offering different ways to calculate forest reserves	Following the lead of the 'main newsmaker'; not questioning state decisions/polices	Highlighting 1990s' emissions drop; highlighting absorptive capacity of Russia's forests	Highlighting Russia's greenhouse gas (GHG) emissions level in relation to 1990s levels; highlighting absorptive capacity of Russia's forests
Climate policy as a Western tool of dominance	Appealing to the lack of a scientific basis for the international agreements	Spreading climate-related conspiracies (e.g. climate weapon; climate change as a myth)	Labelling civil- society groups with foreign links as traitors; doubting anthropogenic climate change; characterizing international policies as 'Western led'	Suggesting climate change is a plot created by the Western profit-driven industries (RE); appealing to Russia's national interests; highlighting international sanctions
Climate change as an opportunity	Highlighting potential benefits from climate change for Russia (agriculture, energy expenses, Arctic access)	Not questioning state and scientist commentary on potential benefits	Highlighting potential investments; overplaying energy efficiency policy; downplaying negative consequences	Promoting the Northern Sea Route (NSR) as an opportunity for greater economic benefits
Climate change as a 'second-order' problem	Offering alternative ways to manage without economic sacrifice (e.g. geoengineering)	Limiting coverage of climate change	Monopolizing public agenda—highlighting the importance of economic development/state security	Branding climate policies as 'uneconomic' or damaging to social stability; portraying fossil fuels as essential to country's survival

integral part of Russian policy documents, strategies, and corporate statements on climate. For example, the founder of the major coal company SUEK, Andrey Melnichenko, stated: 'We are an environmental donor to the planet, including because of the large number of forests rather than a source of emissions. . . . I think we will not need to make global efforts in this direction'.[109] As Nina Tynkkynen[110] observed, the 'Great Ecological Power' approach serves to mask the weaknesses of Russia's climate policy and deflect attention from the country's overdependence on the fossil fuel sector. We see no signs of this changing anytime soon.

Climate policy as a Western tool of dominance

The narrative of climate being a 'Western tool of dominance' feeds off the conspiratorial thinking discussed earlier, as well as the re-emergence of Cold War rhetoric, introducing an antagonistic approach of 'us versus them' to climate politics. During Kyoto deliberations, its antagonist Illarionov stated that climate governance is 'a war, war against the whole world but in this case the first one who got in the way, is our country. . . . It is a total war against our country'.[111] The narrative has remained persistent throughout the past two decades, though has been amplified since Russia's invasion of Ukraine. For example, the pro-government state news agency RIA Novosti published an article in March 2023 on the 'climate weapon': 'This is how the United States wants to fight "Russia's emerging dominance in agriculture"'.[112] This narrative, once again, advances the false assumption that climate change is beneficial for Russia and suggests that it is the United States that has been pushing geoengineering all along (ignoring the early Soviet/Russian role in the field, as discussed earlier).

Within this discursive framework, 'international climate policy is increasingly seen as a Western-led hegemonic project aiming to bypass or overrule the sovereignty of Russia',[113] while Russia's resistance to global climate governance is presented as a sensible and even essential way to defend itself against the West. After February 2022, this theme became even more prominent, with the public agenda monopolized by the 'special military operation's' concerns. Unsurprisingly, after a brief splash of climate-related interest in 2020–2021, the problem has almost disappeared from national discourse. As Doose and Vorbrugg[114] have stated, 'it is undeniable that the economic crisis, sanctions and strengthened anti-Western rhetoric brought on by the war have made it more difficult to pursue decarbonisation plans' as actors that were already trying to obstruct national climate commitments now receive more opportunities to be heard.

For example, the head of the 'Just Russia' party, Sergey Mironov, claimed that 'Russia after Western sanctions must leave the Paris Agreement'.[115] Simultaneously, Igor Sechin, the chair of oil giant Rosneft, claimed that sanctions have ended the green transition as countries try to find alternative sources of hydrocarbons to replace Russian ones, with Europe committing 'energy suicide' in doing so.[116] Ironically, Oreskes and Conway, in their provocative book *Merchants of Doubt* (2011),[117] explain how, in the United States, obstruction narratives tried to frame climate change as something invented by socialists/communists and that, therefore, threatens the prosperity of the capitalist world.

Climate change as an opportunity

Since Medvedev's move during his presidency to give more political prominence to the climate change agenda, he and various other government and business actors have highlighted the potential economic benefits for Russia that are said to emanate from both a changing climate and climate-related policy. For example, the 'National Climate Adaptation Plan to 2022' (signed December 2019) lists the potential negative consequences (for public health, industry, etc.) but also the anticipated positive effects of climate change, including a reduction of energy consumption in winter, greater access for shipping in the Arctic, an expansion of arable land, and the increased productivity of boreal forests.[118]

At the same time, it has been emphasized that Russia meets GHG emission reduction targets without any effort. Indeed, even after the economic recovery from the 2000s onward, and despite the country's remaining one of the most carbon-intensive economies in the world, Russia's emissions did not exceed 2,160 MMT CO_2e—this highest level yet was reached in 2021— and are therefore still well below 1990 levels. However, Russia could still benefit from energy-efficiency plans (to save more fossil fuel for export and reduce national energy expenses). Furthermore, recent documents such as the 'Strategy for the Socio-Economic Development of Russia with a Low Level of Greenhouse Gases to 2050' have placed a stronger emphasis on the opportunities for Russia. These opportunities include the expansion of Russia's nuclear export programme as a core element of its climate agenda, with Russia already being the world's largest exporter of nuclear reactors. Other perceived emerging prospects around hydrogen and renewables have also been emphasized, at least prior to February 2022.

The discourse of 'opportunity' is a complicated one. It can be argued that this 'win-win approach' is a way to overcome climate obstruction because it allows climate advocates to attract the attention of key stakeholders without antagonizing them, something especially valuable in an authoritarian society. At the same time, this positive narrative prevents policymakers from seeing climate change as an environmental problem or threat to the country's wellbeing, thus slowing or limiting the scope of climate policymaking and implementation. Kokorin and Korppoo argue that Russia's leaders follow the 'ostrich approach', persistently delaying climate-related policies. For example, by 2017, despite renewables becoming economically viable in some parts of Russia, Deputy Prime Minister Arkady Dvorkovich suggested waiting until they 'become cost-effective in Russia as a whole'.[119] Furthermore, if climate change is not seen as an existential threat, then it naturally fits into the next narrative of climate as a 'second-order problem': a problem that can be postponed (indefinitely).

Climate change as a 'second-order' problem

The marginalization of the climate change problem in favour of addressing other, seemingly more important difficulties is not unique to Russia, but in fact one of the persistent features of developing societies.[120] As Inglehart argued in his 1995 study,[121] a higher concern for environmental issues is normally accompanied by a 'postmaterialist shift' that goes hand in hand with economic prosperity. While there is evidence suggesting applicability of this argument to Russia,[122] there are also other explanations for the low level of public and state attention to the problem. As discussed earlier, the media overall do not 'see environmental concerns as important compared to political concerns',[123] often resulting in an avoidance of climate change as a topic.

Due to Russia's economic dependency on extractive industries, those with a vested interest (e.g. industry groups, government elites) are more likely to focus on strategies that are not public-facing and seek to influence policy- and decision-makers directly, hence minimizing the public discussion of climate change. Interestingly, in Western countries, especially the United States, a range of stakeholders have contributed to the powerful countermovement that challenged 'the environmental community's definition of global warming as a social problem and blocked the passage of any significant climate policy'.[124] However, Ashe and Poberezhskaya suggest that, in Russia, the need for a countermovement has been negligible because a fully fledged environmental movement never had a chance to flourish due to increased state repression of NGOs. Hence, as we discussed in the media section, there is no need to deny or censor climate-related discussions. Instead, it is much easier to relegate climate to a 'second-order' problem. Because the latest migration wave has forced several top climate correspondents and activists to leave the country, the situation is likely to worsen.[125]

Interested parties also tend to highlight more 'acute' economic and social issues that may affect either a specific region or the country overall. For example, coal companies have been active in leveraging this tactic in corporate communications, arguing that a shift away from coal would have significant implications for employment, energy security, and social stability in major coal regions such as the Kuzbass.[126] The strategy of regarding climate change as something that can be postponed or dealt with superficially became even more fruitful after February 2022. This narrative may be one of the most difficult to overcome in an authoritarian political regime that dominates the public agenda.

CONCLUSION

In their observations on climate obstruction in the Global South, Milani et al.[127] suggested paying greater attention to whether different economies produce 'different types of climate obstruction strategies, discourses, and organizational structures'. We echo these sentiments but add that we also need to understand whether different political systems create different forms of obstruction. The evidence presented in this chapter suggests that climate obstruction in Russia, an authoritarian state dependent on fossil fuels, differs in many ways when compared with other Western countries. For example, according to Brulle,[128] in the United States, 'key opponents to climate action are motivated by private interest in the continuation of the fossil fuel-based economy'. In Russia, there is a less clear-cut distinction between the state and the private sector, which means that some of the more well-known tools and agents of climate obstruction, such as conservative think tanks, do not exist.

Climate is seen as a risk by the Russian state due to its perceived link with foreign influence and, presumably, the challenge that civil society represents to the political system, itself grounded in the fossil fuel economy. Plantan[129] argues that authoritarian governments divide civil society into 'wanted and unwanted elements' to maximize the benefits and minimize risks posed to the regime. For example, Russia's use of labels such as 'foreign agent' and 'undesirable organization' shapes public perception of NGOs and media, thus delegitimizing their work.[130] Indeed, studies have shown limited public demand for climate policy action in Russia.[131] Interestingly, the existing work on climate obstruction beyond Russia emphasizes public opinion as a key focus for actors engaged in climate obstruction, yet in Russia it is irrelevant. Moreover, we argue that public climate disinterest turns into a 'passive' form of climate obstruction, whereas the 'active forms' take place among networks of influential industry groups and state elites.

Regarding possible solutions for these trends, prior research on Russia's climate policy has suggested that support be given to the small but important coalition of national climate change experts and advocates (who have been slowly but surely shaping the country's climate-related agenda).[132] More recent studies have pointed to the potential emergence of influential policy actors within specific areas of the renewable energy industry, including solar photovoltaic manufacturing.[133] Prior to February 2022, it had also been suggested that external actors (including, for example, the European Union, one of Russia's major trading partners) could play an important role in stimulating the development of climate initiatives and projects, leveraging Russia's desire to increase trade and be better integrated within the global community.[134]

Under the ongoing regime of sanctions and Russia's economic, political, and cultural isolation, these strategies have become obsolete, at least for the time being. Since the start of the invasion of Ukraine, the already challenging environment for proactive climate policy has taken another turn for the worse: climate sceptic messages have resurfaced in the major media outlets,[135] business and political actors are capitalizing on hostile relations with the West, and the importance of Russia's international image and engagement in global dialogue has become irrelevant. Hence, national and international stakeholders and researchers need to find new ways to overcome climate obstruction in Russia.

Potential solutions might include a certain degree of depoliticization of climate change by international actors to limit Russia's anti-Western motivated withdrawal from international dialogues and to elevate scientific engagement on climate-related policies. Yet continuing international scientific dialogue with Russia-based climatologists remains a highly controversial topic.[136] Within Russia, climate obstruction could be addressed if there were greater realization among political elites and policymakers that climate-related risks and losses at the national level would surpass any perceived benefits and that assistance for mitigation and adaptation will most likely come only from within Russia itself.

Ultimately, though, given that Russian relations with former Western partners are at their lowest point since the end of the Cold War, a more realistic solution might be to encourage other, non-European/American international partners (e.g. BRICS countries) to take the lead in engaging with Russia on climate for the foreseeable future to ensure it remains on the country's agenda. As we have sought to highlight, climate obstruction is not homogenous globally. We believe there is considerable value in exploring frequently overlooked cases such as Russia to understand how climate obstruction can be overcome in the most difficult political, economic, and social contexts.

NOTES

1. Igor Makarov (2022), 'Does Resource Abundance Require Special Approaches to Climate Policies? The Case of Russia', *Climatic Change*, 170, 3, https://doi.org/10.1007/s10584-021-03280-0.
2. Jonathan Oldfield (2005), *Russian Nature. Exploring the Environmental Consequences of Societal Change*. Hants, UK: Ashgate.
3. Climate Action Tracker (2022, 9 November) 'Russian Federation', https://climateactiontracker.org/countries/russian-federation/.
4. Anna Korppoo (2020), 'Domestic Frames on Russia's Role in International Climate Diplomacy', *Climate Policy* 20, 1: 109–123, doi:10.1080/14693062.2019.1693333.
5. Anna Korppoo and Alex Alisson (2023), 'Russian Climate Strategy: Imitating Leadership', Climate Strategies Report, p. 6, https://climatestrategies.org/publication/russian-climate-strategy-imitating-leadership/.
6. Andy Bruno (2018, July), 'Climate History of Russia and the Soviet Union', *WIREs Climate Change*, https://doi.org/10.1002/wcc.534.
7. David White (2018, January), 'State Capacity and Regime Resilience in Putin's Russia', *International Political Science Review*, 39, 1: 130–143, doi:10.1177/0192512117694481.
8. Dorottya Sallai and Gerhard Schnyder (2021), 'What Is "Authoritarian" about Authoritarian Capitalism? The Dual Erosion of the Private–Public Divide in State-Dominated Business Systems', *Business & Society*, 60, 6: 1312–1348.
9. Thane Gustafson (2012), *Wheel of Fortune*. Cambridge, MA: Harvard University Press.
10. Margarita Balmaceda (2013), *The Politics of Energy Dependency: Ukraine, Belarus, and Lithuania between Domestic Oligarchs and Russian Pressure*. Toronto: University of Toronto Press; Margarita Balmaceda and Andreas Heinrich (2018), 'The Energy Politics of Russia and Eurasia'. In: Kathleen Hancock and Juliann Allison (eds.), *The Oxford Handbook of Energy Politics*, pp. 465–506. New York: Oxford University Press .
11. Marianna Poberezhskaya (2016), *Communicating Climate Change in Russia: State and Propaganda*. Abingdon, UK: Routledge.
12. Barbara Buchner and Silvia Dall'Olio (2005), 'Russia and the Kyoto Protocol: The Long Road to Ratification', *Transition Studies Review*, 12: 349–382.
13. *RIA Novosti* (2003, 7 May) 'Izrael: Kiotskomu Protokolu ne Khvataet Ekonomicheskikh Obosnovaniy', https://ria.ru/20030507/376649.html.
14. Alvin Powell (2004, 2 December), 'Debate over Kyoto Climate Treaty Heats Up at KSG', *The Harvard Gazette*, https://news.harvard.edu/gazette/story/2004/

12/debate-over-kyoto-climate-treaty-heats-up-at-ksg/#:~:text=Illarionov%20w
ent%20on%20to%20say,is%20responsible%20for%20global%20warming.
15. Martin Enserink (2004, 13 July), 'Hot Controversy Over Climate Meeting', *Science*, https://www.science.org/content/article/hot-controversy-over-climate-meeting.
16. *The Irish Times* (2004, 15 April), 'Putin Adviser Likens Kyoto Pact to Auschwitz', https://www.irishtimes.com/news/putin-adviser-likens-kyoto-pact-to-auschwitz-1.1308171.
17. Buchner and Dall'Olio, 'Russia and the Kyoto Protocol'.
18. Jessica Tipton (2008), 'Why Did Russia Ratify the Kyoto Protocol? Why the Wait? An Analysis of the Environmental, Economic, and Political Debates', *Slovo*, 20, 67–96.
19. Laura Henry and Lisa Sundstrom (2007), 'Russia and the Kyoto Protocol: Seeking an Alignment of Interests and Image', *Global Environmental Politics* 7, 4: 47–69.
20. Jonathan Oldfield, Anna Kouzmina, and Denis Shaw (2003), 'Russia's Involvement in the International Environmental Process: A Research Report', *Eurasian Geography and Economics*, 44, 2: 166.
21. For example, see Rosgiromet (2021), 'Doklad ob Osobennostyakh Klimata na Territorii Rossiyskoy Federatsii za 2020 god', https://www.meteorf.gov.ru/upload/pdf_download/doklad_klimat2020.pdf.
22. The development of climate research in the Soviet Union was state-sponsored (like any other research), hence, it presents an interesting comparative case to the private companies-led research and obstruction in the West. The impact of the state over scientific research in the post-Soviet era is a more complex case. As discussed in this chapter, Russian state and industry are closely interconnected, making it extremely complicated to understand whether it is an impact of private or state interests. For more on the Soviet contribution to climate science, see Jonathan Oldfield (2018), 'Imagining Climates Past, Present and Future: Soviet Contributions to the Science of Anthropogenic Climate Change, 1953–1991', *Journal of Historical Geography*, 60: 41–51.
23. There is an ongoing debate on the terminology around climate denialism/scepticism; here we agree with van Rensburg's typology which 'implies that sceptics are disagreeing with parts (rarely all) of the mainstream climate thesis' and more likely to engage in a mainstream discussion on ACC whilst denialists suggest a complete rejection of ACC existence without any engagement with climate science. See Willem Van Rensburg (2015), 'Climate Change Scepticism: A Conceptual Re-Evaluation', *SAGE Open*, 5, 2: 9.
24. A. Gerasimov (2002, 19 July), 'Global'noe Poteplenie Klimata – Eto Mif', *Nauka Izvestiya*, 25, 58: 1.
25. Khabibullo Abdusamatov (2009), 'Solntse Opredelyaet Klimat', *Nauka i Zhizn*, 1: 34–42.
26. Quirin Schiermeier and Bryon MacWilliams (2004, September), 'Crunch Time for Kyoto', *Nature*, 431: 12–13.
27. Dmitry Yagodin (2021), 'Policy Implications of Climate Change Denial: Content Analysis of Russian National and Regional News Media', *International Political Science Review*, 42, 1: 65.
28. Aleksandr Gorodnitskiy (2020, 13 January), 'Devochka i Mif', *Ogonek*, https://www.kommersant.ru/doc/4205270.

29. Elana Wilson Rowe (2012), 'International Science, Domestic Politics: Russian Reception of International Climate Change Assessments', *Environment and Planning D: Society and Space* 30: 711–726.
30. Oldfield, 'Imagining Climates', p. 45.
31. Tamara Kazarina (2015), 'Klimaticheskiy Khaos', *TASS spetsial'nyi proekt*, https://tass.ru/spec/climate.
32. Olga Khrushcheva and Marianna Poberezhskaya (2016), 'The Arctic in the Political Discourse of Russian Leaders: the National Pride and Economic Ambitions', *East European Politics*, 32, 4: 547–566, doi:10.1080/21599165.2016.1231669.
33. President of Russia (2017, 30 March), 'The Arctic: Territory of Dialogue International Forum', http://en.kremlin.ru/events/president/news/54149.
34. Pravitel'stvo Rossii (2020), *Energeticheskaya Strategiya Rossiiskoi Federatsii na period do 2035 goda*, Approved by Government Order 9 June 2020, No 1523-r.
35. Anna Korppoo, Nina Tynkkynen, and Geir Hønneland (2015), *Russia and the Politics of International Environmental Regimes: Environmental Encounters or Foreign Policy?* Cheltenham, UK: Edward Elgar.
36. Van Rensburg categorises this approach as a 'response criticism' which doubts various parts of climate governance and does not necessarily deny ACC (Van Rensburg, *Climate Change Scepticism*).
37. Jonathan Oldfield and Marianna Poberezhskaya (2023), 'Soviet and Russian Perspectives on Geoengineering and Climate Management', *WIREs Climate Change*, e829, https://doi.org/10.1002/wcc.829.
38. Yurii Medvedev (2010, 13 May), 'Yuriy Izrael: Sushchestvuet desyatok stsenariev izmeneniya klimata na Zemle', *Rossiyskaya Gazeta*, https://rg.ru/2010/05/14/izrael-nauka.html.
39. Oldfield, 'Imagining Climates', p. 43.
40. Nikolai Dronin and Alina Bychkova A. (2018), 'Perceptions of American and Russian Environmental Scientists of Today's Key Environmental Issues: A Comparative Analysis', *Environment, Development, and Sustainability*, 20: 2095–2105.
41. Katja Doose (2022), 'Modelling the Future: Climate Change Research in Russia during the late Cold War and Beyond, 1970s–2000', *Climatic Change*, 171, 6: 15.
42. Laura Young and Erin B. Fitz (2021), 'Who Are the 3 Per Cent? The Connections Among Climate Change Contrarians', *British Journal of Political Science*, December 1–20, https://doi.org/10.1017/S0007123421000442.
43. For example, B. Porfir'ev, V. Kattsov, and S. Roginko (2011), *Climate change and International Security*. Moscow: Russian Academy of Science, https://cc.voeikov mgo.ru/images/dokumenty/2016/izmKlim.pdf.
44. Vera Kuzmina (2022, 16 August), 'How Has Russia's Climate Policy Changed Since the Beginning of the War against Ukraine?', *UWEC*, https://uwecworkgroup.info/how-has-russias-climate-policy-changed-since-the-beginning-of-the-war-against-ukraine/.
45. Katja Doose and Akexander Vorbrugg (2022, 23 May), 'Other Casualties of Putin's War in Ukraine: Russia's Climate Goals and Science', *The Conversation*, https://theconversation.com/other-casualties-of-putins-war-in-ukraine-russias-climate-goals-and-science-182995.
46. Rolf Fredheim (2017), 'The Loyal Editor Effect: Russian Online Journalism after Independence', *Post-Soviet Affairs*, 33, 1: 34–48.

47. Anton Troianovski and Valeriya Safronova (2022, 4 March), 'Russia Takes Censorship to New Extremes, Stifling War Coverage', *The New York Times*, https://www.nytimes.com/2022/03/04/world/europe/russia-censorship-media-crackdown.html.
48. Marianna Poberezhskaya (2018), 'Traditional Media and Climate Change in Russia: A Case Study of Izvestiia'. In: Marianna Poberezhskaya and Teresa Ashe (eds.), *Climate Change Discourse in Russia: Past and Present*, pp. 64–79. Abingdon: Routledge.
49. Gazprom Media (2023), https://www.gazprom-media.com/ru.
50. Yagodin, 'Policy Implications'.
51. Marianna Poberezhskaya (2018), 'Blogging about Climate Change in Russia: Activism, Scepticism and Conspiracies', *Environmental Communication*, 12, 7, doi:10.1080/17524032.2017.1308406.
52. Constantine Boussalis, Travis G. Coan, and Marianna Poberezhskaya (2016), 'Measuring and Modeling Russian Newspaper Coverage of Climate Change', *Global Environmental Change*, 41: 99–110.
53. Teresa Ashe and Marianna Poberezhskaya (2022), 'Russian Climate Scepticism: An Understudied Case', *Climatic Change*, 172, 41, https://doi.org/10.1007/s10584-022-03390-3.
54. Liz Koslov (2019), 'Avoiding Climate Change: 'Agnostic Adaptation' and the Politics of Public Silence', *Annals of the American Association of Geographers*, 109, 2, :570, doi:10.1080/24694452.2018.1549472. Please also see a discussion on the lack of climate knowledge in Kari Marie Norgaard, *Living in Denial: Climate Change, Emotions, and Everyday Life* (Cambridge: The MIT Press, 2011).
55. Benjamin Beuerle (2018), 'Climate Change in Russia's Far East'. In: Marianna Poberezhskaya and Teresa Ashe (eds.), *Climate Change Discourse in Russia*, pp. 80–96. Abingdon: Routledge Focus.
56. For example, Marianna Poberezhskaya and Nataliya Danilova (2022), 'Reconciling Climate Change Leadership with Resource Nationalism and Regional Vulnerabilities: A Case-Study of Kazakhstan', *Environmental Politics* 31, 3: 429–452, doi:10.1080/09644016.2021.1920768.
57. Piter.tv (2019, 19 December), 'Putin: 'Nikto ne znaet prichin global "nogo izmeneniya klimata"', https://piter.tv/event/Parizhskie_soglasheniya_Putin/.
58. Nataliya Demchenko (2021, 30 June), 'Putin ob'yasnil pochemu priroda soshla s uma', *RBK*, https://www.rbc.ru/politics/30/06/2021/60dc5f4d9a7947e9e3860da4.
59. Nina Tynkkynen (2010), 'A Great Ecological Power in Global Climate Policy? Framing Climate Change as a Policy Problem in Russian Public Discussion', *Environmental Politics*, 19, 2: 179–195.
60. Ekaterina Postnikova (2021, 22 April), 'Klimaticheski sami: Putin predlozhil chetyre shaga dlya bor'by s potepleniem', *Izvestiya*, https://iz.ru/1155511/ekaterina-postnikova/klimaticheski-sami-putin-predlozhil-chetyre-shaga-dlia-borby-s-potepleniem.
61. For example, Angelina Milchenko (2021, 31 October), 'Osobenno nas bespokoit tayanie vechnoy merzloty', *Gazeta.ru*, https://www.gazeta.ru/social/2021/10/31/14154643.shtml.
62. Boussalis et al., 'Measuring and Modeling'.
63. Agathe Demarais (2023, 13 March), 'Don't Trust Russia's Numbers', *Foreign Policy*, https://foreignpolicy.com/2023/03/13/russia-economy-sanctions-gdp-war-ukraine-disinformation-statistics/.

64. Environmental activist, personal communication, March 2023.
65. Tynkkynen and Tynkkynen, 'Climate Denial Revisited', p. 1116.
66. Steven Lee Myers and Andrew Revkin (2003, 3 December), 'Putin to Reject the Pact on Climate, Putin Aide Says', *The New York Times*, https://www.nytimes.com/2003/12/03/world/russia-to-reject-pact-on-climate-putin-aide-says.html.
67. Tynkkynen and Tynkkynen, 'Climate Denial Revisited'.
68. Liliana Andonova (2008), 'The Climate Regime and Domestic Politics: The Case of Russia', *Cambridge Review of International Affairs*, 21, 4: 483–504; Asya Cooley (2023), 'The Role of the Nonprofit Sector within the Climate Change Discourse: The View Through Russian News Media', *Nonprofit Policy Forum*, 14, 1: 1–23, https://doi.org/10.1515/npf-2022-0002.
69. See Ecodefence (2020, 13 March), 'Pikety protiv dobychi uglya v Kuzbasse', https://ecodefense.ru/2020/03/13/stock-against-coal/.
70. See Ecodefence (2020, 3 February), 'Ugol' – uzhas! Spasem Kuzbass!', https://ecodefense.ru/2020/02/03/ecodefense-kuzbass-statement/; Stephen Fortescue and Ellie Martus (2020), 'Black Jack: Russia's Coal Industry in the Age of Climate Change', *Osteuropa*, 70, 7–9: 103–130, doi:10.35998/oe-2020-0050.
71. Filipp Lebedev (2022, 14 September), 'Russia's First Climate Lawsuit Filed over Greenhouse Emissions', *Reuters*, https://www.reuters.com/article/climate-change-russia-idAFKBN2QE1TO.
72. Jo Crotty, Sarah Marie Hall, and Sergej Ljubownikow (2014), 'Post-Soviet Civil Society Development in the Russian Federation: The Impact of the NGO Law', *Europe-Asia Studies*, 66, 8: 1253–1269.
73. Anna Kireeva and Charles Digges (2019, 25 June), 'Activist from Ecodefense, under Pressure from Russia's 'Foreign Agent' Law, Flees to Germany', *Bellona*, https://bellona.org/news/russian-human-rights-issues/2019-06-activist-from-ecodefense-under-pressure-from-russias-foreign-agent-law-flees-to-germany.
74. Case of Ecodefence and others v. Russia (2022) European Court of Human Rights, https://hudoc.echr.coe.int/fre#{%22itemid%22:[%22001-217751%22]}.
75. Moscow Times (2022, 1 November), 'Climate Activist Arshak Makichyan Stripped of Russian Citizenship', https://www.themoscowtimes.com/2022/10/31/climate-activist-arshak-makichyan-stripped-of-russian-citizenship-a79246.
76. AP (2023, 11 March), 'Russia Lists World Wildlife Fund, Others as Foreign Agents', https://apnews.com/article/russia-world-wildlife-fund-foreign-agents-86b8c97fd44f992264b34de3f7de949e.
77. Reuters (2023, 22 June), 'WWF Russia Cuts Ties with Global Environment Group, Now Labelled 'Undesirable' by Moscow', https://www.reuters.com/world/wwf-russia-cuts-ties-with-global-wildlife-fund-now-labelled-undesirable-by-2023-06-22/.
78. See Ellie Martus and Stephen Fortescue (2022), 'Russian Coal in a Changing Climate: Risks and Opportunities for Industry and Government', *Climatic Change*, 173, 3: 1–21.
79. Tatiana Romanova (2021), 'Russia's Political Discourse on the EU's Energy Transition (2014–2019) and Its Effect on EU-Russia Energy Relations', *Energy Policy*, 154: 112309, https://doi.org/10.1016/j.enpol.2021.112309
80. Tatiana Romanova (2021), 'The 2019 Energy Security Doctrine and Debates around It in Russia'. In: Elizabeth Buchanan (ed.), *Russian Energy Strategy in the Asia-Pacific: Implications for Australia*, pp. 201–217. Canberra: ANU Press.
81. Alexey Kokorin and Anna Korppoo (2017), 'Russia's Ostrich Approach to Climate Change and the Paris Agreement', *CEPS Policy Insights*, 40.

82. Dmitrii Butrin and Aleksei Shapovalov (2019, 17 October), 'Uglerodnye nalogi poshly na vybros', *Kommersant*, https://www.kommersant.ru/doc/4127113.
83. Angelina Davydova (2 June 2021), 'Vybrosy vyzvali voprosy', *Kommersant*, https://www.kommersant.ru/doc/4838307.
84. Balmaceda and Heinrich, 'The Energy Politics of Russia and Eurasia'.
85. Martus and Fortescue, 'Russian Coal in a Changing Climate'.
86. Sergei Tikhonov (2020, 4 September), 'Aleksandr Novak: Dolia VIE v energobalanse Rossii dostignet 4–5%', *Rg.ru*, https://www.Rg.ru/2020/09/04/alksandr-novak-dolia-vie-v-energobalanse-rossii-dostignet-4–5.html.
87. Administratsiya Goroda Kemerovo, Postanovlenie ot 06.06.2022 No 1538, 'ob utverzhdenii toplivo-energeticheskogo balansa goroda Kemerovo na 2021 god i prognoz na 2022–2025 gody', https://kemerovo.ru/sfery-deyatelnosti/gorodskoe-zhkkh/toplivno-energeticheskiy-balans-goroda-kemerovo/.
88. Ellie Martus (2019), 'Russian Industry Responses to Climate Change: The Case of the Metals and Mining Sector', *Climate Policy*, 19, 1: 17–29; Martus and Fortescue, 'Russian Coal in a Changing Climate'.
89. Mathieu Blondeel and Thijs Van de Graaf (2018), 'Toward a Global Coal Mining Moratorium? A Comparative Analysis of Coal Mining Policies in the USA, China, India and Australia', *Climatic Change*, 150: 89–101; Thomas Spencer et al. (2018), 'The 1.5°C Target and Coal Sector Transition: At the Limits of Societal Feasibility', *Climate Policy* 18, 3: 335–351.
90. See Davydova, 'Vybrosy vyzvaki voprosy'.
91. Phoebe Cooke, Rich Collett-White, and Michaela Herrmann (2022, 12 November), 'Sanctioned Coal Barons Among Russia's COP27 Delegates', *DeSmog*, https://www.desmog.com/2022/11/12/sanctioned-coal-barons-among-russias-cop27-delegates/.
92. Stanislav Klimovich, Sabine Kropp, and Ulla Pape (2023), 'Defending Business Interests in Russia: Collective Action and Social Investments as Bargaining Chips', *Post-Communist Economies* 35, 1: 1–24.
93. See Russian Union of Industrialists and Entrepreneurs, https://rspp.ru/about/inform/.
94. Liliana Andonova and Assia Alexieva (2012), 'Continuity and Change in Russia's Climate Negotiations Position and Strategy', *Climate Policy* 12, 5: 614–629; Martus, 'Russian Industry Responses'.
95. See President of Russia, http://www.kremlin.ru/structure/administration/groups#institution-1003.
96. Angelina Davydova (2019, 25 January), 'Parizhskomu soglasheniyu RSPP ne meshaet', *Kommersant*, https://www.kommersant.ru/doc/3862151.
97. Stephen Fortescue (2021), 'Future of Russian Coal Exports in the Asia-Pacific'. In: Elizabeth Buchanan (ed.), *Russian Energy Strategy in the Asia-Pacific. Implications for Australia*, pp. 155–180. Canberra: ANU Press.
98. Anna Korppoo (2018), 'Russian Associated Petroleum Gas Flaring Limits: Interplay of Formal and Informal Institutions', *Energy Policy*, 116: 232–241.
99. Valerii Voronov (2022, 27 April), 'Terpyat otlagatel'stv: v Rossii khotyat otsrochit' sokrashchenie parnikovykh vybrosov', *Izvestiya*, https://iz.ru/1326653/valerii-voronov/terpiat-otlagatelstv-v-rossii-khotiat-otsrochit-sokrashchenie-parnikovykh-vybrosov.
100. Martus, 'Russian Industry Responses'.
101. Martus and Fortescue, 'Russian Coal in a Changing Climate'.

102. For example, Kristoffer Ekberg, Bernhard Forchtner, Martin Hultman, and Kirsti Jylhä (2023), *Climate Obstruction: How Denial, Delay, and Inaction Are Heating the Planet*. London: Routledge, p. 54.
103. Martus, 'Russian Industry Responses'.
104. Tipton, 'Why Did Russia Ratify'.
105. Poberezhskaya, *Communicating Climate Change*.
106. Alexey Kokorin and Darya Lugovaya (2018), 'Pogloshchenie CO2 lesami Rossii v kontekste Parizhskogo soglasheniya', *Ustoichivoe Lesopol'zovanie* 2, 54: 13–18.
107. For example, Felix Light (2021, 7 September), 'Russia Says Its Forests Neutralize Billions of Tons of Greenhouse Gases. Scientists Have Their Doubts', *The Moscow Times*, https://www.themoscowtimes.com/2021/07/05/russia-says-its-forests-neutralize-billions-of-tons-of-greenhouse-gases-scientists-have-their-doubts-a74428.
108. Pravitel'stvo Rossii (2021), 'Rasporyazhenie ot 29 Oktyabr 2021 No3052-r', http://static.government.ru/media/files/ADKkCzp3fWO32e2yA0BhtIpyzWfHaiUa.pdf. Accessed 24 March 2023.
109. RIA Novosti (2017, 2 June), 'Andrei Mel'nichenko: Rossiya odin iz liderov po sokrashcheniyu parnikovykh gazov', *RIA Novosti*, https://ria.ru/20170602/1495700837.html.
110. Tynkkynen, 'A Great Ecological Power'.
111. Vladimir Baburin (2004, 9 July), 'V blizhayshee vremya Rossiya ne ratifitsiruet Kiotskiy Protokol', *Radio Svobody*, https://www.svoboda.org/a/24190936.html.
112. RIA Novosti (2023, 12 March), 'Posledniy argument. Amerika gotovitsya primenit' klimaticheskoe oruzhie', *RIA Novosti*, https://ria.ru/20230312/klimat-1856861090.html
113. Tynkkynen and Tynkkynen 'Climate Denial Revisited', p. 1115.
114. Doose and Vorbrugg, 'Other Casualties of Putin's War'.
115. Spravedlivo (2022, 20 April), 'Sergey Mironov: Rossiya posle sanktsiy Zapada dolzhna vyiti iz Parizhskogo soglasheniya po klimatu', https://spravedlivo.ru/12016310.
116. Quoted in Matvei Katkov (2022, 18 June), 'Igor Sechin rasskazal o pokhoronakh "zelenoi" povestki . . . ', *Vedomosti*, https://www.vedomosti.ru/business/articles/2022/06/18/927288-sechin-rasskazal-pohoronah-zelenoi-povestki.
117. Naomi Oreskes and Eric Conway (2011), *Merchants of Doubt: How a Handful of Scientists Obscured the Truth on Issues from Tobacco Smoke to Global Warming*. New York: Bloomsbury Press.
118. Natsional'nyi Plan Meropriyatii pervogo etapa adaptatsii k izmeneniyam klimata na period do 2022 goda (Natsional'nyi Plan), 3183-r, approved 25 December 2019.
119. Kokorin and Korppoo, 'Russia's Ostrich Approach to Climate Change', p. 9.
120. For example, Ralf Barkemeyer, Frank Figge, and Diane Holt (2013), 'Sustainability-Related Media Coverage and Socioeconomic Development: A Regional and North–South perspective', *Environment and Planning C: Politics and Space*, 31, 4: 716–740.
121. Ronald Inglehart (1995), 'Public support for environmental protection: Objective problems and subjective values', *PS: Political Science & Politics*, 28, 1: 57–72.
122. Poberezhskaya, *Communicating Climate Change*.
123. Cooley, 'The Role of the Nonprofit Sector', p. 16.
124. Aaron McCright and Riley Dunlap (2003), 'Defeating Kyoto: The Conservative Movement's Impact on U.S. Climate Change Policy', *Social Problems*, 50, 3: 348;

see also Aaron McCright and Riley Dunlap (2000), 'Challenging Global Warming as a Social Problem: An Analysis of the Conservative Movement's Counter-Claims', *Social Problems*, 47, 4: 499–522.
125. Environmental correspondent, personal communication, March 2023.
126. Martus, 'Russian Industry Responses'; Fortescue and Martus, 'Black Jack'; Martus and Fortescue, 'Russian Coal in a Changing Climate'.
127. Carlos Milani, et al. (2021), 'Is Climate Obstruction Different in the Global South? Observations and a Preliminary Research Agenda', CSSN Position Paper no. 4, https://cssn.org/wp-content/uploads/2021/10/CSSN-position-paper-4_-Global-South.pdf.
128. Robert Brulle (2021), 'The Structure of Obstruction: Understanding Opposition to Climate Change Action in the United States', CSSN Briefing, https://cssn.org/wp-content/uploads/2021/04/CSSN-Briefing_-Obstruction-2.pdf.
129. Elizabeth Plantan (2022), 'Not All NGOs Are Treated Equally: Selectivity in Civil Society Management in China and Russia', *Comparative Politics*, 54, 3: 501–524.
130. Maria Tysiachniouk, Svetlana Tulaeva, and Laura Henry (2018), 'Civil Society under the Law 'On Foreign Agents': NGO Strategies and Network Transformation', *Europe-Asia Studies*, 70, 4: 615–637.
131. For example, Oleg Anisimov and Robert Orttung (2019), 'Climate Change in Northern Russia through the Prism of Public Perception', *Ambio*, 48: 661–671, https://doi.org/10.1007/s13280-018-1096-x
132. For instance, Aleksey Kokorin and Anna Korppoo listed NGOs (WWF, the Social & Ecological Union, Greenpeace-Russia), environmental committees' members from major business associations, certain civil servants from the presidential administration, the MED, and prominent academics who, one way or another, contributed to the adoption of the Climate Doctrine and its implementation plan, raised climate awareness, supported Russia at UNFCCC events, encouraged transition to the low carbon economy, and so on. (Aleksey Kokorin and Anna Korppoo (2013, May), 'Russia's Post-Kyoto Climate Policy: Real Action or Merely Window-Dressing?' *FNI Climate Policy Perspectives*, 10: 1–8).
133. Anatole Boute and Alexey Zhikharev (2019), 'Vested Interests as Driver of the Clean Energy Transition: Evidence from Russia's Solar Energy Policy', *Energy Policy*, 133: 110910.
134. For example, Igor Makarov, 'Does Resource Abundance Require Special Approaches?'
135. For example, Anna Urmantsev (2023, 16 January), "CO2 tut ni pri chem". Inoy vzglyad na global'noe poteplenie', *Gazeta.ru*, https://www.gazeta.ru/science/2023/01/16/16088101.shtml?updated.
136. Eric Piaget, Luk Van Langenhove, and Luc Soete (2022, 31 May), 'Science Diplomacy in Times of War: To What Extent Should Western Countries Distance Themselves from Russian Science?', *Impact of Social Science blog*, https://blogs.lse.ac.uk/impactofsocialsciences/2022/05/31/science-diplomacy-in-times-of-war-to-what-extent-should-western-countries-distance-themselves-from-russian-science/.

10
Climate Obstruction in the Czech Republic
Winning by Default

MILAN HRUBEŠ AND ONDŘEJ CÍSAŘ

INTRODUCTION: CLIMATE OBSTRUCTIONS IN CZECHIA

Despite the recent development of renewable energy sources, Czechia, like Poland, remains one of Europe's most coal-dependent economies, with coal accounting for approximately 50% of the national energy mix.[1] In terms of greenhouse gas (GHG) emissions, coal-fired power plants produce nearly 90% of emissions in the country's energy sector; coal- and gas-fired power plants together accounted for 96% of the sector's emissions in 2018. In the wider economy, also in 2018, the energy sector produced 40% of all emissions. Transportation was second, with 16%, and industrial production third, with 13%.[2]

From the beginning of the modern era, the Czech lands (territory that, until 1918, was part of the Austro-Hungarian monarchy) have been characterized by their concentration of energy-intensive industrial production. At the end of the nineteenth century, the Czech lands accounted for more than 2% of all global CO_2 emissions (they emit less than 0.25% today). Per capita CO_2 emissions peaked in 1978 at 18.39 tonnes, and, in the twenty-first century, have oscillated between 12.5 and 8.72 tonnes.[3]

The country's total 1990 emissions were 198 million metric tonnes of carbon dioxide equivalents (MMT CO_2e). By the early 2000s, they had dropped to 150 MMTCO_2e due to the shuttering of many energy-intensive industrial facilities after the fall of communism. This post-transformation shock resulted in a sharp drop in emissions in the last decade of the twentieth century, which continued even during the first decade of the new millennium, though it was not as steep. As a result, the country reported 118 MMT CO_2e in 2021, 60% of its CO_2 production in 1990 (Figure 10.1).[4]

As in Poland (Chapter 8), these transformation-related changes made it possible for Czechia to meet the reduction targets set by the Kyoto Protocol without any explicit policy measures aimed at protecting the climate. The country's projected future emissions (based on nationally determined contributions [NDCs]) are to be reduced by at least 55% compared with 1990 by 2030. As part of the European Union's 'Fit for 55' climate package, Czechia's commitment to reduce its GHG emissions increased from 14% to 26% by 2030 compared with 2005.[5]

This track record sets the stage for reconstructing the country's climate story. Here we aim to provide a better understanding of climate obstruction in Czechia, particularly the high direct involvement of actors from the sphere of politics rather than from business; the significant role of fossil fuel companies, which is often hidden; and the low level of public interest in discussing climate mitigation and policies. We follow Kristoffer Ekberg and colleagues' definition of climate obstruction (Chapter 5), which describes the concept as an umbrella term covering 'complex ways in which the status quo is reproduced, be it in the dimension of science, politics, culture or the economy'.

We begin by introducing our argument and follow with a brief contextualization of the development of climate obstruction in Czechia. Then we focus on describing the most important actors and the strategies and tactics they deploy. Here we differentiate between 'hard' strategies and tactics and 'soft' (discursive) ones, showing how these actors put their words into practice. Last, we analyse the specific meanings the actors construct within different obstruction discourses.

The Czech climate story

The Czech climate story begins in the early 1990s, when Czechoslovakia was undergoing its transition to democracy, which in turn brought a significant improvement in emissions levels (see above), making the issues of environmental protection and climate change seem less important to

Czech Republic Greenhouse Gas Emissions

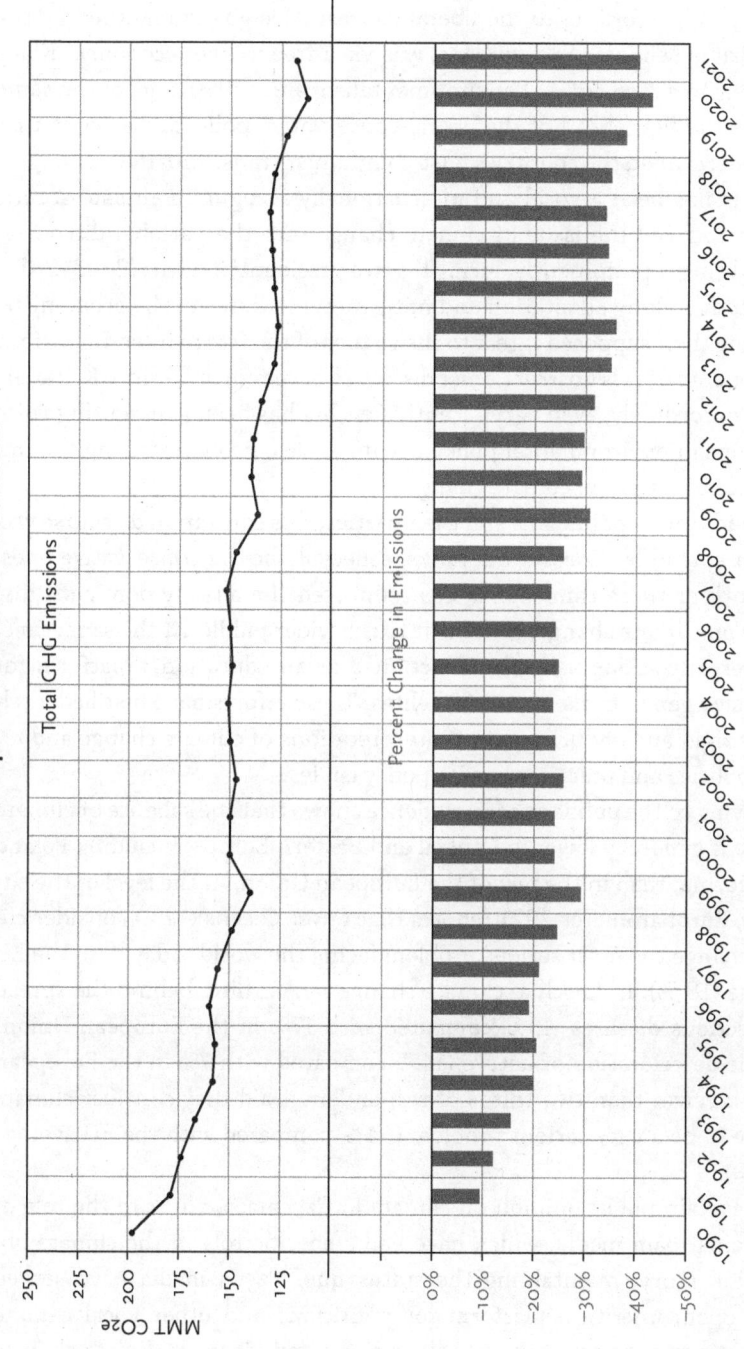

Figure 10.1 Total greenhouse gas (GHG) emissions (in MMT CO_2e) and percentage change in emissions in Czechia between 1990 and 2021, inclusive.

Source: Total GHG emissions based on data provided by Gütschow and Pflüger (2023) for Kyoto Six Greenhouse Gas Totals.

mainstream society than others. The nation's main political goal was to catch up economically with Western Europe, while the only way to effectively do so, according to the liberal-conservative government led by the then-Prime Minister Václav Klaus, was via a free market economy.[6] Klaus has been an active critic of environmentalism since 1990, and, at the same time, is the figurehead of the liberal-conservative political discourse that has underpinned the country's post-communist transformation strategy.

Klaus has been a vocal and internationally recognized climate sceptic who introduced the issue of climate change into the national discourse, spotlighting it prominently during his two presidential terms (2003–2013) by directly linking adaptation and mitigation measures to the economy by stressing their supposed threat to the virtues of the free market. Important components of this obstructionist discourse have remained since Klaus left office. Accordingly, even very recently Czechia has been a dissenting voice in European environmental policy debate circles, de-emphasizing the importance of action on climate change.

In addition to politicians and their parties, this contrarian discourse was further spread by Czech think tanks, especially liberal-conservative ones. The work of these think tanks has reinforced the already dominant discourse on climate change, offering it to the wider public. At the same time, these organizations have also functioned as an educational platform for successive generations of political elites. These efforts have instilled a relatively rigid and enduring set of interpretations of climate change and related policies and practices in elite policy circles.

Moving to the public sphere, evidence shows that the salience of climate change is generally lower in Central and Eastern Europe, including Poland and Czechia, than in the rest of the European Union. At the level of the citizenry, Eurobarometer 2021 reports that fewer Czechs (12%) consider climate change the most serious problem facing the world today than the EU average (18 %). In Czechia, climate change ranks third, behind the spread of infectious diseases (15%, compared with 17% in the European Union) and the deterioration of nature (14%, compared with 7% in the European Union). Fewer than two-thirds of respondents said they consider climate change to be a very serious problem (64%, compared with the EU average of 78%).[7]

To explain public opinion on climate in Czechia, we look to the role of the mainstream media, which have had a specific role in the climate obstruction story: maintaining the status quo. Czech media have served as an open, uncritical platform for politicians and other vocal climate obstructionists to communicate their views and ideas on climate change, failing even to encourage an exchange of various viewpoints on the nature of climate change or debate on climate-related policies. As wealthy coal and energy production company owners have also owned media companies, their role in this story perfectly fits their needs.

Surprisingly, coal and energy producers have not played a visible role in the climate obstruction story. Rather, they have remained in the background because most of their job has been accomplished for them by politicians and related think tanks. Important politicians not only push the agenda of climate obstruction, but also seem to accept the demands of the coal and energy producers (as much as they can given the demands of the international environmental arena, which limits room for more radical political moves).

For all of these reasons, we interpret the Czech story of climate obstruction as 'winning by default': climate denial and scepticism along with opposition to, delay of, and dismissal of effective climate policies is the established mind-set of the Czech political mainstream. As such, obstructionists need not do much additional lobbying to make an impact. To put it metaphorically: much effort is needed to make a fire, but once the fire ignites, one need only throw a small log on it from time to time to keep it burning.

A HISTORICAL DESCRIPTION OF CLIMATE OBSTRUCTION

As noted, Czechia has a well-documented history of broad climate scepticism,[8] which is also reflected in quantitative indicators such as the Climate Change Performance Index.[9] Czechia is home to one of the most famous climate deniers—its former president, Václav Klaus—and is regarded by researchers as 'one of the most sceptical countries in Europe'.[10] This designation is due not only to the general differences between Eastern and Western Europe, but also to the fact that the sceptical position was articulated in the country relatively early on and, more importantly, came from the top.

Already prime minister (1992–1998), Klaus contributed significantly to the closed political climate surrounding environmental issues in Czechia because he bundled environmentalism together with feminism and Europeanism, labelling them collectively as a new form of communism in disguise that threatened human freedom.[11] Miloš Zeman, who served as prime minister after Klaus (1998–2002), was similarly militant against all types of environmental activists and their political messages. In 2003, Klaus was elected Czech president, ascending to the most influential position in terms of symbolic importance. Research indicates that it was in this role that he made significant progress in spreading climate scepticism in the country, legitimizing it in the eyes of important political agencies and even among parts of the population.[12]

At the same time, our recent research on Czech media has demonstrated that open climate denialism and scepticism are now playing a relatively minor role in the Czech media; climate obstruction rather exists on a spectrum. The voices of climate deniers, even though relatively strong in the past and representing the political elite, in more recent years are becoming marginal and slowly fading away.

AN ANALYSIS AND DESCRIPTION OF THE MAJOR ACTORS AND TYPES OF INSTITUTIONS INVOLVED

The role of former President Klaus cannot be understated in the mainstreaming of climate scepticism in Czechia, but it was not only he who helped to legitimize this type of discourse. Most important were political parties whose representatives actively denied the human origin of climate change and challenged coordinated action to protect the climate. The most important of these was the Civic Democratic Party (ODS; *Občanská demokratická strana*), established by Klaus in 1991. Next has been an active and visible network of obstructionist think tanks that produced various cultural products such as publications, commentaries, and media appearances. Some of them, such as the Centre for Economics and Politics and the Civic Institute, received direct support from their US-based partners and are personally linked to Klaus and/or a political party, most often ODS. Finally, there are businesses and media, owned or potentially owned by oligarchs and/or important investors in the fossil fuel industry, that offer a platform for types of climate obstruction. At present, two important media corporations are owned by Andrej Babiš (until 2023), a Czech oligarch and former prime minister, and Daniel Křetínský, an internationally significant investor in the energy sector.

In the following section, we examine these types of actors, presented in order of their importance based on their explicit and nationally visible involvement in climate obstruction. (It is impossible to base the criteria on their actual influence, which we are unable to measure at present.)

Political parties

The Civic Democratic Party

The ODS was established as a liberal-conservative political party modelled on British conservatism in the Thatcherite tradition, which formed the core of the political programme advocated by Klaus, the party's lead figure for the first decade of its existence. Regarding the climate and environmental protection, the party mirrored the position of its founder, traditionally downplaying their importance. For Klaus, environmentalism since the 1990s is a dangerous ideology, a belief he would consistently tie to his explicit denial of the human origin of climate change later on. Through its overlapping membership and cooperation, the party (including Klaus and other important members) has been closely linked with other organizations, mostly libertarian think tanks (discussed later).[13]

ODS has historically formed an important part of the anti-climate network of political organizations (Figure 10.2). As one of the most important political forces in the country until 2013, when the political spectrum began to shift—the hegemon of the centre-right and a senior member of several coalitional cabinets—it was undoubtedly influential in shaping public opinion on climate change. Currently, although there are some active deniers among the more visible ODS politicians, including MPs, the party itself now pragmatically accepts the reality of climate change and the need to decarbonize the European economy.

ANO 2011 (Action of Dissatisfied Citizens 2011)

Action of Dissatisfied Citizens 2011 (ANO 2011) is the most significant new populist party, usually classified as managerial or anti-elitist.[14] It was part of governing coalitions during two terms between 2013 and 2021. Its leader, Babiš, served as prime minister between 2017 and 2021, when ANO was the senior member of the governing coalition. The party was founded in 2011, by Babiš, the second-richest Czech entrepreneur and owner of the country's largest agricultural and food processing holding, Agrofert (which is also active in multiple business sectors, including the news media). ANO and its founder have regularly declared their intention directly to help 'the people' and have attacked elite professional politicians. Although the party has at times characterized policy measures for climate protection as politically dangerous, it has also pragmatically accepted the international mainstream consensus, seeing climate change as a business opportunity, and it supports Czechia's conformity with international climate agreements.

The Far Right

Currently, only far-right parties directly attack measures to protect the climate and/or deny the human origin of climate change. Except for the

Freedom and Direct Democracy party (SPD, represented in the parliament), these groups (such as Trikolóra and Svobodní) are marginal, although the latter is linked to think tanks involving the Klaus family, as discussed later). SPD wants to radically transform Czechia's political system by introducing mechanisms of direct democracy, such as general referenda on fundamental political issues and political mandates that can be directly revoked by the public. The SPD's first priority is to call a referendum on leaving the European Union. Accordingly, it is against international cooperation on climate protection and Czechia's participation in the process.[15]

Think tanks

Drawing partly on our past research on Czech think tanks,[16] we have identified those that are actively involved in the issue of climate change.

The Klaus family think tanks: Centre for Economics and Politics (CEP) and the Václav Klaus Institute (IVK)

An advocacy think tank, the Centre for Economics and Politics (CEP, or *Centrum pro ekonomiku a politiku*) was founded as a nonpartisan association in 1998. It has been seen as an institutional umbrella for associates and followers of Klaus. The CEP's main goal was to promote the principles of a free market economy, limited government, and individual freedom and to formulate and further public policies based on these principles. A climate agenda has been part of this ideologically libertarian organization. The CEP is considered the most important and also most resourceful think tank among climate sceptical organizations in the country, at least historically.[17]

The CEP has closely cooperated with the Václav Klaus Institute (*Institut Václava Klause*, or IVK) which has declared the same goals, even using the same words. Many of the CEP's activities were performed jointly with the IVK and generously supported by the PPF company, established and controlled until his death by the wealthiest Czech businessman, Petr Kellner, whose activities traditionally relied on political backing.[18] Based on the volume of their current public output, the IVK has coordinated the groups' main activities since 2013. That year, Klaus retired from office, and his public activities (as well as those of the network of his associates) found their institutional home in the IVK. IVK is very active in publishing and otherwise informing the public on many aspects of political and social life including climate change.

The Liberal Institute

The Liberal Institute (LI, *Liberální institut*) is an advocacy think tank, established as an independent association in 1990. Its main goal is to spread, develop, and apply classical liberal ideas as well as to promote programmes based on the principles of classical liberalism. In terms of climate scepticism, it has not been particularly active, but it did help to publish the book *The Sceptical Environmentalist* by Bjørn Lomborg and organized his first visit to the country; both of these efforts were supported by Czech Coal, among other companies.[19]

The Civic Institute

The advocacy think tank Civic Institute (OI, *Občanský institut*) was founded in 1990. It is an independent association originally focused on promoting a free market economy. Since the mid-1990s, the OI has moved to a more conservative position, stressing cultural and social issues, mainly what the organization understands to be the moral, religious, and pre-political foundations of a free society: the traditional family. This orientation may explain why the organization is sympathetic to the recent success of nationalists and populists in Poland and Hungary and accepts broadly nationalist politics as the right response to all current problems, including climate change.[20]

Businesses

As noted, the major mainstream Czech media outlets are or were owned by two oligarchs, one of whom is a major European player in the energy sector. The first is Babiš, who is generally regarded as the most important Czech populist politician and whose holding, Agrofert, owned the biggest Czech media company, MAFRA. Currently, the company has been sold to a former PPF manager and Kellner associate Karel Pražák, the owner of the investment company Kaprain Holdings. Originally, negotiations were under way with two businessmen, both very active in the fossil fuel industry and the production of obstructionist content: Pavel Tykač and Daniel Křetínský. Both Tykač and Křetínský are major investors in the energy sector and fossil industry. Tykač is the owner of the Sev.en AG group; Křetínský owns the Energetic and Industrial Holding company (EPH) and is also currently the majority owner of the Czech News Centre, one of the biggest media companies in the country. Both also invest internationally, especially Křetínský, who is involved in the energy sector, including fossil fuel sources and the building of new power plants.

Due to the typical business strategy of secrecy and the lack of transparency in industrial lobbying in Czechia, concrete evidence of the ways big businesses obstruct climate change mitigation is unavailable. The information that is available comes from publicly accessible media content, which itself qualifies as climate obstruction, in the outlets owned by these investors. The prevailing interpretation is that both Křetínský and Tykač, both heavily invested in fossil businesses, also invest in the media to gain leverage over public opinion. Because the future of fossil business depends in part on government regulation, wealthy executives acquire media in hope of influencing policymaking through their own influence over the public.[21] In addition, some relationships between fossil businesses and specific politicians have been documented, including former Czech President Zeman and factions of ODS. Also, energy businesses have sponsored public events for prominent guests and opinion leaders.[22]

In their public statements, these business owners do not see themselves as climate change deniers, and, for example, Křetínský has explicitly stated that his EPH does a 'tough job' in keeping unpopular assets viable, which at the same time provide needed energy. According to him, for example, the German economy cannot currently do without fossil fuel sources, but he has also stressed the need for an energy transformation and a future in which fossil fuels have been replaced with renewable energy sources.[23] Needless to say, Křetínský claims non-interference in, and the editorial independence of, the media he owns and even views his acquisition of traditional media as a service to a liberal democracy currently under siege from the boom in Internet-based social media and media platforms.[24]

Media and Internet-based platforms

Evidence from our research on the media formerly owned by Babiš (MAFRA) reveals some trends in media content related to climate obstruction in Czechia. We categorize these media as part of the mainstream and contrast them with alternative media, represented by smaller, web-based leftist media. A significant difference between different newspapers is evident in their framing of climate change. Whereas the mainstream media focus on adaptation measures or understand the climate crisis as an opportunity for business, the alternative media stress the importance of mitigation. The discourse of adaptation is also represented in the media owned (until 2023) by former Czech Prime Minister Babiš.

Media owned by Křetínský were not covered by our research, and other systematic research on the topic is sparse. However, critical journalists have repeatedly identified open attacks against demands for climate protection and their advocates, including activists and international organizations. As the critical web-based papers Referendum.cz, Alarm.cz, and other critical sources have pointed out, reporters from papers and journals owned by Křetínský often frame climate activists and the European Union alike as the 'green Taliban' or eco-terrorists.[25] At the same time, these critical sources stress the urgent need for further research to track the influence of the fossil fuel industry on the content of the Czech media. For example, in the case of Tykač, investigative journalists were able to uncover a direct connection between his company and Facebook trolls ridiculing and attacking climate activists.[26]

Apart from the traditional media, we can also identify some Internet-based sceptical platforms spreading obstructionist content, such as reformy.cz and D-Fens, which are rather limited in their scope and resonance.[27] At the same time, their content creators are embedded in the aforementioned organizations, mostly far-right parties, and some of them identify themselves as part of the Czech climate science community.

CLIMATE OBSTRUCTION STRATEGIES AND TACTICS

Politicians, think tanks, media, and other actors are involved in many activities to promote their own perceptions of climate change and related policies. Although their strategies and tactics differ, a closer look reveals these efforts to be interconnected and complementary.

Political strategies and tactics

In her research on think tanks, Diane Stone defines the core of what they can achieve in policy transfer: 'Their prime importance is in the construction of legitimacy for certain policies and in agenda-setting. They transfer the ideas and ideologies, the rationalisations, and legitimations for adopting a particular course of action. . . . However, to see policy transfer occur, these organizations are dependent on formal political actors'.[28] Formal political actors not only push their agenda to get voter support, but also tend to prioritize policies that are salient to their voters.[29] Applying Stone's definition to the practice of climate obstruction in Czechia can help in understanding the logic of the specific strategies and tactics the major actors have applied to achieving their climate obstruction (and other) goals. The logic of each sector, and how it has interacted with the other actors, is as follows.

The topic of climate change was first raised in the 1990s by political parties and politicians with a liberal conservative-leaning ideology. They collaborated with allied think tanks to spread the information they wished to emphasize to establish a specific hegemonic discourse in Czech society, or particular system of practices and interpretations,[30] around climate change. The discourse advances the claim that climate change is an ideology hostile to freedom rather than a phenomenon proven by (proper) science. Mainstream media helped to spread this discourse by reporting on these politicians and think tanks without directly promoting the agenda of climate obstruction themselves. Rather, these outlets provided a platform for climate obstruction actors and did not challenge their claims, nor did they provide any forum to discuss the topic of climate change more broadly. As noted, the Czech public has historically not considered climate change to be a very important public policy issue. This media passivity and citizen indifference have been useful to the coal and energy magnates who own some of the biggest media houses in Czechia because their agenda of climate obstruction had already been successfully advanced by politicians and related think tanks. In other words, the coal and energy industries have had to do little if anything on this front because the politicians have always done it first.

Taking a closer look at the think tanks involved in climate obstruction, it is important to note that they are not oriented solely around the topic of climate change. This is evident from their focus on and activities promoting liberal-conservative values. Thus, while they might sponsor activities dedicated to climate change, such as a 'Global Warming – Facts and Myths' scientific community meeting (organized by CEP in 2007), they might also concurrently organize a seminar on the performance of the Czech economy, such as the one CEP organized around Klaus's book on the topic in which climate policies were criticized as a form of regulatory overreach that harms free market economies.

To show how such logic works in practice, we have used Stone's list of think tank policy diffusion tactics[31] and applied it to climate obstruction in Czechia. Table 10.1 provides an overview of the type of information the major Czech think tanks collect and produce, the topics within which climate change is discussed, the target groups to whom the information is directed, and the various ways the information or discourse is disseminated.

The table shows that there is much in common among the think tanks analysed here. First, all of them conduct their own research (mostly on

Table 10.1 CZECH THINK TANKS' 'HARD' STRATEGIES AND TACTICS

	Type of information collected/produced	Topic (within which climate change is discussed)	Target groups	Methods of information/discourse dissemination
CEP	Research Publications Good practice (e.g. discussions/debates with the policy makers on various policies applied in different countries)	Free market Democracy European Union	Policy makers (politicians and policy experts) Academia Media editors	Education (summer schools, lectures, conferences) Meeting with policy makers (roundtables) Publishing books and translations Public discussions
IVK	Research Publications	Free market Democracy European Union	Policymakers (politicians and policy experts) Academia Media editors Public	Education (summer schools, lectures, conferences) Library service Publishing books and translations Public discussions
LI	Research Publications	Economic policies Macroeconomic transition	Policymakers (politicians) Academia (students) Public	Education (summer schools, lectures, conferences) Meeting with policy makers (roundtables) Publishing books and translations Public discussions
OI	Research Publications	Cultural issues Social issues Free market Christian values	Policymakers (politicians and policy experts) Media editors Diplomats Academia (students)	Education (summer schools, lectures, seminars, conferences) Meeting with policy makers (roundtables) Library services Publishing books and translations

'Hard strategies' refer to any activity that is not a discursive framing (e.g. organizing seminars is a hard strategy, while the way climate change is discussed at a seminar is a discursive framing).

Source: Based on Císař, O., Hrubeš, M. 'Think Tanks and Policy Discourses in the Czech Republic' in Veselý, A., Nekola M, and Hejzlarová E. M. (eds). *Policy Analysis in the Czech Republic*. Bristol: The Policy Press, 2016 and data collected from think tank web pages and annual reports.

economics) and collect other information via various publications (desk research), focusing mostly on economics and liberal-conservative values. For example, LI concentrates on translating classical texts by liberal philosophers. These think tanks do not embrace pluralism but advance a one-sided, ideological perspective intended to shape their audiences' understanding of climate change and related topics. Climate change itself has been a focus of these think tanks, especially during the period of Klaus's presidency. It has also been discussed within the context of economics, especially in relation to free markets. CEP and later IVK (its successor) have discussed their opposition to various proposed measures to address climate change in the context of politics, describing it until very recently as a threat to democracy and democratic development in Czechia.

In terms of their targeted groups, think tanks focus on influencing the triad of policy makers (politicians and policy experts), media employees (mainly editors), and academics (scholars and students). To do so, these think tanks organize educational programmes and meetings with policymakers, publishing bulletins, newsletters, original research/policy papers, and books as well as translations of books and other texts on climate denialism. CEP, IVK, and OI have released the largest number of obstructionist texts, books, and book translations, with OI publishing since the early 1990s and CEP's (now IVK)'s publishing programme active since 2005.[32] Beyond publishing, think tanks have built their influence through networking. The boundaries between the think thanks and the triad of groups they target are porous: for example, we see individuals who are simultaneously affiliated with a think tank and also active in academia. This is the case for controversial economist Miroslav Ševčík, who was one of the cofounders of LI and is also a professor at Prague University of Economics and Business.[33] As such, he serves as a 'bridge' between these two networks, enabling information and discourses to move from one to the other.

Besides making use of this system to exchange information,[34] climate obstruction actors use it to keep climate sceptic discourse visible at various levels of society. Politicians and policy experts have been trained to understand the data related to climate change, including how to read graphs and Intergovernmental Panel on Climate Change (IPCC) research reports. For example, such trainings occurred at a meeting of the Senate Committee on EU Affairs in September 2010[35] and at a conference at Czech National Bank where Lomborg gave another speech about his book.[36] Political leaders have also been instructed on how to interpret the EU response to climate change according to neoliberal ideology: students and scholars have listened, for example, to Kutílek's[37], Singer's[38], and Klaus's[39] speeches alleging 'no convincing evidence of global warning'

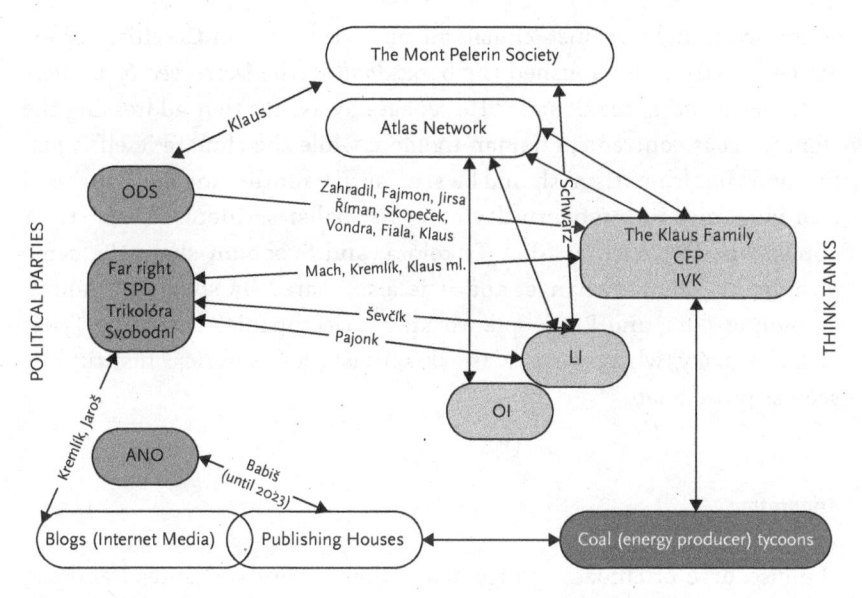

Figure 10.2 Czech climate obstruction activists' relationships/networks.

during seminars and lectures at universities. In addition, most of these activities have been actively promoted to gain media attention. In turn, the media found such controversial conversations compelling, especially during Klaus's presidency[40] as the debate over climate change would attract public attention.

Figure 10.2 summarizes the material just discussed. Although it is designed to capture the relationships among the actors involved in climate obstruction, it also illustrates how interrelated the field is.

Communication strategies: Types of discursive obstruction tactics and who uses them

We have identified four broad discourses related to climate change in the Czech media, three of which may be interpreted as forms of climate obstruction.[41] Besides providing context for the specific frames used, the four discourses also guide obstructionists' discursive obstruction strategies.

Open denialism

Though only marginally present, open denialism continues. Klaus, the founder of ODS and the leading figure of the climate sceptic camp,

THE CZECH REPUBLIC [257]

continues to fight against 'climatism' and 'alarmism' in Czechia. For example, in 2017, he published the book *Shall We Be Destroyed by Climate or by Our Fighting the Climate?* He repeatedly argues that addressing the climate crisis contradicts human freedom while the climate itself is just fine and that 'climatism' should be seen as yet another ideology that will lead humanity into modern-day crypto-socialist serfdom. The far-right populist parties such as SPD, Trikolóra, and Svobodní share the same position on climate change, and it is also shared by some journalists, opinion leaders, and PR people working for companies owned by Tykač and Křetínský, who regularly attack activists and political institutions seen as pro-climate.

Adaptation

The discourse of climate change adaptation is very widespread and can also be found in the discourse of ANO and the media formerly owned by Babiš. Here, climate change-related problems, such as dealing with the consequences of droughts repeatedly affecting some parts of the country, are presented as challenges and puzzles to be solved through public investment, technology, and capable management. In this discourse, society is expected to adapt to its new reality by employing new technical solutions, with no significant attention paid to the root cause of the problem (emissions); the possibility of lifestyle or structural changes is never seriously considered. To use a simple but telling metaphor, here the engineers are expected both to achieve the desired technical solutions to climate change and continue to drive their SUVs.

Business opportunity

Here, climate change is framed as an opportunity for business. This discourse is found mostly in the media targeting entrepreneurs and investors. Economics journalists present new technologies for combating drought, such as green roofs and vertical gardens, and mitigation activities, such as producing electric cars and solar panels, not as climate solutions but as new business opportunities. Although climate change is also viewed as a problem that will bring costs due to natural disasters and changing temperatures, in this discourse it can be transformed relatively easily into a good investment—a sentiment that the current ODS and ANO also embrace. The climate change adaptation and business opportunity discourses can be seen in the products of some of the think tanks discussed earlier. For example, the LI publishes articles on the search for solutions in technology and entrepreneurial activity, or as they themselves put it: 'Entrepreneurs are the solution'.

Mitigation

The mitigation discourse is the only one whose theme is the root cause of the climate crisis, emissions. It is nearly absent from Czechia's mainstream media and located mainly in critical, alternative outlets with limited reach and influence. Our research found that the debate on reducing carbon emissions is muted, with one exception: mitigation is occasionally mentioned in the corporate press when the article deals with countries other than Czechia.[42] At home, however, the topic of mitigation is conspicuously missing, thus contributing to the generally more sceptical climate change 'climate' in the media. As we can see, different versions of climate obstruction form the prevailing discourse on climate change in the Czech media.

With important exceptions, such as Klaus and his allies, the problem of climate change is no longer widely denied in Czechia, a situation that corresponds to the global situation generally.[43] Still, the issue tends to be depoliticized by the mainstream media and, in that sense, is a form of obstruction. In particular, audiences' attention is turned to political activities only in the form of shallow adaptation measures without an explanation of the need to protect the climate for the future, which in turn would mean raising the issue of significantly reducing carbon emissions. The mainstream media do not deny climate change as such, but they do deny the public a forum to discuss mitigation measures and tend to avoid arguments criticizing state energy policy.[44] This situation has worsened since the outbreak of the Russian war against Ukraine, when the country's continued dependence on coal began to be considered a realistic policy option. This type of (non)reporting may be one of the reasons for the scepticism regarding effective policy responses to climate change demonstrated among significant parts of the Czech population.[45]

In Table 10.2, we capture the combinations of types of discursive obstruction and the actors involved. Depending on their strategies, some actors can be included in multiple categories.

Table 10.2 TYPES OF DISCURSIVE OBSTRUCTION STRATEGIES AND THE ACTORS INVOLVED

		Type of actors involved		
		Politicians/ Parties	Think tanks	Business/ Media
Type of discursive obstruction strategy	Open denial of the human origin of climate change and/ or explicit attacks on activists, the EU, and climate scientists	Klaus, ODS (in the past) SPD Trikolóra Svobodní	Klaus family (CEP and IVK)	Czech News Center (Křetínský) Sev.en AG (Tykač)
	'Soft' type of obstruction through distraction and focus on particular problems and business opportunities	ANO (Babiš), Current ODS	Liberal Institute Civic Institute	Agrofert (Babiš) Czech News Center (Křetínský) Sev.en AG (Tykač)

DISCURSIVE FRAMINGS

In perhaps his best-known book, *Blue Planet in Green Shackles: What Is Endangered: Climate or Freedom?* Klaus states,

> Global warming has recently become a symbol and, in fact, a prototype of the truth vs. propaganda problem. A single, politically correct truth has been established and it is not easy to oppose it, even though a significant number of people, including top scientists, see the problem of climate change and its causes and consequences quite differently. . . . The advocates and promoters of those hypotheses are mostly scientists who profit from their research, both financially and in the form of scientific recognition, and also politicians (and their fellow travellers in academia and in the media) who see it as a political issue attractive enough to build their careers on.[46]

These lines offer a rich illustration of the framing that the dominant portion of Czech climate obstruction actors use to convey a specific and simplified meaning to a complex phenomenon (climate change).[47] These actors use this framing to attack climate activists, the European Union, and climate scientists. Although this group comprises politicians/political parties, think tanks, and media, they have all framed climate change in a similar way using a few master frames that have been bridged, amplified, and extended.[48]

At first, in the late 1990s, climate change was described as 'global warming', a problem addressed by Klaus and his think tank CEP. They raised uncertainty not only about the data scientists had gathered but also about the way these data were interpreted. Over this first, relatively brief period, a master frame we call Science (*proper science versus biased/ ideological science*) was constructed. The key factor here was that the science was said to be not just biased but also ideologically driven. This was an important moment in the evolution of climate obstruction as the master frame was bridged, bringing ideology into the meaning-making process related to understanding global warming/climate change. The broad meaning of the subsequent Ideology frame (*communism/totalitarianism versus classical liberalism*) opened a path for the development of many frame extensions and amplifications, which we witnessed during Klaus's presidency. While Klaus (and CEP) framed concern about climate change as an ideology based on values in opposition to freedom (liberalism), others (politicians and some regular newspaper columnists) explicitly allied those concerned with climate change progressivists, communists, and adherents of the prior Soviet regime. Such was the case for blogger Petr Jaroš.

> However, it came as if on cue to the totalitarian parasites, who had just hastily completed the cutting of the red base with the green top layer and who urgently needed a new enemy in order to reunite the scattered hordes. And no matter how hard I say it, I have to say it—they thought it up brilliantly. Or would it occur to any of you to take one chemical element and put it in the place of Trotsky, Kamenev, Tito the Bloodhound, or any other deviants from the valid party line drawn by the last Politburo meeting? From the ordinary C in the periodic table to the new class enemy Carbon—isn't that just breath-taking?[49]

Research[50] shows that the Ideology framing, which explicitly links unpopular actors or events with the previous nondemocratic Czechoslovak regime, resonated with the public, particularly when freedom was emphasized, a key factor said to help us distinguish between democracy and totalitarianism. Emphasizing freedom when discussing climate change politics—that is, constructing its meaning by comparing it to regulation—is itself a frame that supports the master frame of Ideology. In combination, Science (*proper science versus biased/ideological science*) and Ideology (*communism/totalitarianism versus classical liberalism*), with additional support from the Freedom frame, form a powerful and complementary set of meaning-making tools. Consider the framing in the following quote from one of Klaus's texts:

> In my speech here—in Erice—in 2012, I said: 'this doctrine, as a set of beliefs, is an ideology, if not a religion. It lives independently on the science of climatology. Its disputes are not about temperature, but are a part of the conflict of ideologies. . . . This doctrine is a loosely connected cascade of arguments, not a monolithic concept which—because of its structure—escapes the scrutiny of science.' I don't have any reason to change this seven years' old statement of mine.[51]

Although referring to (the concept of) science and ideology, the current political elite,[52] mainly the politicians of ODS and ANO, construct their frames a bit differently. In both parties, a significant number of politicians still deny climate change or oppose climate change policymaking. Those who favour the soft form of obstruction tend to question the science and the role of society in causing climate change but in a subtler way, as can be seen in the following autumn 2021 quote from Prime Minister Petr Fiala:

> As a scientist who does not do this professionally, I try to follow the various debates. I think the answer is not entirely clear. But I don't think that's the most important part of it. We have to perceive that some change is taking place and we have to be careful to some extent not to cause worse consequences.[53]

Most important, these leaders acknowledge that the climate is changing, but usually do not discuss who is responsible for that fact. They admit that some type of action needs to be taken, just in case, to ensure that society will be able to adjust to a new situation some time in the future. This framing—'better safe than sorry'—is relatively recent, appearing in public discourse only after 2017 and anchored in the discourse of adaptation.

In the case of think tanks, this type of soft obstruction is rare. Such think tanks, particularly LI and sometimes OI, typically draw on the open denial discourse, in this case using the Ideology frame. Here, the frame is used not to shift the conception of climate change from a scientific problem to an ideological one, but to oppose the measures taken to fight it on the grounds that they are state-driven and thus contravene the logic of the free market. In practice, this framing manifests in discourse pointing to the high costs of the transition to a carbon-neutral economy. As economist Dominik Stroukal stated in his article 'As a climate leader, we will be poorer':

> The rest of the world will run away from us economically and we will have to justify being relatively poorer, which will make us the climate leader of the world.

Table 10.3 TYPES OF DISCURSIVE OBSTRUCTION STRATEGIES AND THEIR FRAMES

		Frames used		
		Politicians/ Parties	Think tanks	Business/ Media
Type of discursive obstruction strategy	Open denial of the human origin of climate change and explicit attacks on activists, the EU, and climate scientists	Master frame: Science (*Proper science versus biased science*) Master frame: Ideology (*Communism/totalitarianism versus classical liberalism*) Frame: Freedom		
	'Soft' obstruction through the use of distraction and focusing on particular problems and business opportunities	'Better safe than sorry' Pragmatism	'Irrational' 'Cost versus benefit'	'Cost versus benefit'

Personally, I have no problem with that, I'm happy to reduce my own wealth for the sake of higher goals. The question is how the Europeans, who are deep in their pockets, will view this, and we will have a better environment at the expense of their standard of living. I'm already doing it voluntarily myself, but it will bother a lot of people, especially those for whom it's an expensive trade-off. Will it hold up politically then? A greener world is the ultimate good, but doing good is not free.[54]

Table 10.3 summarizes the discursive framings Czech obstructionist actors use, placing them in the context of the obstruction strategies they employ. As the table shows, the open denial group has used identical master frames, while the soft obstruction group has used more diverse frames while still drawing on the dominant meanings already used in this discourse.

CONCLUSION

This overview of the climate obstruction landscape in Czechia raises a question: How influential have obstructionist actors been to date? It is not easy to answer because, unlike many other countries with extensive fossil fuel production, Czechia has been limited in terms of activities both aimed directly at policy change and visible to the public, such as the publication of policy papers, the staging of debates on existing policies, the preparation of bills, or the organization of national issue campaigns. The Czech obstruction scene also has a relatively small number of actors involved. Several think tanks, tens of politicians, and a few journalists/bloggers represent the core players in climate obstruction, while wealthy coal- and energy-producing industrialists and certain publishing/media houses periodically appear in greater or lesser roles.

As we have demonstrated, there has been little need for concerted or visible lobbying campaigns against climate action by businesses or other nominally extra-political forces as mainstream Czech politicians have maintained friendly relations with fossil businesses in any case. However, more research is needed to tease out the interactions between business and politics on climate. Probably due to the indifferent stance of the political elite on climate change, only 39% of the Czech population state that they are currently interested in the problem. Furthermore, 37% of the population believe climate change's effects on Czechia will be half negative, half positive, while only 42% believe it will be all negative. These opinions exist even though most of the population—92%—believe climate change has been caused by humans at least to a certain degree, and 71% say climate change can be influenced if we change our behaviour.[55]

The results presented do not reveal the reasons why Czechs are not interested in climate change and are rather restrained in their concern over its impacts. However, we may assume that the country's relatively small climate obstruction enterprise, in cooperation with mainstream politicians and via its close relationships and networks with business and media, has succeeded in promoting its own interpretation of climate change. Moreover, climate obstructionists have established certain relationship structures and discourses that have come to define what is taken for granted in Czech society; they have become part of common political logic. In other words, perhaps the biggest success of Czechia's climate obstructionists has been their mastery of discursive framing to create an environment that limits opportunities for 'opening windows to let the fresh air in'.

NOTES

1. O. Císař worked on the chapter within the framework of the research project 'Under Pressure: Crisis, Emotions, and Political Transformations around Climate Change' (Czech Science Foundation, GA22-00800S).

2. Fakta o klimatu (2023), https://faktaoklimatu.cz/. 21.3.2024.
3. Hannah Ritchie and Max Roser, Czechia: 'CO2 Country Profile', https://ourworldindata.org/co2/country/czech-republic. 21.3.2024.
4. Gütschow, J.; M. Pflüger (2023), 'The PRIMAP-hist National Historical Emissions Time Series v2.4.2 (1750-2021)'. zenodo, doi:10.5281/zenodo.7727475.
5. Climate Watch (2023), Czech Republic, https://www.climatewatchdata.org/countries/CZE?end_year=2020&start_year=1990#climate-commitments. 21.3.2024.
6. Martin Myant and Jan Drahokoupil (2010). *Transition Economies: Political Economy in Russia, Eastern Europe, and Central Asia*. New York: Wiley.
7. Eurobarometer (2021). Climate Change, https://europa.eu/eurobarometer/surveys/detail/2273. 21.3.2024.
8. For example, Petr Vidomus (2018), *Oteplí se a bude líp: Česká klimaskepse v čase globálních rizik*. Praha: Slon; and Daniel Čermák and Věra Patočková (2020). 'Individual Determinants of Climate Change Scepticism in the Czech Republic', *Sociológia*, 52, 6: 578–598.
9. Climate Change Performance Index (2023), 'Climate Change Performance Index 2023: Ranking', https://ccpi.org/ranking/. 21.3.2024.
10. Čermák and Patočková, 'Individual Determinants of Climate Change Scepticism', at 591.
11. Ondřej Císař (2007), 'Between the National and Supranational? Transnational Political Activism, Conflict, and Cooperation in the Integrated Europe', *Contemporary European Studies*, 2, 1: 21–36.
12. Vidomus, *Oteplí se a bude líp*.
13. Ibid.
14. Sean Hanley and Milada Vachudová (2018), 'Understanding the Illiberal Turn: Democratic Backsliding in the Czech Republic', *East European Politics*, 34, 3: 276–296 and Vlatimil Havlík (2019), 'Technocratic Populism and Political Illiberalism in Central Europe', *Problems of Post-Communism*, 66, 6: 369–384.
15. Ondřej Císař and Jiří Navrátil (2019), 'For the People, by the People? The Czech Radical and Populist Right after the Refugee Crisis'. In: Manuela Caiani and Ondřej Císař (eds.), *Radical Right Movement-Parties in Europe*, pp. 184–198. London: Routledge.
16. Ondřej Císař and Milan Hrubeš (, 'Think Tanks and Policy Discourses in the Czech Republic'. In: Arnošt Veselý, Martin Nekola and Eva Hejzlarová (eds.), *Policy Analysis in the Czech Republic*, pp. 273–290. Bristol: Policy Press.
17. Vidomus, *Oteplí se a bude líp*.
18. Vojtěch Pecka (2023), *Továrna na lži. Výroba klimatických dezinformací*. Praha: Alarm and Utopia Libri, pp. 181–183.
19. Vidomus, *Oteplí se a bude líp*.
20. Obcinst. 'Orbán jako klimašampión', http://www.obcinst.cz/madarsko-orban-jako-klimasampion/. 21.3.2024.
21. Pecka, *Továrna na lži*, pp. 181–183.
22. For example, Eliška Hradilková Bártová (2023, 23 May), Tykačovo impérium na výsluní. Žralok z devadesátek zažívá nečekaně tučné období a expanduje. *Deník N*, https://denikn.cz/1143010/tykacovo-imperium-na-vysluni-zralok-z-devadesatek-zaziva-necekane-tucne-obdobi-a-expanduje/; see alternative perspectives at https://www.alternativeperspectives.info/alter-eko/probehla-fora/forum-2023-jaro. 21.3.2024.

23. The Times (2023, 27 May), 'The Czech Sphinx Speaks: Daniel Kretinsky on the Future of Royal Mail and Sainsbury's', *The Sunday Times*, https://www.thetimes.co.uk/article/the-czech-sphinx-speaks-daniel-kretinsky-on-the-future-of-royal-mail-and-sainsburys-hm2l32sm3.
24. Petr Šimůnek and Zdravko Krstanov (2023, 7 July), Daniel Křetínský (interview). *Forbes*, Czech ed., https://forbes.cz/exkluzivne-nejbohatsi-cech-daniel-kretinsky-poprve-o-byznysu-za-devet-miliard-euro/.
25. Radek Kubala (2022, 25 November), Křetínského fosilní hyena. EPH vydělává sabotáží řešení klimatické krize, https://denikreferendum.cz/clanek/34658-kretinskeho-fosilni-hyena-eph-vydelava-sabotazi-reseni-klimaticke-krize; and Vojtěch Pecka (2022, 31 October), 'Špinavý byznys: Tři zdroje české klimaskepse', https://a2larm.cz/2022/10/spinavy-byznys-tri-zdroje-ceske-klimaskepse/.
26. Truchlá, Helena (2020, 13 January), 'Stránku urážející ekology spravuje tvůrce facebookové image Tykačovy elektrárny', https://zpravy.aktualne.cz/ekonomika/profil-elektrarny-chvaletice-spravuje-stejny-clovek-jako-str/r~edf388ba346511ea858fac1f6b220ee8/.
27. Vidomus, *Oteplí se a bude líp*, p. 219; Pecka, 'Špinavý byznys'.
28. Diane Stone (2000), 'Non-Governmental Policy Transfer: The Strategies of Independent Policy Institutes', *Governance: An International Journal of Policy and Administration*, 13, 1: 66.
29. Heike Klüver and Iñaki Sagarzazu (2016). 'Setting the Agenda or Responding to Voters? Political Parties, Voters and Issue Attention'. *West European Politics*, 39, 2: 380–398.
30. Hendrik Wagenaar (2011), *Meaning in Action: Interpretation and Dialogue in Policy Analysis*. Armonk, NY: M. E. Sharpe, p. 223.
31. Stone, 'Non-Governmental Policy Transfer', p. 45.
32. Vidomus, *Oteplí se a bude líp*, p. 254.
33. Liberální institut (2021, 3 May), Miroslav Ševčík opět dokazuje, proč nemůže být členem Liberálního institutu. https://libinst.cz/miroslav-sevcik-opet-dokazuje-proc-nemuze-byt-clenem-liberalniho-institutu/
34. Vidomus, *Oteplí se a bude líp*, p. 211.
35. Senát ČR (2010, 8 September), https://www.senat.cz/cinnost/galerie.php?aid=8049.
36. 'Pražská podzimní přednáška' (2006, 12 October), https://libinst.cz/prazska-podzimni-prednaska-2006/; and 'Lomborg změnil podobu diskuse o životním prostředí a globálních problémech' (2006, 14 October), https://libinst.cz/lomborg-zmenil-podobu-diskuse-o-zivotnim-prostredi-a-globalnich-problemech/.
37. 'Globální oteplování—Fakta a Mýty' (2007, 15 November), http://cepin.cz/cze/pozvanka.php?ID=122.
38. 'Global Warming: Man-Made or Natural?' (2007, 10 October), http://cepin.cz/cze/pozvanka.php?ID=116.
39. 'Rekordní návštěva přednášky prezidenta Václava Klause' (2007, 13 May), https://www.vse.cz/zpravodaj/38/.
40. Pecka, 'Špinavý byznys' and Petr Vidomus (2018, 20 May), 'Dlouhý stín Václava Klause: Češi dál pochybují, že změny klimatu jsou vážný problém. Klimatologové jim to nevyvracejí, nechtějí si pošpinit prestiž expert', https://domaci.hn.cz/c1-66143120-dlouhy-stin-vaclava-klause-cesi-dal-pochybuji-ze-zmeny-klimatu-jsou-vazny-problem-klimatologove-jim-to-nevyvraceji-nechteji-si-pospinit-prestiz.

41. Ondřej Císař, Marta Kolářová, and Tomáš Profant (2023). 'Climate Reframed: Economic Discourses on Climate Change in Czech Newspapers'. MS.
42. Ibid.
43. See Michael Mann (2021), *The New Climate War: The Fight to Take Back Our Planet*. New York: PublicAffairs; see also Chapter 8, on Poland, in this volume.
44. Císař, Kolářová, and Profant, 'Climate Reframed'.
45. For example, Daniel Čermák and Věra Patočková (2020), 'Individual Determinants of Climate Change Scepticism in the Czech Republic', *Sociológia* 52, 6: 578–598.
46. Václav Klaus (2008), *Blue Planet in Green Shackles: What Is Endangered: Climate or Freedom?* Washington: Competitive Enterprise Institute, p. 12.
47. See David A. Snow, and Robert D. Benford (1992), 'Master Frames and Cycles of Protest'. In A. D. Morris, C. M. Mueller (eds.), *Frontiers in Social Movement Theory*, pp. 133–155. New Haven, CT: Yale University Press.
48. By 'bridging', we refer to the linking 'of two or more ideologically congruent but structurally frames regarding a particular issue or problem'; by 'amplification', we refer to the 'idealization, embellishment, clarification, or invigoration of existing values or beliefs', and by 'frame extension', we refer to 'extending [actors frames] beyond its primary interests to include issues and concerns that are presumed to be of importance to potential adherent'. See R. D. Benford and D. A. Snow (2000), 'Framing Processes and Social Movements: An Overview and Assessment', *Annual Review of Sociology*, 26, 1: 611–639 and David A. Snow, et al. (1986), 'Frame Alignment Processes, Micromobilization, and Movement Participation', *American Sociological Review*, 51, 4: 464–481.
49. Petr Jaroš (2019, 1 April), Klimajugend, marsch! https://dfens-cz.com/klimajugend-marsch/
50. See, e.g., Adéla Gjuričová (2011), *Rozděleni Minulostí: Vytváření Politických Identit v České Republice Po Roce 1989*. Praha: Knihovna Václava Havla; Jiří Koubek, et al. (2012), 'ČSSD a KSČM v Perspektivě Stranickopolitické'. In: M. Polášek, M. Novotný, and M. Perottino (eds.), *Mezi Masovou a Kartelovou Stranou: Možnosti Teorie Při Výkladu Vývoje ČSSD a KSČM v Letech 2000–2010*. Praha: Sociologické Nakladatelství; and Milan Hrubeš and Jiří Navrátil (2017), 'Constructing a Political Enemy: Intersections', *East European Journal of Society and Politics*, 3 (Suppl.), 3. 3, 3: 41–62.
51. Václav Klaus (2019, 20 August), 'The 2019 Climate Alarmists´ Offensive is Exclusively Politically Driven', https://www.klaus.cz/clanky/4422.
52. By 'current political elite', we refer to parliamentary politicians starting their office in 2021 elections.
53. A journalist citing Fiala's opinion on climate change, see Prokop Vodrážka (2021, 6 October), 'Experti: Fialovo zpochybňování vlivu člověka na klima je manifestace populismu. Cílí na bojovníky za „normální svět"',https://denikn.cz/716973/experti-fialovo-zpochybnovani-vlivu-cloveka-na-klima-je-manifestace-populismu-cili-na-bojovniky-za-normalni-svet/.
54. Domik Stroukal (2020, 24 June), 'Jako „klimatický lídr" budeme chudší', https://iuhli.cz/jako-klimaticky-lidr-budeme-chudsi/.
55. Radka Hanzlová (2022), 'Postoje české veřejnosti ke změně klimatu na Zemi – září až listopad 2022', https://cvvm.soc.cas.cz/media/com_form2content/documents/c2/a5585/f9/oe221208.pdf. 21.3.2023.

11

Climate Obstruction in Italy

From Outright Denial to Widespread Climate Delay

MARCO GRASSO, STELLA LEVANTESI, AND
SERENA BEQJA

INTRODUCTION: A HOTSPOT FOR CLIMATE CHANGE BUT LITTLE ACTION

Italy is a hotspot for climate change due to its combination of multiple major risk factors and high vulnerability.[1] According to the European Severe Weather Database, the country experienced 3,191 extreme weather events in 2022, compared with 2,072 the year before, and 380 in 2010. Because Italy is particularly exposed to climate impacts, it should follow that the country would have very ambitious mitigation objectives and work hard to adapt to these inevitable impacts. However, Italy's political and institutional commitment to decarbonization and the energy transition has been weak. For example, Italy approved the 2018 National Adaptation Plan (Piano Nazionale di Adattamento al Cambiamento Climatico, or PNACC) in December 2023, and its already tepid transition seems to have been further diluted by the right-wing coalition currently leading the country. No long-term strategy has been submitted to the United Nations Framework Convention on Climate Change (UNFCC), and the country lacks a national, economy-wide emissions reduction target. Figure 11.1 includes the absolute values and percent change of Italian greenhouse gas emissions from 1990 to 2021.

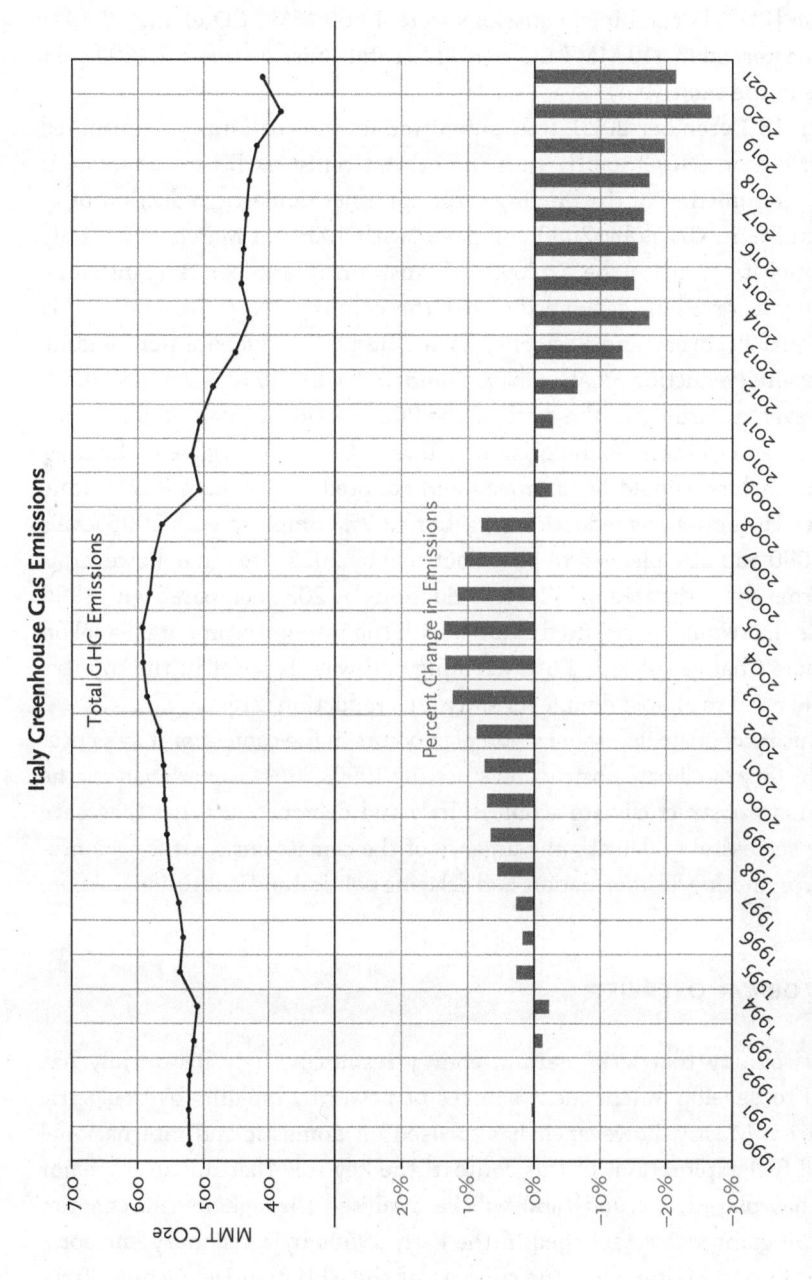

Figure 11.1 Total greenhouse gas (GHG) emissions (in MMT CO_2e) and percentage change in emissions in Italy between 1990 and 2021, inclusive.

Source: Total GHG emissions based on data provided by Gütschow and Pflüger (2023) for Kyoto Six Greenhouse Gas Totals.

As Figure 11.1 shows, Italy's total greenhouse gas (GHG) emissions (excluding LULUCF) in 1990 amounted to 522 million metric tonnes of carbon dioxide equivalents (MMT CO_2e) (for comparison, the European Union [EU27]'s combined emissions were 4,860 MMT CO_2e) and by 2021 had decreased to 410 MMT CO_2e (EU27's total emissions were 3,460 MMT CO_2e in the same year).[2]

On 18 December 2020, Italy submitted its first nationally determined contribution (NDC) jointly with the other twenty-six EU member countries, committing to the binding target of a net domestic reduction of at least 55% in GHGs by 2030 compared with 1990. In March 2022, Italy adopted its Plan for the Ecological Transition (Piano per la Transizione Ecologica, or PTE), developed under the country's Next Generation EU National Recovery and Resilience Plan. The plan included a non-binding emissions reduction goal of 51% compared with 1990 levels by 2030. However, according to the draft of the 2023 National Integrated Plan for Energy and Climate (Piano Nazionale Integrato per l'Energia e il Clima, or PNIEC, which should be approved and adopted before June 2024), Italy set a GHG emissions reduction target of 43.7% compared with 2005 levels by 2030 and also pledged to phase out coal by 2025. This goal, however, is far from the estimated 61–71% in reductions by 2030 compared with 1990 levels that would be required to align with the Intergovernmental Panel on Climate Change (IPCC)'s 1.5°C warming pathway. To meet it, the country would need to almost double its emissions reduction target.[3]

This inadequate decarbonization plan occurs in the context of Italy's peculiar history of climate obstruction since the 1990s. This chapter analyses the efforts to obstruct climate action in Italy and demonstrates that they have been successful in denying the urgency of the climate crisis, creating confusion, promoting disinformation, and delaying political and institutional action.

HISTORICAL OVERVIEW

As a country that lacks natural energy resources, Italy historically has tried to develop independent sources of power by building hydroelectric capacity. Mostly, however, it has focused on domestic and international fossil fuel exploration. In this context, the key role that Italian oil major Eni has played is crucial and will be analysed throughout the chapter. The oil giant was established in the early 1950s by a visionary entrepreneur, Enrico Mattei, with the support of the Christian Democratic Party (Democrazia Cristiana, or DC) by merging several entities working in the exploration, refinement, transport, and distribution of oil and gas. Since the beginning and throughout the 1960s, the company's business was directed at breaking the country's dependence on the international oil industry led by American and British companies. Among other things, Eni developed fossil fuel projects in the Po Valley and established autonomous relations with North African and Middle Eastern oil- and gas-producing countries, thereby challenging other interests in the regions. On a symbolic yet eminently practical level, Eni aligned the international energy technocracy with its industrial strategy through its Graduate School for the Study of Hydrocarbons (Scuola di Studi Superiori sugli Idrocarburi), located in Milan, which prepared a global managerial elite for the fossil fuel business.[4]

On the national level, the company has held close ties to all Italian governments, and today, the country's Ministry of Economy and Finance and the development bank Cassa Depositi e Prestiti hold the state's one-third ownership stake in Eni. The company is currently among the world's largest fossil fuel companies and operates in more than sixty countries at all stages of the oil and gas business. From 1950 to 2018, Eni ranked twenty-fourth among global oil and gas majors for cumulative CO_2 and methane emissions.[5] The company also runs extensive advertising, sponsorships, and partnerships with multiple Italian academic institutions.[6]

Against this backdrop, dominated by the pervasiveness of Eni in the country's economic, social, political, and cultural spheres, the recent history of climate obstruction in Italy can be divided into five periods characterized by the development of consistent obstructionist narratives that emerged in response to both earlier and contemporary events. For analytical purposes, these periods are considered separately, albeit in practice their features and dynamics substantially overlap.

Period 1 (1990–2000)

As the science of anthropogenic climate change became more certain and consistent, directed efforts to counter climate action began to take shape in Italy in multiple ways.[7] They consisted mostly of fully fledged forms of denial that disputed that the climate was changing due to human causes, asserting falsely that these changes were natural and have always occurred.

In the early 1990s, a rudimentary yet effective form of obstruction emerged based on 'instrumental realism', it largely used cherry-picked information, and redirected the responsibility for climate change to non-anthropogenic causes.[8] Additionally, the increasing scientific consensus and the rising but still limited public awareness of climate change were mostly ignored or downplayed in mainstream media, which erased the issue from public discourse. In the decade under scrutiny, there was a clear trend of decreasing coverage of environmental issues by Italian newspapers.[9] For example, over the 1989–1994 period, the two leading dailies, *Corriere della Sera* and *Repubblica*, published 272 articles about climate change compared with the leading US papers, *The New York Times* and *The Washington Post*, which published a total of around 1,000 articles on the topic.[10]

Period 2 (2001–2007)

From 2001 to 2007, an outright hostility to and denial of the available science on climate change took root. In this period, despite that individuals and groups within the Roman Catholic Church held differing positions on climate change, the Church—which had had a long-standing, prominent role in the country's cultural, political, and socioeconomic debates—emerged as a crucial player in climate obstruction efforts. The politicization of Catholicism has a long history in the country: for more than forty years, the Italian Republic was dominated by the DC party.[11] During the first decade of the 2000s, however, such politicization processes went further and exploited religion for political gain on a variety of issues, turning it into a media commodity.[12] In terms of climate and environmental discourse, right-wing populists in particular saw the Catholic Church as an ally in their efforts to dispute climate science and resist action.[13]

Period 3 (2008–2013)

Climate obstruction during these years served mostly to shift attention away from climate change by consistently diverting the public's attention from the problem.[14] This period was dominated by the libertarian narrative, fabricated by the centre-right government led by Silvio Berlusconi (May 2008–November 2011) when the country's attention was focused mostly on other issues perceived as far more urgent (e.g. tax reform, judicial reform). During this time, the climate change question was relegated to the margins by efforts to deny its relevance and the need for action.

Period 4 (2014–2018)

In the fourth period, the previous modes of obstruction gave way to something new in the Italian context: the right-wing ideologization of climate change. This form of obstruction was rooted in previous attempts to dispute climate science through faith-based arguments, which, as noted, helped to fuel anti-scientific perspectives.

Politicians from both sides of the spectrum took advantage of the politicization and ideologization of climate change to implement climate obstruction. Paradoxically, in the almost three years of its mandate (February 2014–December 2016), the governing centre-left coalition led by Matteo Renzi waged war against renewables by introducing new incentives for the construction of major biomass plants and incinerators but reducing incentives for photovoltaics and by favoring extractive activities and the underground storage of gas.[15]

Period 5 (2019–Present)

The dominant role of Eni, the Italian oil and gas major, marks the fifth period. According to Rino Formica, an Italian politician who has played a prominent role in multiple administrations since the 1970s, 'the weakness of [Italian] parties and politics has allowed Eni to capture the state'.[16] While Eni's role is considered central to the fifth period, the company's influence had been significant and widespread throughout the periods described. Experts argue that this key role was also promoted through the company's efforts to guarantee energy security during and after various sociopolitical and diplomatic crises through forging energy deals with fossil fuel-producing countries and developing fossil fuel infrastructure. Its reputational capital endures to this day.[17]

ITALY'S MAJOR ACTORS AND TYPE OF INSTITUTIONS

Climate obstruction in Italy is employed by numerous political, institutional, media, commercial, and financial actors. Most have close ties to one another while operating through different and sometimes overlapping modes of obstruction.

Fossil fuel companies and industry groups

Fossil fuel companies and fossil fuel-adjacent companies, such as pipeline operators and energy distributors, have engaged in climate obstruction through oil and gas expansion activities, lobbying, political influence peddling, advertising, and sponsorships prior to and since 1990. Engaging in documented lobbying activities at both the national and European levels, oil and gas multinational Eni and pipeline operator Snam (owned by Eni until 2012) are among the leading actors in this space.[18]

Between January and June 2021, Eni and Snam met more than one hundred times with Italian ministers including Roberto Cingolani, former minister of the 'Transizione Ecologica' (ecological transition) under former Prime Minister Mario Draghi.[19] The companies wanted to ensure that Italy's COVID-19 recovery funds would be used to 'tie us to gas for the next decades'.[20]

The degree to which Eni in particular is entrenched in the political, social, and cultural life of the country, as noted earlier, also translates to influence in the decision-making process on the national, European, and global levels. Eni has portrayed itself as a crucial facilitator of the energy transition while also promoting reliance on 'silver bullet' technological solutions and offsetting as the best ways to address GHG emissions and climate change.[21] While the company rebranded its utility services division as 'Plenitude', represented by Eni's traditional logo of a six-legged dog rendered in shades of green, Eni's business plan to 2025 remains focused on gas.[22]

In 2020, the Italian Antitrust Authority fined Eni €5 million ($5.5 million) for its misleading advertising messaging.[23] The company's Eni Diesel+ promotional campaign had made clear references to environmental sustainability although, according to the country's Antitrust Authority, 'the product is a diesel fuel for automotive use that by its nature is highly polluting and cannot be considered green'.[24] In the course of the authority's proceedings, Eni discontinued the campaign and, according to a statement, pledged to stop using the word 'green' with reference to its automotive fuels.[25]

On 9 May 2023, on the eve of Eni's annual meeting, Greenpeace Italy and the advocacy group ReCommon publicly announced Italy's first climate lawsuit against Eni,[26] which began in February 2024. The suit rests, in part, on documents unearthed by the two environmental groups that show Eni had known of the risks posed by burning its products since 1970.[27] Further research[28] by nonprofit climate news service DeSmog showed that Eni's company magazine *Ecos* made repeated references to climate change during the late 1980s and 1990s while simultaneously running advertising campaigns promoting gas, composed of methane, as a 'clean' fuel. Eni said it would prove the lawsuit is 'groundless' and, if necessary, demonstrate in court that it has taken the correct approach to decarbonization.[29] Despite these efforts to hold the company to account, the national climate debate is driven in large part by Eni and its counterparts, with the direct result that the industry narrative on climate appears across many sectors, including politics.

Eni is also involved with the European and global network of associations and groups tied to the oil and gas industry.[30] The company and many more in the oil and gas industry in Italy are represented by the Italian Chamber of Commerce, Confindustria, which, as documented by InfluenceMap and others,[31] routinely lobbies against regulatory legislation for fossil fuels at the European level and exercises political influence on the national level. Industry groups also have a history of weaponizing their political and economic influence to act in the interest of the companies they represent. For example, according to Influence Map, Confindustria has been lobbying the European Union to back new fossil gas projects while opposing policies to limit demand.[32]

Beyond direct lobbying and the types of greenwashing noted earlier, actors in the oil and gas industry implement climate obstruction through a number of tactics that will be discussed later in the chapter.

Politicians and political parties

Prominent actors in the political sphere, many in leadership roles, also participate in climate obstruction by hindering the development of climate policies and environmental protection legislation; promoting oil and gas through a number of different approaches; spreading climate denial, disinformation, and delay; minimizing the urgency and effects of climate change; and delegitimizing climate activists, as will be discussed later in the strategies and tactics section.

These actors are mostly on the right of the Italian political spectrum,[33] as right-wing ideology overlaps, at least in part, with climate denial and delay. Members of the Brothers of Italy (Fratelli D'Italia), the League (Lega), and Berlusconi's Forza Italia parties have (1) prioritized advancing industry over developing climate policies; (2) promoted and facilitated fossil fuel infrastructure, national drilling, gas deals, and fossil fuel subsidies; (3) voted against climate policies at the European level; (4) promoted climate disinformation and denial online and through their social media accounts; (5) made instrumental use of discourses of climate delay in the public debate[34]; and (6) attacked climate activists and movements.

These parties' position is to back the oil and gas industry publicly while facilitating fossil fuel companies' access to the public sphere. Right-wing Italian politicians have lobbied at the EU level, voting against European environmental and renewable energy policies. In the European Parliament in 2018, for example, the League voted against all climate and sustainable energy policy proposals except for a directive on energy conservation in buildings.[35] While in the Italian Parliament, the League, including member Giancarlo Giorgetti, head of Draghi's government Ministry of Economic Development, abstained from ratifying the Paris Agreement.[36]

Along with other populist parties of the European right, the League boasts of its 'green patriotism', geared toward supporting environmental conservation on the surface but without any real political impact on climate.[37] Ultra-nationalist parties like the League support renewable energy in their programmes and public statements because 'they are perceived to benefit domestic industries and people'.[38] More recently, such politicians, along with the Italian and European gas lobbies, have voiced their support for renewable energy in order to foster the perception that gas and renewables are roughly equivalent in terms of sustainability and can work together in a decarbonized energy system.[39] Under the leadership of the far-right Brothers of Italy party, statements by politicians in Giorgia Meloni's government, elected in September 2022, have commonly featured outright climate denial and climate disinformation. These statements have included decades-old arguments against action, such as pointing out colder temperatures to promote the idea that global warming is a 'hoax'.

In 2022, a new centrist coalition called Third Pole (Terzo Polo), led by politicians Carlo Calenda and former Prime Minister Matteo Renzi, entered the political arena. This coalition promoted oil and gas expansion while advocating for nuclear energy and also denying anthropogenic responsibility for climate change and the urgency of action to curb emissions.

This obstructionist trend could be seen during the September 2022 snap elections, when politicians referenced the climate crisis in fewer than 0.5 percent of their statements on Italian TV talk shows, online, and on their Facebook accounts.[40] Above all, legislation for environmental and climate protection has increasingly been quashed. In July 2023, under Meloni's government, nearly €16 billion directed toward nine environmental regulations within the Next Generation EU National Recovery and Resilience Plan, including those aimed at fighting the country's hydrogeological vulnerability, were tabled.[41]

Think tanks

Groups in the neoliberal camp have also been active in promoting climate change denial and, more recently, delay. The most prominent among them has been the Istituto Bruno Leoni (IBL), an Italian think tank that supports free markets and a non-interventionist state policy with close ties to the United States' climate denial machine. IBL is a member of the Cooler Heads Coalition, whose website is paid for and run by the Competitive Enterprise Institute (CEI), a major think tank with a key role in the US withdrawal from the Paris Agreement[42] As early as 2010, in an article in the newspaper *Il Foglio*, Carlo Stagnaro, then senior fellow and now research director at IBL, praised the outcomes of the fourth International Conference on Climate Change, held in Chicago by the Heartland Institute, and aligning the narrative promoted by the IBL to that of the American think tank.[43] This narrative was grounded mostly in the 2009 'Climategate' controversy—the hacking of an email server at the Climatic Research Unit at the University of East Anglia in the UK—which was subsequently weaponized by climate deniers: first, to 'prove' that global warming was a conspiracy just weeks before the COP 15 summit on climate change in Copenhagen and, second, to attack climatologist Michael E. Mann's famous 'hockey stick graph', where the 'blade' of the stick represented the rapid warming of the late twentieth century.

While presenting itself as a supporter of science, IBL imported an ideological framework from its American counterparts that fueled the politicization of climate change. On one side stands the libertarian, pro-market ideology that promoted opposition to any public 'interference' in climate action and on the other side stand the IPCC, mainstream scientists, and pro–government intervention environmentalists, who the IBL has portrayed as irrational, anti-modernist, and hostile to innovation, technology, and progress.[44] Francesco Ramella,[45] an IBL research fellow, for example, had long deployed the Heartland Institute's false rhetoric on the 'positive' effects of climate change, a myth associated[46] with the 'realism' narrative—a term used in opposition to 'alarmism' to delegitimize those who warn about the catastrophic impacts of the climate crisis.[47]

Throughout its recent history, Italian climate obstruction, particularly in its institutional and political contexts, is in part traceable back to the US denial machine. On 26 and 27 April 2007, the Pontifical Council for Justice and Peace (Pontificio Consiglio della Giustizia e della Pace) held an international conference on 'Climate Change and Development', which was attended by well-known denialist US think tanks directly or indirectly funded by ExxonMobil and the Western Fuels Association.[48] One of the messages of the conference was to discourage the use of birth control—and therefore also the promotion of abortion and distribution of contraceptives—as fewer people on the planet would allegedly not reduce the quantity of climate-changing emissions.

This reactionary position was eventually superseded by the new approach of Pope Francis, elected on 13 March 2013. In 2015, Francis published *Laudato Sì*, his first original encyclical, on humans' responsibility to act on climate, which was followed in October 2023 by the *Apostolic Exhortation Laudate Deum*, a call for action against the climate crisis that strongly condemns climate denial.

Individual climate deniers

A large part of the misinformation around climate change in Italy's public sphere has emanated from well-known figures who deny the existence of, human responsibility for, or urgency of the issue. Franco Battaglia, a professor of chemistry at the University of Modena, was probably among the first outspoken climate change deniers in Italy, whose work since the beginning of the twenty-first century has provided some of the basis for denial messaging in the Italian media. Battaglia falsely argued that human activity had a negligible effect on climate change, which, according to him, was due to natural causes and was part of an endless pattern of natural climate modifications.[49] In the years following the publication of the IPCC report of 2007, for example, Battaglia attacked the report, arguing that it was funded and staffed by politicians motivated purely by hope of political gain unrelated to science. Similarly, Adriano Mazzarella, a professor of atmospheric physics at the University Federico II in Naples, accused the IPCC of missing the complexity of climate change: according to Mazzarella, humans were responsible only for what he called 'local warming' but not for 'global warming'.[50]

In the same period, making arguments similar to Bjørn Lomborg's, which view poverty and climate change as mutually exclusive, individuals such as Antonino Zichichi, a professor of physics at the University of Bologna, focused on the idea that the most serious environmental problem humanity faced was not climate change but poverty. Such claims were also leveled at the IPCC reports, with their advocates postponing the so-called possible impacts of climate change to a distant future and land, far away from Italy.

Battaglia also contributed to the work of Galileo 2001 for the Freedom and Dignity of Science, an organization he established in 2001 with engineer and 'futurologist' Roberto Vacca and Renato Angelo Ricci, the organization's president and a professor of physics at the University of Padua. In 2001, Ricci was appointed as the 'government's commissioner' at the National Agency for the Protection of the Environment (then Agenzia Nazionale per la Protezione dell'Ambiente, or ANPA, a technical organization that supported the Ministry of the Environment), with Battaglia named as coordinator of its scientific committee. In 2002, ANPA published a report, 'Science and the Environment, Scientific Knowledge, and Environmental Priorities' ('Scienza e Ambiente, Conoscenze Scientifiche e Priorità Ambientali'), which tried to weaken the credibility of the IPCC's science and, in particular, the 2001 Third Assessment Report. ANPA created confusion by making misleading comparisons and discussions demonstrating the alleged inconsistency between the IPCC's reports and its summaries for policymakers and by instrumentally emphasizing and distorting the uncertainties, controversies, and disagreements within climate science.[51] As Oreskes and Conway put it in their book *Merchants of Doubt*,[52] climate denial has been allowed to develop thanks to such contrarians being treated as 'experts' regardless of the reliability of their records and publications. Italy is a perfect example of this phenomenon, especially considering that several of the individuals promoting climate change denial in the early 2000s were based at respected Italian academic institutions.

The majority of these denialist perspectives are still present in the public debate on climate, often recurring among the same well-known individuals. They might have remained mostly at the margins if it had not been for some enabling actors, particularly media platforms. In this arena, one of the most vocal denialists currently is Franco Prodi, a former professor of physics at the University of Ferrara and brother of Romano Prodi, the two-time Italian prime minister and former president of the EU Commission between 1999 and 2004.

Media

The media in Italy work as an echo chamber and, as mentioned, also constitute a significant vehicle for denial and obstruction messaging. Media platforms including newspapers, online magazines and outlets, television shows, and their respective social media accounts engage in obstruction by spreading climate disinformation, promoting climate deniers' views and arguments, advancing discourses of climate delay, diverting responsibility for the climate crisis, signing partnership deals with polluting industries, hosting advertising and sponsorships with and by the oil and gas industry, and attacking and delegitimizing climate activists.

The public receives confusing messages via these media platforms, which then fuel the denialist and delayer perspectives. These platforms include traditionally right-wing newspapers such as *Il Giornale* and *La Verità*, as well as the *Il Foglio* daily newspaper and others. For example, in the summer of 2022, while Italy was experiencing its worst drought of the last seventy years, two interviews were published in the paper *Il Mattino* during the last week of June. In one, the interviewee stated, among other things, that the UN climate data are 'wrong and exaggeratedly warm to begin with', that scientific information is 'spread in a propagandistic way', and that Earth is warm because of 'millennial cycles and a lot of speculation'.[53] The other interview stated that 'record heat is nothing new' and is affected by the 'influence of solar cycles'.[54] In another article published in *Il Foglio* on 24 June 2022, it was stated that 'other than drought, the real water crisis in Italy is ideological'.[55] In 2021, when the cyclone Qendresa hit Sicily, a climate denier claimed on a prominent television show that human activity 'has nothing to do' with climate change.[56]

Climate change denial, delay, and obstruction are also still promoted on mainstream Italian television talk shows. These include popular programmes such as *Otto e Mezzo* and *Carta Bianca* as well as widely followed radio shows such as *La Zanzara*, broadcast daily by Confindustria's Radio 24. Messaging on these platforms generally follows a pattern whereby a climate denier is invited to debate climate change, energy, or adjacent issues with a climate scientist, climate campaigner, or environmental activist.

It is important to note that climate denialism, disinformation, and obstruction are widespread not only in politically right-leaning newspapers but also on platforms, channels, and broadcasts that the Italian public consider more progressive. The result is that the public receives contradictory messages that feed the perspective of the denialists, who have continued to leverage doubt about climate science, creating confusion on the causes and effects of climate change and disseminating political propaganda to obstruct climate policies.

Leading national newspapers also often engage in these modes of obstruction while downplaying the role of renewables in the energy system, diverting responsibility for the climate crisis, presenting inaccurate information about the actors contributing to the climate crisis, and promoting disinformation on extreme weather events and their connection to climate change. Media platforms including leading national newspapers and media groups also engage in minimization of the urgency to act on climate and the effects of climate change.

The 'we will adapt' argument is also commonly heard in the news media and aims to downplay the impacts of the climate crisis. Moreover, this argument implies that efforts to mitigate the effects of climate change are futile and frames adaptation as the only possible response.

Finally, the media echo chamber engages in climate obstruction by allowing fossil fuel companies, oil and gas industry actors, and other polluting companies to buy advertising that employs greenwashing and sometimes false or misleading claims. On 25 January 2023, for example, an article in *Il Corriere della Sera*, Italy's leading newspaper, claimed that an expert 'with a lifetime in the fossil fuel industry, a past in Eni and also in Russia's Lukoil, is really one of the best-equipped people imaginable to assess what is happening in the composite world of the energy transition'.[57]

Analysis of media coverage in Italy shows that the climate crisis is often on the sidelines and not a priority in the news, with a general lack of attention devoted to investigating the underlying causes and the responsible actors.[58]

Financial institutions and banks

Financial institutions and banks also promote climate obstruction by funding fossil fuel projects, infrastructure, and expansion. Unicredit and Intesa Sanpaolo are the main banks financing carbon-intensive industries in Italy, and globally, they rank within the top 100 banks that fund fossil-fuel industries. Between 2016 and 2022, Intesa Sanpaolo invested US$21,031 (€19,228) billion in fossil fuels, ranking forty-fifth globally, and Unicredit invested US$42,801 (€39,131) billion in fossil fuels, ranking thirty-ninth.[59] Intesa Sanpaolo also spent US$6,294 (€5,745) billion in fossil fuel expansion between 2016 and 2022, ranking forty-first globally, and, over the same period, Unicredit spent US$8,846 (€8,088) billion in fossil fuel expansion, ranking thirty-seventh globally.[60] Unicredit also ranked second among global banks for Arctic oil and gas financing between 2016 and 2022.[61]

SACE, the Italian export credit agency, also finances oil and gas operations worldwide and guarantees carbon-intensive industries and activities with public money.[62] After the IPCC published its Sixth Assessment Report Summary for Policymakers in March 2023 and issued a 'final warning' that global emissions must fall, the Italian government published a policy for SACE that promised continued fossil fuel support past 2022, which is 'at odds with IPCC fossil fuel phase-out trajectories'.[63] According to an analysis by Oil Change International,[64] SACE is the biggest public financier of fossil fuels in Europe. Between 2016 and 2021, SACE supported €13.7 billion (US$15.3 billion) in fossil fuels, and it is considering financing for international fossil fuel projects with projected emissions equivalent to more than three times Italy's entire annual emissions.[65]

THE STRATEGIES AND TACTICS UTILIZED

Italy's industrial, political, and media actors have adopted multiple tactics and strategies to obstruct climate action, from blatantly anti-scientific narratives to scare tactics such as weaponizing energy insecurity to flagrant misinformation and disinformation campaigns to discourses of climate delay. They typically deploy their full armamentarium of climate obstruction techniques when climate policies and legislation are at the centre of the public debate and when extreme weather events contribute to visible evidence of climate change's impacts on the country.

The tactics and strategies described in this section and used by the actors described above are not mutually exclusive and, in some cases, overlap. Their ultimate, common objective is to delay or hinder climate action.

Outright climate change denial

As mentioned earlier, outright climate denial is still present and widespread in both politics and the media in Italy. Its main objectives in these contexts are to fuel the perception that the debate on the existence, causes, and urgency of climate change is still ongoing and to create confusion. Prominent climate deniers who employ decades-old arguments are hosted on major TV shows and interviewed by mainstream newspapers. These arguments include denying the existence and urgency of, and anthropogenic responsibility for, climate change, as well as falsely attributing the causes of climate change to other phenomena such as the sun. As described earlier, politicians toward the right end of the political and ideological spectrum also veer into outright climate change denial during their public statements.

Aside from these still present but more isolated episodes of outright climate denial, most strategies of climate obstruction make use of more subtle tactics.

Greenwashing and climate washing

One of the most common obstruction tactics used in Italy is greenwashing. Its main objective is to promote the perception that a business or organization is part of the solution to climate change, operates in a sustainable manner, and engages in clean and non-polluting activities—all in service of maintaining a social license to operate.

Greenwashing is used by polluting industries, especially oil and gas companies, and by politicians through advertising, sponsorships, and political influence peddling. It is expressed mainly through (1) misleading language and visuals, (2) use of selective facts, (3) stating outright falsehoods, (4) factual omissions, and (5) rhetorical distortions.

Greenwashing is developed through the use of language and visual tools with positive associations—for example, words such as 'eco-friendly', 'green', and 'sustainability', or images of nature with green and blue palettes. Greenwashing is used mostly in the creation of misleading advertising to induce consumers to buy a product the company wishes to promote as sustainable or renewable although it is not. Although greenwashing is a decades-old strategy, it is also part of a new climate denialism, widespread online and on social media, that allows companies to continuously mislead the public and evade accountability.

'Climate washing' is a common form of greenwashing that is visible in the wide gap between an organization's public statements and tangible climate commitments.[66] Fossil fuel companies and major polluters thus adopt communication strategies to create the perception that their activities are part of the solution to climate change rather than being a root cause of it.[67]

Greenwashing is also a common political tool in Italy, used by leaders and parties with the aim of obstructing climate action by deceptively promoting the perception that their commitment to creating effective climate change policies is concrete.

The idea of falsely portraying gas as a clean energy source and the message that it is crucial to the energy transition also fall under this strategy.[68] Terms such as 'renewable gas' or 'lower-emissions fuels' appear in online messaging from gas lobby groups and political actors as well as in politicians' public statements and media interviews and articles. These communications amount to greenwashing because they downplay the industry's climate impact.[69]

Redirecting responsibility

Redirecting responsibility from business and industrial production to individuals is a common strategy of climate obstructionists.[70] Both fossil fuel companies and media engage in this strategy to divert attention from the industrial, political, and institutional responsibility for the climate crisis. The main objective of redirecting responsibility is to shift attention and accountability from production to consumption, from industry to individuals, and from systemic to secondary causes.

In 2021, one of Eni's promotional campaigns claimed, 'To change things, we need Silvia who is always careful at home not to waste water. Because Eni + Silvia is better than Eni'. The reference to individual responsibility is evident: Silvia, like Eni, is also responsible for the environment, and if Silvia is not environmentally aware, Eni cannot change things.

Because of its opaque and nuanced nature, this strategy is deeply internalized in numerous social, political, and cultural dimensions of Italian society and has been weaponized by bad faith actors when, for example, media or politicians have emphasized positive individual agency in solving the climate crisis.

Delaying action

More recently, 'discourses of climate delay'[71] have entered the public debate on climate policy and action. The main objectives of this group of tactics include delaying climate action and denying its urgency while promoting the perception that something is being done. In Italy, these tactics (which may overlap) include but are not limited to (1) technological optimism, (2) fossil fuel solutionism and saviourism, (3) appeals to social justice, (4) policy perfectionism, and (5) 'doomism'.[72]

Discourses of climate delay are used mainly by politicians, political institutions, the fossil fuel industry, and the media. Some right-wing and centre-right wing politicians, for example, recur to technological optimism by holding that technological breakthroughs such as nuclear fusion, for example, are real solutions to climate change and 'right around the corner'.

The media echo chamber has often reiterated this discourse of delay through the energy security narrative: the idea that, beyond the legitimate need to secure energy sources, fossil fuel companies like Eni have 'saved' the country during the energy crisis and in the wake of the war in Ukraine by providing alternatives to gas from Russia. The fossil fuel lobby, meanwhile, had similarly leveraged fears of energy insecurity in wartime, promoting gas as a means to maintain energy security, as a bridge fuel, and as a short-term fix for energy crises, all with the ulterior motive of ensuring fossil fuel lock-in for years to come.[73]

Politicians also appeal to social justice and policy perfectionism to obstruct climate policies,[74] framing such policies as too costly or burdensome to the country. Social justice appeals promote the perception that there are other, more important political priorities to address (e.g. energy issues) and that these priorities are separate from and unconnected to the climate crisis and environmental protection. Policy perfectionism, in turn, postpones action by setting unrealistic policy ambitions. Similarly, politicians also use discourses of delay to postpone the phaseout of oil and gas by promoting the perception that it is both too costly and essentially impossible.

Discourses of doomism and defeatism[75] are also used by political leaders and fueled by the media echo chamber. Policy statements also fall under this category of discourse when they raise doubts whether mitigation is possible, pointing to seemingly insurmountable political, social, or technological challenges. Defeatism also argues that any action we take will not be enough and that, in any case, it is too late. Like other discourses of climate delay, this strategy discourages climate action and any commitment to developing effective solutions.[76]

Additional tactics

Italian obstructionist forces have used many additional tactics to curtail or delay climate action, including: spreading misinformation on renewables; using pseudo-religious or religious terms when referencing climate issues whereby 'ecology' becomes 'a religion to replace canceled Christianity' and switching to an electric car is 'fanatical'[77]; scare tactics, such as engaging in direct attacks on climate campaigners and using words such as 'environmentalist' in a derogatory manner, as well as advancing ad hominem arguments to delegitimize individuals who fight for climate action; creating confusion[78] around climate issues by inaccurately portraying the drivers and effects of climate change or promoting the idea that the debate on the existence of human-caused global warming is still ongoing; and engaging in sponsorships, advertising, and partnerships associating polluting companies with highly regarded institutions, events, and social, cultural, and sports initiatives to promote the false perception that these actors are part of the solution to climate change as well as social benefactors.[79]

DISCURSIVE FRAMINGS

During the five periods described at the beginning of this chapter, Italian climate obstruction has been enacted using five dominant discursive framings. There has been considerable overlap of these narratives over the years, but, together, they have shaped and reinforced the complex Italian climate denial machine.

Period 1 (1990-2000): Defensive obstruction

During this period, climate obstruction had the objective of maintaining the status quo and was thus 'defensive', focused on the alleged nonexistence of climate change and/or the false notions that the phenomenon was natural and that there was still widespread scientific uncertainty around the issue. These arguments were advanced mostly by supposed experts in the orbit of the Italian academic world who lacked relevant expertise in the climate change issues on which they commented; they built their narratives on cherry-picked data or studies and focused on unexplained and anomalous details in the research while ignoring more comprehensive findings about the issue.[80]

Their distorted narratives were repeated often in the national media and took advantage of the journalistic norm of balance, which assumes that every story has two equally valid sides and thus deserve the same level of coverage.[81] During a time when climate change was largely unfamiliar to the public, such false balance promoted the perception that those who warned about climate change and those who rejected climate science (and thus spread climate misinformation) had equal standing.[82] This tradition prevented the public from being properly informed about the nature and seriousness of climate change and ultimately favored the enduring climate obstruction the country is still experiencing.[83]

Due to the often-inadvertent support of mostly complacent media, Italy's defensive obstruction blurred the lines between facts and opinions, real and fake news, accurate information and misinformation. This stream of misleading discourses converged into a false narrative about the nonexistence of anthropogenic climate change.[84]

Period 2 (2001-2007): Oppositional obstruction

Following the path led by the George Marshall Institute—a now-defunct conservative US think tank funded by the fossil fuel industry that conducted campaigns to undermine the credibility of the IPCC[85]—climate obstruction in Italy became 'oppositional' during this period. Opponents of climate action focused on attacking scientists directly, using false narratives of corruption and/or incompetence. In particular, those adopting oppositional obstruction accused the IPCC of political biases supposedly hidden in its reports, claiming that the UN body deleted and distorted evidence. They also continued to emphasize the 'uncertainty' of climate science findings and promoted several unsubstantiated theories (e.g. that solar activity is the main or only cause of climate change). This type of resistance to the scientific consensus on climate change resembles the framings used by the Marshall Institute's most prominent climate deniers, Fred Singer and Fred Seitz, who accused the IPCC of 'scientific cleansing' and of unauthorized changes to parts of the report.[86]

A prominent feature characterizing this discursive framing is the presence of 'logical fallacies':[87] the presentation of invalid conclusions achieved by oversimplifying and misinterpreting data, graphs, statistics, and the broader arguments in the IPCC reports. These logical fallacies still involved cherry-picking techniques and were further developed to support conspiracy theories, such as inaccurate claims of fabricated scientific data and corrupted scientific processes.

The central role of the Catholic Church in the life of Italians was ripe terrain for the unique turn oppositional obstruction took toward the end of this period. Climate obstruction actors intentionally used religious influence to shape discourses, cultural imagery, and behaviors. Through oppositional obstruction, religion and politics intersected in a mutually reinforcing manner: in the 2000s, Catholicism represented and advocated cultural and identity values linking religion, people, places, and the natural world, arguments that were then employed in the nationalist and antiscientific rhetoric of Italy's right-wing parties.[88]

Period 3 (2008–2013): Dismissal obstruction

Climate obstruction during this period did not confront climate science or scientists directly but centred on the obscuration and/or minimization of the implications of climate change to deny its urgency and, ultimately, discourage action. Interestingly, this discursive framing continued to identify the IPCC as the epicentre of fabricated climate science and falsely portrayed its reports, especially those regarding climate impacts, adaptation, and vulnerability, as the mainstream narrative on climate. To a degree, the dismissal obstruction frame is consistent with science denialism expert Mark Hoofnagle's FLICC framework: *f*ake experts, *l*ogical fallacies, *i*mpossible expectations, *c*herry picking, and *c*onspiracy theories.[89]

A quintessential example of dismissal obstruction often reiterated by disparate fake experts in Italy was the need to devote greater attention to local environmental issues than to 'abstract' global climate change.

Period 4 (2014–2018): Ideological obstruction

In this period, the driving narrative was based on the premise that climate action is ideologically driven and threatens 'our way of life'. It often pitted the need for environmental protection efforts at the domestic level against 'globalist' climate change policymaking at the international level. The tactics used to shape and justify this view were taken from the arsenal of 'climate scepticism' which, like outright denial, seems to reject the evidence of climate change. They included sowing doubt on the increase in global temperatures, rejecting the link between global warming and human activities, and denying the consequences of climate change.

Period 5 (2019–Present): Greenwash-and-delay obstruction

Far from outright climate denial or skepticism, the current form of climate obstruction is sophisticated and highly effective in eroding political and public support for climate policies, as well as in burnishing the image of fossil fuels and promoting the false image of gas as a 'clean' fuel necessary to the energy transition. 'Greenwash-and-delay' involves promoting the narrative that something is being done about climate change, with conventional fuels and their supporters positioned as the heroes. This narrative adheres to the notion of 'fossil fuel saviorism'[90] and involves the advancement of arguments favoring 'non-transformative solutions'[91] in which technological optimism, fossil fuel solutionism, and 'all talk-little action' discourses are dominant (e.g. long-term net zero commitments and short/mid-term sustained fossil expansion). These arguments have also included promoting a greater reliance on gas as a matter of energy security, as mentioned earlier. Greenwashing strategies, such as emphasizing companies' offsetting practices and using nature-evoking visuals in advertising, have also been central during this period. All of these tactics serve to delay truly transformative action on climate. Above all, this discourse has served to strengthen oil and gas companies' hegemony and dictate a future entrenched in fossil fuel use.[92]

CONCLUSION

This chapter has explored the intentional efforts by specific actors to obstruct climate action in Italy through a range of tactics, strategies, and discourses. The weak—and in many instances nonexistent—climate commitments on the political, institutional, and corporate levels in the country show that these obstruction efforts have been at least partly successful in either delaying or hindering effective climate policies and tangible progress toward the national decarbonization targets necessary to meet international climate goals. In addition, whenever the country is hit by extreme weather events or climate change policies are under the spotlight, vested interests, individual climate deniers, and the media continue to fuel, circulate, and ramp up climate obstruction strategies and misinformation.

The problematic dynamics that have entrenched climate obstruction in the sociopolitical and cultural fabric of the country warrant further research. The interrelationships between the corporate world and politics in the Italian climate policy context is ripe terrain for further investigation, as are those between polluting industries and the media. Further research into how the processes of climate obstruction affects academic research into climate mitigation and adaptation, public perception of the climate question, climate legislation, and the decarbonization of polluting sectors is also recommended. Finally, further exploration of the best avenues for combatting current climate obstruction efforts will be necessary. These efforts may be constituted by a range of different strategies and processes, including social science studies, climate litigation, grassroots climate action, nationwide educational initiatives, and bans on misleading advertising and polluting industry sponsorships, among others.

NOTES

1. International Panel on Climate Change (IPCC) (2022), *Climate Change 2022: Impacts, Adaptation, and Vulnerability. Contribution of Working Group II to the Sixth Assessment Report of the Intergovernmental Panel on Climate Change.* Cambridge: Cambridge University Press.
2. J. Gütschow and M. Pflüger (2023), 'The PRIMAP-hist National Historical Emissions Time Series v2.4.2 (1750-2021)', zenodo, doi:10.5281/zenodo.7727475.
3. https://1p5ndc-pathways.climateanalytics.org/countries/italy/.
4. E. Bini (2014), 'A Transatlantic Shock: Italy's Energy Policies between the Mediterranean and the EEC, 1967–1974', *Historical Social Research/Historische Sozialforschung*, 145–164.
5. Climate Accountability Institute (2020, December) Carbon Majors 2018 Data 2020, https://climateaccountability.org/carbon-majors-dataset-2020/.
6. S. Levantesi (2022, 28 October), 'Professor Resigns from Research Center Over Partnership with Oil and Gas Major Eni', *DeSmog*, https://www.desmog.com/2022/10/28/unimib-eni-research-marco-grasso-resigns/.

7. Reteclima (2021, 30 April), 'Le Bufale sul Clima: Una Storia di Consapevoli Menzogne e di Disinformazione', *Reteclima*, https://www.reteclima.it/bufale-clima/.
8. G. Liva (2019, 21 November), 'Tra Politica, Media e Accademia: Com'è Fatta la Rete dei 'Negazionisti' Climatici Italiani', *Pagella Politica*, https://pagellapolitica.it/articoli/tra-politica-media-e-accademia-come-fatta-la-rete-dei-negazionisti-climatici-italian.
9. Reteclima, 'Le Bufale sul Clima'.
10. S. Caserini (2008), *A Qualcuno Piace Caldo: Errori e Leggende sul Clima che Cambia*. Milano: Edizioni Ambiente.
11. L. Ozzano and A. Giorgi (2016), *European Culture Wars and the Italian Case: Which Side are You On?*. Abingdon: Routledge.
12. M. Lövheim (2011), 'Mediatisation of Religion: A Critical Appraisal', *Culture and Religion*, 12, 2: 153–166.
13. N. Marzouki, D. McDonnell, and O. Roy (2016), *Saving the People. How Populists Hijack Religion*. London: Hurst Publishers.
14. Liva, 'Tra Politica, Media e Accademia'.
15. Greenpeace (2016), 'Rinnovabili nel mirino', http://www.greenpeace.org/italy/Global/italy/report/2016/clima/Rinnovabili_nel_mirino.pdf
16. R. Formica (2023, 28 March), 'La Debolezza dei Partiti ha Consentito all'Eni di Prendersi lo Stato'. *Domani*, https://www.editorialedomani.it/idee/commenti/la-debolezza-dei-partiti-ha-consentito-alleni-di-prendersi-lo-stato-ajwny1yh.
17. A. Aresu (2020), 'Enrico Mattei una Figura di Manager Pubblico', *Rivista Edita da Studiare Sviluppo Srl*, 3, 1: 41–63.
18. Fossil Fuel Climate Lobbying (2023, February), 'Update #5', *InfluenceMap*, https://influencemap.org/report/Fossil-Fuel-Climate-Lobbying-Update-December-2022-January-2023-21039.
19. ReCommon (2021, 28 June), 'Ripresa e Connivenza', *ReCommon*, https://www.recommon.org/ripresa-e-connivenza/.
20. Ibid.
21. M. Grasso and S. Vergine (2020), *Tutte le Colpe dei Petrolieri. Come le Grandi Compagnie Ci Hanno Portato sull'Orlo del Collasso Climatico*. Milano: Piemme.
22. Oil Change International (2022, 24 May), 'Big Oil Reality Check: Updated Assessment of Oil and Gas Company Climate Plans', *Oil Change International*, https://priceofoil.org/2022/05/24/big-oil-reality-check-2022/.
23. Autorità Garante della Concorrenza e del Mercato (2019, 20 December), *Autorità Garante della Concorrenza e del Mercato*, https://www.agcm.it/dotcmsdoc/allegati-news/PS11400provv.pdf.
24. Ibid.
25. Autorità Garante della Concorrenza e del Mercato (2020, 15 January), 'PS11400 – Sanzione di 5 Milioni di Euro a ENI per Pubblicità Ingannevole nella Campagna ENI diesel+', *Autorità Garante della Concorrenza e del Mercato*, https://www.agcm.it/media/comunicati-stampa/2020/1/PS11400.
26. Greenpeace (2023, 9 May), 'Portiamo ENI in tribunale per #LaGiustaCausa. Vogliamo giustizia per le persone e il Pianeta!', *Greenpeace Italy*, https://www.greenpeace.org/italy/storia/17735/portiamo-eni-in-tribunale-per-lagiustacausa-vogliamo-giustizia-per-le-persone-e-il-pianeta/.
27. S. Levantesi (2023, 9 May), 'Italian Oil Firm Eni Faces Lawsuit Alleging Early Knowledge of Climate Crisis. *The Guardian*, https://www.theguardian.com/envi

ronment/2023/may/09/italian-oil-firm-eni-lawsuit-alleging-early-knowledge-climate-crisis.
28. S. Levantesi (2023, 9 May), 'Italy's Eni Faces Lawsuit Alleging Early Knowledge of Climate Change', *DeSmog*, https://www.desmog.com/2023/05/09/italys-eni-faces-lawsuit-alleging-early-knowledge-of-climate-change/.
29. Ibid.
30. S. Levantesi and T. Lewton (2023, 24 February), 'Europe's Gas Lobby Exploits Energy Security Fears in Year Since Ukraine War', *DeSmog*, https://www.desmog.com/2023/02/24/european-gas-lobby-tweets-ukraine-war/.
31. InfluenceMap (2021, July). 'Industry Associations and European Climate Ambition', https://influencemap.org/report/Industry-Associations-and-European-Climate-Ambition-fdaeeb57dc404c90aaf2f82bbd729733.
32. S. Levantesi (2022), 'Italy's Lurch to the Right Raises Risk of Fossil Gas Lock-In', https://www.desmog.com/2022/09/21/italys-lurch-to-the-right-raises-risk-of-fossil-gas-lock-in/
33. S. Levantesi (2021), *I Bugiardi del Clima*. Roma: Laterza.
34. William F. Lamb, et al. (2020), 'Discourses of Climate Delay', *Global Sustainability*, 3: 1–5, doi:10.1017/SUS.2020.13.
35. Adelphi (2019), 'Convenient Truths', https://adelphi.de/de/publikationen/convenient-truths.
36. Levantesi, *I Bugiardi del Clima*.
37. Adelphi, 'Convenient Truths'.
38. Ibid.
39. Levantesi and Lewton, 'Europe's Gas Lobby'.
40. Greenpeace Italy and Osservatorio di Pavia (2022, September), 'Monitoraggio Dei Temi Ambientali Nei Primi 15 Giorni Di Campagna Elettorale', *Greenpeace Italy and Osservatorio di Pavia*, https://www.greenpeace.org/static/planet4-italy-stateless/2022/09/e724748f-report-greenpeace-campagna-elettorale-def.pdf.
41. S. Cantarini (2023, 28 July), 'Governo Italiano Propone una Maxirevisione del PNRR, Fuori dal Piano Miliardi Destinati al Dissesto Idrogeologico'. *Euractive*, https://euractiv.it/section/economia-e-sociale/news/governo-italiano-propone-una-maxirevisione-del-pnrr-fuori-dal-piano-miliardi-destinati-al-dissesto-idrogeologico/.
42. DeSmog, 'Competitive Enterprise Institute', https://www.desmog.com/competitive-enterprise-institute/
43. A. Scalari (2019, 2 December), 'Quelli che Fanno le Battaglie Pro Scienza e Poi Negano il Cambiamento Climatico', *Valigia Blu*, https://www.valigiablu.it/cambiamento-climatico-negazionisti/.
44. M. Lockwood (2018), 'Right-Wing Populism and the Climate Change Agenda: Exploring the Linkages', *Environmental Politics*, 27, 4: 712–732; K. M. Jylhä and K. Hellmer (2020), 'Right-Wing Populism and Climate Change Denial: The Roles of Exclusionary and Anti-Egalitarian Preferences, Conservative Ideology, and Antiestablishment Attitudes', *Analyses of Social Issues and Public Policy*, 20, 1: 315–335.
45. F. Ramella (2019, 24 September), 'Catastrofismo e Scienza: Un Approccio Equilibrato'. *Aspenia Online*, https://aspeniaonline.it/catastrofismo-e-scienza-un-approccio-equilibrato/.
46. Scalari, 'Quelli che Fanno'.

47. S. Levantesi and G. Corsi (2020, 6 August), 'Climate "Realism" Is the New Climate Denial', *The New Republic*, https://newrepublic.com/article/158797/climate-change-alarmism-greta-thunberg-naomi-seibt.
48. Caserini, *A Qualcuno Piace Caldo*.
49. Ibid.
50. Ibid.
51. Ibid.
52. N. Oreskes and E. M. Conway (2010), *Merchants of Doubt: How a Handful of Scientists Obscured the Truth on Issues from Tobacco Smoke to Global Warming*. London: Bloomsbury Press.
53. S. Levantesi (2022, 6 July), 'Chi Fa da Megafono ai Negazionisti Climatici', *Internazionale*, https://www.internazionale.it/opinione/stella-levantesi/2022/07/06/informazione-negazionisti-climatici
54. Ibid.
55. Ibid.
56. S. Levantesi (2021, 4 November), 'Italy's First Climate Lawsuit Seeks Bold Emissions Target in Effort to Protect the Planet and Human Rights', *DeSmog*, https://www.desmog.com/2021/11/04/italy-climate-lawsuit-giudizio-universale-human-rights/
57. G. Mercuri (2023, 25 January), 'Ma Dopo Algeri, Meloni Dovrebbe Fare Yrgentemente un Salto a Piombino', *Il Corriere della Sera*, Il Punto, https://www.corriere.it/rassegna-stampa/2023/01/25/piano-mattei-nodo-piombino-parole-papa-romanzo-cremlinale-mele-veleni-49f45606-9c02-11ed-b717-184306d51af5.shtml.
58. Greenpeace Italy (2022), 'L'Informazione sulla Crisi Climatica in Italia', https://www.greenpeace.org/static/planet4-italy-stateless/2023/04/3eac43b3-report-2022-linformazione-sulla-crisi-climatica-in-italia-odp-gp-def.pdf.
59. Banking on Climate Chaos (2023), 'Banking on Climate Chaos, 2023', https://www.bankingonclimatechaos.org/
60. Ibid.
61. Ibid.
62. D. Finamore (2022, 1 July), 'SACE's Opacity on Climate Will Not Guide Us Toward the Ecological Transition', *ReCommon*, https://www.recommon.org/en/saces-opacity-on-climate-will-not-guide-us-toward-the-ecological-transition.
63. V. Stackl (2023, 21 May), 'Italy Breaks Climate Promise to End Public Financing for International Fossil Fuel Projects, Publishing 'Worst-In-Class' Climate Policy', *Oil Change International*, https://priceofoil.org/2023/03/21/italy-breaks-glasgow-statement-climate-promise/.
64. Ibid.
65. Ibid.
66. L. Benjamin, A. Bhargava, B. Franta, K. Martínez Toral, J. Setzer, and A. Tandon (2022), 'Climate-Washing Litigation: Legal Liability for Misleading Climate Communications'. Policy Briefing. The Climate Social Science Network. https://cssn.org/wp-content/uploads/2022/01/CSSN-Research-Report-2022-1-Climate-Washing-Litigation-Legal-Liability-for-Misleading-Climate-Communications.pdf
67. Ibid.
68. Climate Reality Project (2018, 6 July), '3 Big Myths About Natural Gas and Our Climate', https://www.climaterealityproject.org/blog/3-big-myths-about-natural-gas-and-our-climate.

69. S. Levantesi (2022, 16 June), 'Climate Deniers and the Language of Climate Obstruction', *DeSmog*, https://www.desmog.com/2022/06/16/climate-deniers-fossil-fuel-language-obstruction/
70. M. Grasso (2022), *From Big Oil to Big Green. Holding the Oil Industry to Account for the Climate Crisis*. Cambridge, Ma., MIT Press.
71. Lamb et al., 'Discourses of Climate Delay'.
72. Ibid.
73. Levantesi and Lewton, 'Europe's Gas Lobby'.
74. Lamb et al., 'Discourses of Climate Delay'.
75. Ibid.
76. Ibid.
77. Levantesi, 'Climate Deniers and the Language of Climate Obstruction'.
78. Oreskes and Conway, *Merchants of Doubt*.
79. S. Levantesi (2022, 11 November), 'How the Fossil Fuel Industry Buys Goodwill', *DeSmog*, https://www.desmog.com/2022/11/11/fossil-fuel-industry-sponsors-art-sports-universities-greenwashing/.
80. J. Cook, S. Lewandowsky, and U. K. Ecker (2017), 'Neutralizing Misinformation Through Inoculation: Exposing Misleading Argumentation Techniques Reduces Their Influence', *PloS One*, 12, 5: e0175799.
81. Oreskes and Conway, *Merchants of Doubt*.
82. S. Lewandowsky (2021), 'Climate Change Disinformation and How to Combat It', *Annual Review of Public Health*, 42: 1–21.
83. Cook, Lewandowsky, and Ecker, 'Neutralizing Misinformation Through Inoculation'; and A. M. McCright, M. Charters, K. Dentzman, and T. Dietz (2016), 'Examining the Effectiveness of Climate Change Frames in the Face of a Climate Change Denial Counter-Frame', *Topics in Cognitive Science*, 8, 1: 76–97.
84. J. S. Damico and M. C. Baildon (2022), *How to Confront Climate Denial: Literacy, Social Studies, and Climate Change*. New York: Teachers College Press.
85. Oreskes and Conway, *Merchants of Doubt*.
86. K. Ekberg, B. Forchtner, M. Hultman, and K. M. Jylhä (2022), *Climate Obstruction: How Denial, Delay and Inaction are Heating the Planet* (1st ed.). London: Routledge.
87. G. T. Farmer, J. Cook, J., G. T. Farmer, and J. Cook (2013), 'Understanding Climate Change Denial'. In: Kristoffer Ekberg, Bernhard Forchtner, Martin Hultman, Kirsti M. Jylhä (eds.), *Climate Change Science: A Modern Synthesis: Volume 1 – The Physical Climate*, pp. 445–466.
88. P. Consorti (2020), 'The Role of Law and the Political Use of Religion. The Italian Experience from «Mani Pulite» to the Rise of the Lega', *Quaderni di Diritto e Politica Ecclesiastica, Rivista Trimestrale*, 2: 309–320.
89. M. Hoofnagle (2007, 30 April), 'Hello Scienceblogs', *Denialism Blog*, http://scienceblogs.com/denialism/about/.
90. G. Supran and N. Oreskes (2021), 'Rhetoric and Frame Analysis of ExxonMobil's Climate Change Communications', *One Earth*, 4, 5: 696–719.
91. Lamb et al., 'Discourses of Climate Delay'.
92. C. Wright, R. Irwin, D. Nyberg, and V. Bowden (2022), 'We're in the Coal Business': Maintaining Fossil Fuel Hegemony in the Face of Climate Change', *Journal of Industrial Relations*, 64, 4: 544–563.

12

Climate Obstruction in Spain

From Boycotting the Expansion of Renewable Energy to Blocking Compassion Toward Animals

JOSE A. MORENO AND NÚRIA ALMIRON

INTRODUCTION: MORE EFFORT IS NEEDED

Spain has a long track record of environmental pollution tolerated by the political elites. Despite EU pollution directives dictating that the 'polluter pays', the public sector has taken responsibility for the costs and protected the polluters.[1] This principle has been applied not only to toxic discharges but also to greenhouse gas (GHG) emissions. For example, while most EU countries reduced their GHG emissions and complied with the Kyoto framework, Spain was one of the few that fell further short of its targets.[2] Specifically, GHG emissions in Spain have followed this trend, rising to almost 300 million metric tonnes of carbon dioxide equivalents (MMT CO_2e) by 1990, peaking in 2007 at around 450 MMT CO_2e, and then falling to around 300 MMT CO_2e at the beginning of the 2020s, as shown in Figure 12.1.

Since 2007, decarbonization in Spain has been led by the progressive dismantling of fossil fuel sources such as coal-fired power plants and circumstantial factors such as the economic crisis that put the brakes on the nation's industry. The EU National Energy and Climate Plans (NECPs)

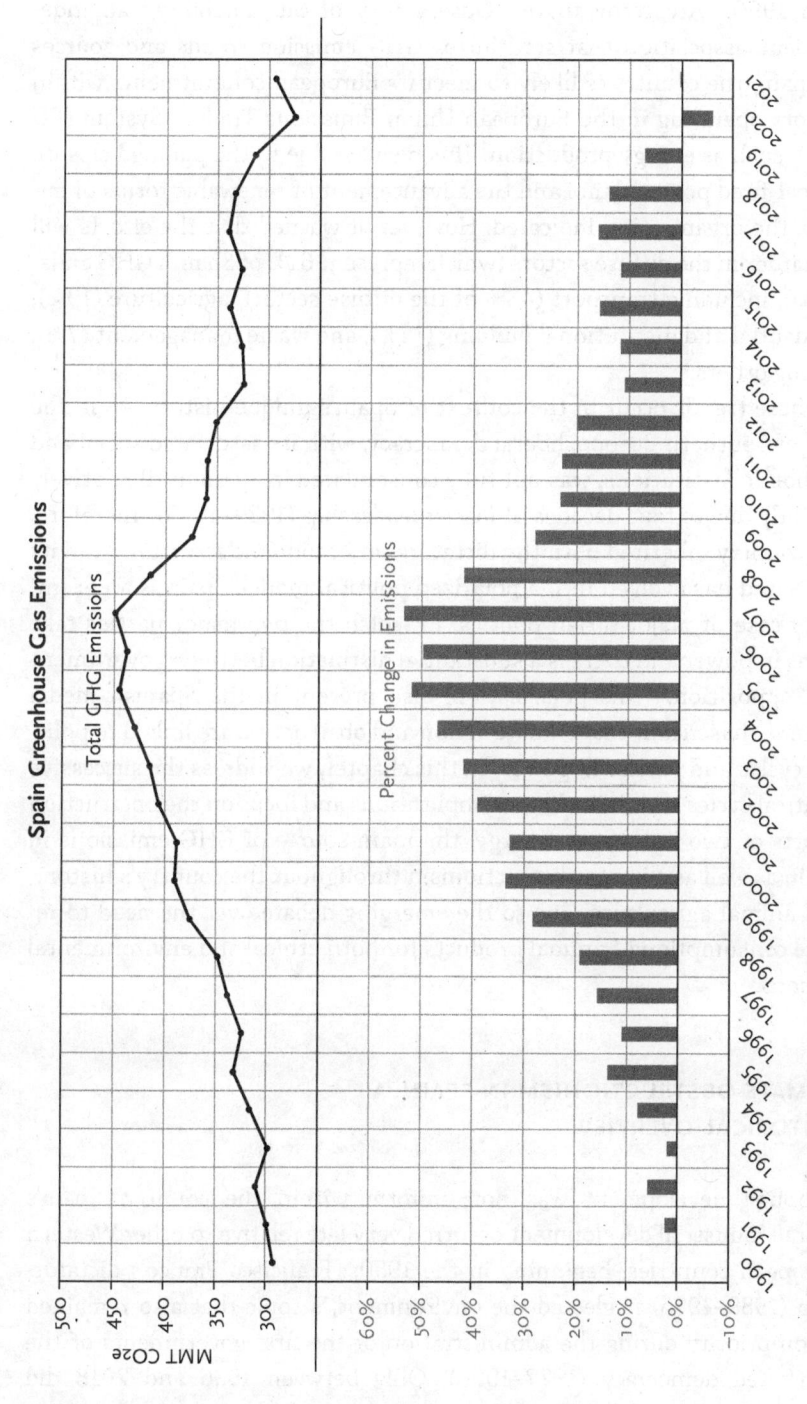

Figure 12.1 Total greenhouse gas (GHG) emissions (in MMT CO_2e) and percentage change in emissions in Spain between 1990 and 2021, inclusive.
Source: Total GHG emissions based on data provided by Gütschow and Pflüger (2023) for Kyoto Six Greenhouse Gas Totals.

stipulate that Spain must reduce its GHG levels by 23% by 2030 compared with 1990.[3] According to the Observatory of Sustainability,[4] an independent association that scrutinizes GHG emission trends and sources in Spain, the country is likely to meet its European commitments within sectors operating in the European Union Emissions Trading System (EU ETS), such as energy production. This result is due to the planned closure of coal-fired power plants and the advancement of renewable forms of energy, the organization indicated. However, it warned that the efforts will be harder in the diffuse sectors (which represent 61% of Spain's GHG emissions), including transport (43% of the diffuse sector), agriculture (19%), residential and institutional building (14%), and waste management (7%), among others.[5]

These trends occur in the context of Spain's unique history. As in the rest of Southern Europe, liberal democracy, with its associated social and economic institutions, was not fully consolidated in Spain until relatively recently due to four decades of fascist leadership (1939–1975). Therefore, the country inherited both the dictatorship's economic structures and its elites and has evolved into a polarized political model. Spain is an exemplary case of majoritarian politics, in which the two major parties take turns in power and there is a substantial distinction between government and opposition.[6] This polarization, also present in the Spanish media system,[7] has meant that climate action and obstruction are linked to political cycles and parties in power. In this chapter, we address the successive political cycles and their climate implications and focus on the obstruction efforts of two industries: energy, the main source of GHG emissions in Spain as well as climate obstructionism throughout the country's history, and animal agriculture, due to the emerging debate over the need to reduce consumption of animal products for both ethical and environmental reasons.

CLIMATE OBSTRUCTIONISM IN SPAIN: A HISTORICAL OVERVIEW

Although development was not uniform within the country, Spain's overall industrial development occurred very late relative to other Western European countries, beginning in the 1950s. Francisco Franco's dictatorship (1939–1975) neglected the environment, a topic that also remained a non-priority during the administration of the first governments of the reinstated democracy (1977–1996). Only between 1996 and 2018 did climate awareness emerge in the country, together with explicit climate obstructionism. And, from 2018 to 2023, climate action has been marked by culture wars.

Genealogy of environmental obstructionism in Spain (1939–1996)

Franco's fascist administration applied tactics to ignore, deny, hide, and absolve polluting industries of the environmental impacts they produced, as described by environmental historian Pablo Corral-Broto.[8] The Franco government used experts to generate doubt about the evidence of environmental pollution, a technique that Corral-Broto notes is in line with the thesis of Oreskes and Conway's book *Merchants of Doubt*.[9] The dictatorship sought to portray an image of modernity and industrial prosperity, a goal to which pollution complaints were an obstacle. This authoritarian past helps to explain Spain's late implementation of ambitious climate policies. Spain is one of several European countries that later freed themselves from fascism, a feature that has affected the quality of contemporary democracies in Southern Europe.[10] Francoism's sociological and structural legacy reflects what Franco biographer Paul Preston[11] has called the 'institutionalized pillage' of the dictatorship. When liberal institutions and democracy eventually developed in Spain, they did so under the auspices of neoliberal capitalism, which placed former non-democratic elites in control of key sectors of the economy—what some authors have defined as *authoritarian liberalism*.[12] Thus, Spanish elites' ideological alignment toward the old regime's value of environmental neglect was the starting point for delay on climate action in this country.

The first democratic administrations in Spain managed key sectors, such as energy, with the economy rather than the environment as a priority. An example of this trend was the speculative bubble in nuclear energy generated during the Franco regime, which ended with a costly moratorium on the construction of new plants during the government of the social democrat Felipe González (1982–1996). The privatization of the state-owned energy company Endesa (founded in 1944) is another case of interest in this period due to its role as one of the largest GHG emitters in Spain. Part of the company's stock was sold to private investors in 1988, with the Socialist Party (PSOE) still in power, while a conservative government (Popular Party, or PP) completed Endesa's privatization in 1997.[13] This company grew and expanded thanks to state support but was then progressively transferred to private investors, a paradigmatic example of the impact of neoliberal policies on the energy sector in Spain.[13]

Pioneering renewables and the penalization of self-generated electricity (1996–2018)

Between 1996 and 2004, the PP governed Spain under President José María Aznar, whose administration signed the Kyoto Protocol in May 2002. Later, during José Luis Rodríguez Zapatero's PSOE government (2004–2011), former Aznar administration members positioned themselves as climate sceptics. The ideological hub of the PP was, at the time, the Foundation for Analysis and Social Studies (FAES), a think tank founded by Aznar in 2002. Although this think tank has not been an overt source of climate obstruction, it published in Spanish some climate obstructionist books by relevant European politicians, such as *Planeta Azul (No Verde)* (Blue Planet in Green Shackles) by Vàclav Klaus (2008) and *Una Mirada Fría al Calentamiento Global* (An Appeal to Reason: A Cool Look at Global Warming) by Nigel Lawson (2009).

For a short period during the 2000s, Spain was a pioneer in installing renewable forms of energy in the European Union.[14] Aznar's outgoing conservative government had established the first policy to incentivize renewable energies in 2004, in hope of promoting private investments in this area,[15] and, during the next socialist administration (Zapatero's, from 2004–2011),[16] incentives grew. As a result, in 2008, Spain found itself amid a renewables speculative bubble that coincided with the eruption of a global economic crisis that also affected the electricity market; consequently, the PSOE government adjusted the renewable energy subsidies. In 2012, Mariano Rajoy's PP administration introduced a moratorium on the construction of new installations and, in 2013, ended the policy mechanism designed to accelerate investment in renewables. The reasons for the dismantling of renewable energy promotion varied from poor policy design to the political economy of the sector[14] but also included obstructionist strategies: 'Lobbying by the incumbent energy utility companies played a major role in driving the policy dismantling process' because 'they perceived renewables as unwanted competition and as a threat to their business models'.[14]

The 2010s in Spain were characterized by the austerity government of Mariano Rajoy's conservative party (2011–2018). Rajoy had a poor understanding of climate science, as shown by his previous statements. For example, in 2007, when his party was the political opposition, he said that he trusted his cousin—a physicist sceptical about anthropogenic climate change—and that if humans can barely predict the weather, how much more difficult must it be to predict the climate?[17] Once in office, Rajoy's government instituted a tax on solar photovoltaic self-generation (the 'sun tax') in 2015. The 2012 moratorium on renewable energy development combined with the brake on self-generation paralyzed the development of renewables, especially photovoltaics, in Spain.[18] Critics argued[19] that this decision favoured the interests of the energy oligopoly—a market controlled by a few companies, such as Endesa, Iberdrola, Naturgy, Repsol, and Energias de Portugal (EDP)—which could continue to operate their existing power plants profitably and for whom household PV self-generation could be a threat.

The climate culture wars (2018–2023)

One of the first decisions taken by Pedro Sánchez's first PSOE administration when it came into power in 2018 was to repeal the 'sun tax'. This period has been marked by the diversification of climate measures at various political levels, which in turn increased opposition and delay-based obstructionism. Although Sánchez's government can be considered the most ambitious administration on climate issues in Spain's history, it wasn't without controversy. An example of this was the poor reception of COP 25, organized by Chile in Madrid in December 2019. On this occasion, Spain presented itself as an international flag-bearer in the fight against climate change while allowing the country's most prominent GHG-emitting energy companies, such as Endesa and Iberdrola, to sponsor the summit. COP 25 did not produce any substantial results, negotiations between countries were unsuccessful, and the summit received criticism for greenwashing.[20]

Nevertheless, Spain eventually acted during this period. One of the milestones of the second Sánchez administration, in coalition with the left-wing party Unidas Podemos (2020–2023), was the approval in 2021 of the Climate Change and Energy Transition Bill. This law set the goal of achieving a 23% reduction in Spain's emissions levels compared with 1990 by 2030, as required by the European Union. The law aroused both enthusiasm and mistrust from environmental groups, who considered it necessary but insufficient. During the legislative process, the conservative PP abstained from supporting the law, and the far-right party Vox emerged as an obstructionist stakeholder. In 2020, the latter presented a controversial amendment to the entire law, which repeated the denial, delay, and policy sceptic arguments disseminated for years by the climate countermovement.[21]

In addition to the central government, other Spanish governments, including autonomous communities and city councils, implemented strategies to tackle climate change during this period. A relevant case was the creation of low-emission zones, restricted traffic areas based on vehicle pollution levels in the urban centres of large cities, including Madrid and Barcelona. This measure also became a politically polarized cultural flashpoint. In Barcelona, Mayor Ada Colau (2015–2023) of the left-wing Barcelona en Comú party focused on pedestrianizing streets and banning high-emissions vehicles from entering the city. In Madrid, left-wing Mayor Manuela Carmena and her Ahora Madrid coalition pursued the same strategy between 2015 and 2019. However, since the 2019 elections, conservatives used the growing fight against car bans as a marker of right-wing identity. For example, in the 2023 municipal elections, the campaign of right-wing parties in some cities turned the elimination of bike lanes into an ideological issue.[22]

Thus, climate action in Spain became a polarized cultural issue from which all sides sought political capital. This process occurred with the help of equally polarized media and interest groups. Unsurprisingly, the conservatives (PP) and the far right (Vox) pushed the culture war narrative most actively. For example, PP's president of Madrid's region, Isabel Díaz Ayuso (2019–), positioned herself against the upcoming ban on combustion cars in 2035, arguing that the environment 'cannot be an excuse' to control the population.[23] Ayuso also used retrograde tropes of climate obstructionism, such as the idea that climate change is natural or that climate action favours communism, as her discourse in Madrid's regional parliament on 10 November 2022 showed.[24]

In short, although climate awareness grew in Spain during the period, socioeconomic privileges, political polarization, and entrenched habits still played a major role in a country where neoliberalism had merged with a fascist inheritance. As might be expected to be the case in the Southern European context, the history of climate obstructionism in Spain has been interlinked with politics and neoliberal authoritarianism. Over time, power distribution among obstructionist stakeholders and their strategies have varied. However, the obstructionists all share an outdated notion of progress that places economy and ideology ahead of the environment.

THE FIVE TYPES OF OBSTRUCTIONIST STAKEHOLDERS IN SPAIN

Politicians

The first group of obstructionist stakeholders in Spain is politicians. The conservative PP stands out for rejecting climate policies and even casting doubt on the scientific evidence on climate change, as noted in this chapter's history section. This is the party that inherited the values of the previous authoritarian regime, including corruption and use of administration and state at the service of the country's economic elites—though the latter trait is also shared by the socialist party, PSOE. As described earlier, the conservative party rode the sceptic wave in the 2000s, obstructed electricity self-generation in the 2010s, and made political capital from the culture wars on climate change in the 2020s.

In this vein, the emerging far-right party Vox stands out for its populist climate obstructionist discourse. Vox's electoral platform in the national elections of July 2023 reaffirmed the party's position of the harshest form of obstructionism: it included the abandonment of the Paris Agreement and the derogation of the Climate Change Law, arguing with nationalist rhetoric that climate measures are against the interests of Spain and especially of its rural areas. However, Vox's climate discourse was contradictory until 2023,[25] the last period studied, so it remains to be seen whether it will moderate its position to capitalize on eco-nationalism as other similar parties have done in Europe.

Although not comparable with the aforementioned parties, the socialist party PSOE can also be counted among the obstructionist stakeholders in Spanish politics. Despite its climate policymaking efforts during the two Sánchez administrations, including the first Climate Law, the outcomes were mixed. These efforts seemed to be more a response to the need to 'jump on the EU bandwagon'—including the push and funding from the European Union for climate action—than to a strong commitment.[26] They were described as unambitious by environmental activists; this lack of ambition can be seen in the PSOE's support for an expansion of Barcelona's airport despite the need to decrease aviation due to its climate impact and the possibility that it could affect protected land.[27] Indeed, the coalition government was sued for climate inaction in 2020 by three green nongovernmental organizations, although that lawsuit was dismissed by the Supreme Court in 2023.[28] In Spain, unlike in other European countries, there is no standalone Green Party. Factions of this type tend to form coalitions with left-wing parties such as Podemos, which has had some influence as part of the PSOE coalition government with Sánchez, but not enough to implement strong and decisive climate policies.

Industry

A second set of obstructionist stakeholders comprises the most polluting industries, usually organized into trade groups. Among the most influential are the Spanish Electricity Industry Association (AELEC, formerly UNESA), which holds a dominant position in the market by representing the interests of major companies such as Endesa, Iberdrola, Naturgy, and EDP; the Nuclear Forum, which unites the interests of the eight active nuclear plants in Spain; Sedigás, which represents the interests of the gas sector for all transportation, storage, and marketing phases; Carbunión, which covers the coal mining sector and lobbies to delay the closure of coal-fired plants; and the Spanish Association of Oil Product Operators, which represents oil companies such as Repsol, Cepsa, and BP. Through these trade organizations and their individual public relations strategies, the energy sector has historically lobbied for energy regulation in Spain to be more favourable to their businesses than to environmental criteria.

Animal agriculture is another significant emissions-intensive sector with extensive lobbying activity. In this area, Interporc stands out in defending the interests of the pig farming sector, one of the most prominent in both revenues and pollution generated.[29] Another relevant organization is ANICE, the National Association of Meat Industries in Spain, which represents slaughterhouses and meat-processing plants. The lobbying success of such animal trade groups is demonstrated by the fact that they continue to be subsidized by the Spanish state and the European Union despite the environmental, ethical, and health impacts of their business. To maintain direct and indirect economic EU aid from the Common Agricultural Policy (CAP) is one of their main goals.[30]

Think tanks

A third set of climate obstructionists can be found among think tanks. These organizations have had less relevance in Spain than in other countries but still have been able to influence or align with political parties and industry to distract the climate debate. Most think tanks are neoliberal and libertarian, with a focus on the dissemination of anti-state interventionist ideas. Notable ones include FAES, the think tank linked to the conservative PP; the Juan de Mariana Institute, the most prominent in Spain for generating explicitly climate-obstructionist content; and the Civismo Foundation. The latter two are linked to the Atlas Network, a US-based network of conservative international think tanks. There are also think tanks directly related to industry. One such is Enerclub, which spans different companies from the Spanish energy sector, including the most polluting ones such as Repsol, Endesa, Iberdrola, Naturgy, and EDP. Enerclub prepares reports and analyses and makes recommendations to advance these companies' positions and increase their profits in the electricity market. Another case is that of InLac, the Interprofessional Dairy Organization, which not only lobbies policymakers to serve their interests but also strives to convince the population of the need to consume dairy products daily.

Media

The fourth obstructionist stakeholder in Spain is the news media. In the Spanish press, climate scepticism has been limited, at least until the 2020s (i.e. it was concentrated within conservative or economics-oriented publications and within the opinion genre).[31] In general, political parallelism—in which the media system reflects the structure of the political system—has resulted in the topic of climate change being subject to a polarization of views in the press similar to that found in politics. The media landscape has become a breeding ground for obstructionism, with some media content depicting climate action as an ideological issue to legitimize inaction when discussing, for example, private car use versus bicycle lanes, or meat consumption. Additionally, Spanish media coverage has followed the same cycles of attention and neglect seen in other countries of the Global North.[32] Political summits, mobilizations, and scientific breakthroughs have dominated the media's focus. This event-driven coverage has avoided deeper explorations of the systemic nature of the climate crisis and possible solutions, such as the connection between dietary choices and the condition of the environment. This link has been a commonly shared omission in all media until recently. The *carnist* mindset—defined by psychologist Melanie Joy as the cognitive bias that makes us think of meat-eating as normal, natural, and necessary[33]—has produced climate coverage in some Spanish media in which the environmental impact of animal agriculture has been neglected and, at times, even ridiculed.[34]

Citizens

The fifth sphere of obstructionism in Spain is its citizenry. According to Eurobarometer, 81% of Spaniards see climate change as a fundamental problem (3% above the EU average), but the proportion of people who change their habits accordingly is much lower.[35] The most significant climate action Spaniards perform is recycling (78% say they do so, compared with 75% for the EU average). In contrast, more impactful activities such as reducing meat consumption are undertaken by only 15% in Spain (31% in the European Union); sustainable mobility is practiced by 26% (30% in the European Union), and only 3% practice sustainable tourism when planning a trip (11% in the European Union). The lack of cultural deconstruction of their polluting privileges—acknowledging that car use, meat consumption, air travel, and even faith in economic growth as the definition of progress can be harmful to the environment—makes Spanish citizens believe they are engaging in climate action when they are actually delaying decisive changes.

STRATEGIES TO OBSTRUCT CLIMATE POLICIES

In this section, we summarize the main strategies used in Spain by the top obstructionist stakeholder, polluting industries. These include influencing regulation and/or public opinion directly or indirectly through the media. The main tactics used by these climate-obstructionist industries include lobbying ('in house' by the company itself, externally by hired firms, and in coalition with other organizations in the sector), gaining access to politicians' influence via 'revolving doors', giving policy briefings; sponsoring favourable scientific research, disseminating favourable information (by, e.g., generating it through think tanks), media ownership infiltration, threatening legal action, and jumping on the climate bandwagon through greenwashing, corporate advertising, and publicity stunts. Though their tactics have been varied, the following are examples of some of the more prominent and effective of these tools.

Lobbying

The main strategy used by industry is lobbying policymakers to create the best regulatory atmosphere for their business needs, including preventing regulations from being passed and promoting or amending certain regulations. To help ensure success, companies hire former legislators or regulators and/or promote some of their corporate members to run for office or work in the administration. This close relationship between the private sector and the government—almost nonexistent among green non-governmental organizations (NGOs) lobbying for climate action—grants companies privileged access to the administration and its functioning. This revolving-door reality has been described as a common practice among the most polluting industries in Spain.[36]

Since the restoration of democracy, the two major parties that have taken turns leading the government have assumed power positions in energy companies before and after holding office. For example, the first president of the PSOE government ended up on the board of Gas Natural (Naturgy), and the first president of the PP government became an advisor to Endesa after leaving public office. Many cabinet members of various governments since 1977 have been directors, board members, or advisors of Endesa, Iberdrola, Naturgy, Repsol, and Enagás, dominant companies in the fossil fuel sector in Spain. According to criminologist Mònica Pons-Hernández,[36] 'the lack of experience, the high remunerations, and the previous connections with the private companies lead to questioning whether individuals involved in revolving out are selling their access or expertise.'

Legal action

A second industry tactic, important to mention because of its novel use in Spain, is legal action as a form of climate obstruction. Litigation is not as ingrained in Spanish culture as it is in English-speaking countries, but it is emerging amid the culture wars over issues such as meat eating. Recently, the animal agriculture lobbies have begun to use legal action to prevent widespread dietary changes and thereby reinforce the meat culture in Spain. In 2021, for example, the meat lobbies Provacuno and Interporc brought[37] a vegan food company, Foods for Tomorrow, to court over its campaign promoting vegan 'hamburgers' as a more climate-friendly dietary option via the slogan 'A meat hamburger pollutes more than your car'. The meat trade groups denounced the campaign, saying that it constituted unfair and misleading advertising, and, in 2022, a court ruled in its favour, arguing that the slogan met the criteria for misleading advertising and constituted a denigration of meat products. Foods for Tomorrow was forced to withdraw the campaign (although it was no longer running) and ordered never to repeat similar statements. The court also stated that the evaluation of the environmental sustainability of production processes is too complex for a vegan company to be able to state definitively that their products have a smaller environmental impact than meat products.[37] The Spanish meat lobbies thus successfully prevented a brand from communicating to the public the environmental externalities of meat and created a threat against anyone trying to do the same.

Media co-optation

A third industry tactic involves the media sector. Through ownership, sponsorship, and advertising, the Spanish media system has become vulnerable to corporate interference. This problem has been compounded by deteriorating working conditions due to the 2008 economic crisis, resulting in precarious jobs and a lack of expert staff in newsrooms, including a lack of expertise in issues such as the environment. This situation has made the media vulnerable to the power of vested interests, including those of the polluting industries, either directly or through their lobbies and funded think tanks. For example, Grupo PRISA, one of the largest media groups in Spain, has long incorporated stakeholders with fossil fuel interests among its shareholders.[38] Also, during the COP 25 event hosted by Madrid and Chile, Endesa and Iberdrola were both sponsors and dominated many media outlets by, for example, purchasing ads on the front pages of the country's newspapers and generating coverage marked by greenwashing.[18]

Another example of a tactic used by obstructionists' industrial interests in Spain is attempting to pass responsibility for GHG emissions on to the public. While the citizenry, through changing consumption patterns and other habits, can play a key role in climate protection, most of the responsibility and power to act rest with industry. However, under the guise of knowledge dissemination and climate education, some industries try to delegate this responsibility to individual citizens. For example, the Spanish oil company Repsol, one of the biggest GHG emitters in Spain, in 2022 launched the Green Engine emissions compensation project.[39] Under this initiative, citizens can calculate their GHG emissions and offset them by supporting reforestation projects. This type of project seeks to instill in citizens the responsibility to act while the company continues to profit from the extraction, refining, and commercialization of fossil fuels.

Greenwashing

A final major obstruction tactic is greenwashing. Many industries have jumped on the bandwagon of climate concern in their corporate rhetoric and public discourse while also lobbying against degrowth and sustainability—the very things needed to protect the environment—behind the scenes. For example, Spain's tourism sector officially holds a very pro-climate-action position and has partnered with the United Nations Framework Convention on Climate Change (UNFCC) to combat climate change through its World Travel and Tourism Council (WTTC), the industry's largest international lobby. However, some of WTTC's members continue to lobby for increased tourism despite its large environmental footprint. A remarkable example of the success of this trend is the millions in subsidies Ryanair, the ultra-low-cost Irish tourist airline, has received from Spanish authorities.[40]

THE DISCURSIVE FRAMINGS DISTORTING THE CLIMATE DEBATE

Overt obstructionist discourse has been deployed in Spain mostly through the use of discursive frames similar to those used in other European countries and the United States. We focus here on two cases,[41] one involving the right-libertarian think tank Juan de Mariana Institute (JMI) and the other the animal agriculture lobbies, to exemplify these tactics. The first occurred during the period of direct opposition to climate action in the 2000s and 2010s; the second, during the climate culture wars of the 2020s.

Juan de Mariana Institute: Right libertarianism and climate obstruction

Because of its international networking, the JMI has played a role in the global climate countermovement. In 2009, with Universidad Rey Juan Carlos (Madrid), the JMI published a report 'Study of the effects on employment of public aid to renewable energy sources', arguing that the 'green employment' policies of Zapatero's socialist administration had been a fiasco. These policies encouraged industries such as energy, construction, and transportation with subsidies to meet certain environmental criteria. The US Institute for Energy Research (IER), a think tank whose work focuses on fossil fuel interests, sponsored the study.[42] Obstructionist organizations in the United States, such as Heritage Foundation, American Enterprise Institute, and the American Energy Alliance, used the report to oppose then-President Obama's environmental policies.[42] The 'Spanish case' also inspired other obstructionist think tanks to produce similar reports in Germany, Denmark, Italy, and the United States. The circulation of these European reports generated tensions in these countries' embassies in the United States as they tried unsuccessfully to manage this flow of misinformation.[42]

The JMI has become the most overt and prolific think tank in Spain in disseminating obstructionist messages about climate action. As a right-libertarian organization, the JMI supports free market capitalism, civil liberties, individual rights, and minimal government intervention. Its concern over the concentration of political power is evident in its first criticism of climate action: a 2005 opinion piece (op-ed) about the Kyoto Protocol, which criticized climate science on the grounds that it would lead to a global government. Between then and April 2022, the institute published approximately 180 articles concerning climate change, mostly op-eds. Through an inductive analysis[43] of the focus on climate issues in JMI's blog posts, we found recurrent themes: undermining climate science (33% of the texts), questioning climate policies (22%), and criticizing climate action supporters (22%). The publication of these texts peaked in 2007 and again in 2009, with another, smaller rise in 2019.

As for its critique of climate science up to the 2020s, the institute has maintained almost the same frames it has used since its inception, though it published many fewer texts after 2010. For example, in 2005, the JMI was featured at the 'Global warming: myth or reality' seminar held by the Rafael del Pino Foundation on 16 May in Madrid. There, the JMI president co-moderated a session with the president of the Science and Environmental Policy Project, a US advocacy group that aims to undermine climate science and helped to organize the Nongovernmental International Panel on Climate Change (NIPCC), a group of climate science contrarians. As recently as 2015, their blog[44] featured a counterargument appealing to the myths they promote, namely that the planet has been warm in other past periods; CO_2 is not the most common GHG; extreme weather events are less common now than in the past, not more; CO_2 is not a toxic environmental pollutant; and scientists' agreement on the anthropogenic cause of climate change is not true. In the discourse of this think tank, climate science is a religion, as expressed in a 2019 blog post.[45]

The frames the JMI has used to criticize climate policies (the main focus, as noted, of 22% of the 180 blogs we analysed) show a recurrent pattern of opposition to such measures on the grounds that they would involve state intervention—a theme that repeats across time. The texts range from a 2007 article 'The fight against climate change costs a lot of money'[46] to a 2019 op-ed, 'Taxes on sin and climate change',[47] in which taxes in general, including climate taxes, are defined as 'robbery' and thus harmful to the economy.

JMI's targeting of activists and politicians promoting climate action (the main approach of another 22% of the blog posts studied) includes questioning the climate actions of members of the PSOE government, activists from environmental organizations such as Greenpeace, and international figures such as Greta Thunberg and Al Gore, the latter of whom the think tank severely criticized[48] in the mid-2000s, the period during which his documentary *An Inconvenient Truth* was in movie theaters.

Other frames (the remaining 23% of the blogs analysed) included 12% that mentioned climate change governance without a specific obstructionist focus, such as commentaries on social and political events. Also, 4% of the texts reflected nuclear expansionism and techno-optimism as their primary framing, despite the unfeasibility or unavailability of these technologies. In addition, 3% of the texts question renewable energy because of its cost or alleged non-viability—especially during the 2000s, when the speculative bubble in renewables emerged in Spain. Similarly, another 3% of the texts used frames centred on the idea that current lifestyles require fossil fuels.

Although the JMI is not highly visible in the media, it has gained some popularity in the niche world of the liberal economic press. This pattern can be demonstrated by the fact that up to 61% of the blog posts on climate change on the JMI website were reprints of articles first published in other media outlets by think tank members. Here, the online newspaper *Libertad Digital* stands out for having published 51% of the 180 texts in question. Several members of the JMI are regular columnists for *Libertad Digital*, which could explain why they have been able to introduce their discourse in the newspaper. This online outlet is another proponent of neoliberal libertarianism that adheres to criticism of climate science. For example, after the controversy over the emails from the Climate Research Unit of the University of East Anglia ('Climategate'), *Libertad Digital* published a column by a JMI member stating that 'In Spain we have been alone for many days informing readers of this pseudo-scientific scandal subsidized by politicians'.[49]

Provacuno and Interporc: Fuelling meat culture in Europe with public funds

The meat industry is Spain's largest revenue sector within the agri-food business[50] in a country with the highest meat consumption per capita of all EU member states.[51] The animal-food industry promotes and protects the meat culture that sustains this high consumption in Spain. Animal-rich diets everywhere are the outcome of technological changes and marketing efforts in the twentieth century allowing for widespread production of animal-based foods and triggering a 'meatification' of diets.

In Spain, meat consumption has been in a slow but steady decline since 2008 (except for 2020). Industry lobbies have been making considerable efforts to counteract the increasingly negative image of their products among consumers in what has already become another culture war. This industry has been spreading a counter-discourse to undermine the scientific evidence connecting meat consumption to health problems[52] as well as to climate change: the meat industry accounts for 70% of all agricultural emissions in Spain.[53] A relevant episode occurred in July 2021, when the Ministry of Consumer Affairs, under the control of the Unidas Podemos political coalition, launched a campaign on reducing meat consumption. Industry associations tried to distort the debate using arguments about the alleged environmental sustainability and social importance of meat production[54] while the prime minister himself, Pedro Sánchez (PSOE), ridiculed his government partners' campaign, declaring that steaks are 'unbeatable'.

A case of particular interest in the greenwashed reframing of the meat industry's role in Spanish society is the discourse used in the 'Proud of EU beef' campaign, promoted by Provacuno in Spain and APAQ-W in Belgium under the financial sponsorship of the European Commission.[55] In 2019, the EU Directorate General for Agriculture and Rural Development granted €3.6 million to this project, of which more than €2 million were allocated to the Spanish lobby 'to incite the consumers not to have a stereotyped idea about red meat and to enable them to be again confident about their consumption decision'. The campaign aimed to cleanse the industry's image in the wake of the criticism the meat sector had increasingly received during the last decades for its animal exploitation, the human health impacts of its products, and, more recently, its contribution to GHG emissions. The campaign targeted Spain, France, Portugal, Germany, and Belgium. As the organizers stated, their aim was not only to show the 'benefits of the product but to make the consumer feel identified and supported in its choice regarding it'. In their description, they acknowledged that information about meat consumption had affected consumers and that they wanted to make people feel comfortable again about eating animals.

The controversial nature of this campaign prompted thirty-four MEPs to issue in 2020 a letter of protest[56] led by Portuguese MEP Francisco Guerreiro, a member of the Greens/European Free Alliance. The letter pointed out the campaign's incompatibility with the Commission's objectives in its Farm to Fork and Green New Deal strategies, albeit it was approved shortly before the launch of these green agriculture programmes. In particular, the letter noted that the campaign was against the interests of the EU because Spain 'has the highest per capita meat consumption per year (over 100kg), meaning national consumption of meat would need to drop by 76% by 2030 to keep our climate goals within reach'.

The campaign also received harsh social criticism, prompting the organizers to change their approach slightly. Previously, the campaign included advertising content that reflects the thesis of Carol J. Adams in *The Sexual Politics of Meat*,[57] which discussed the objectification of the female body in meat-oriented communication. For example, one ad had featured a close-up of a woman wearing lipstick biting into a piece of meat; another pictured a slim, feminine-looking waist to try to sell the idea that a meat-based diet is balanced and healthy—images that Provacuno later withdrew.[58] In another example, a poster shows a woman with cooking utensils and a picture of a salad with meat on top, on one side, and, on the other side, a seated man being served a big steak accompanied by a few vegetables.[59]

The strategy of obtaining public funds to promote meat consumption was also followed by Interporc, which, in coalition with other meat industry lobbies in Portugal and France, obtained more than €2 million from the European Commission.[60] In this case, the campaign 'Let's talk about pork' warned that pig flesh consumption is decreasing for sustainability and animal compassion reasons. Accordingly, the campaign sought to inform the public 'about the reality of production in the entire pork chain'. Specifically, it aimed to show 'the conditions of production in the farms with scrupulous respect for the highest standards of animal welfare, and then proving the sustainability of the production process'.[60] The campaign's target age group was people under 35 years old, and Interporc based their strategy on a display contrasting 'true and fake messages', jumping on the bandwagon of purportedly debunking false claims. Inspired by the news media trend of fact-checking and debunking actual misinformation, Interporc's public relations team took this trend as an opportunity to 'debunk' any information that harmed its reputation.[60]

Provacuno and Interporc have been very active in different arenas, from political lobbying to media and educational advocacy including, as we have shown in a previous example, taking legal action against vegan companies that dare to compare plant-based foods' GHG emissions with those of animal-based foods. The Spanish meat industry has not been alone in framing its products as sustainable; the PP and PSOE governments have lent their full support to these companies' efforts.[61] However, and most revealingly, these two lobbies have gone from competitors to allies within the meat market, both trying to debunk facts that may engender climate action and thus disrupt their business.

CONCLUSION

Climate obstruction in Spain involves politicians, industry, the media system, and everyday citizens and has continued to evolve. In Spain, obstructionism directed at any major shift in the energy model, particularly the rise of solar power, was substantial in the 2000s and 2010s, while today (the early 2020s) the advances in deployment of renewable energy have been remarkable.[62] The greatest controversy during this later period has been over dealing with diffuse (non-energy) GHG emissions, especially in areas such as transport, diet, and tourism because they involve important business interests and entrenched ideological stances.

To unpack the Spanish climate obstructionism context, we have here identified three main factors: ideology, industry, and privilege.

Regarding ideology, we have shown how the Spanish political system of alternating majorities and political parallelism has produced a polarization of stances on important issues. Climate concern, like environmental concern generally, was first neglected and then instrumentalized by the country's political and economic elites via greenwashing and to fuel culture wars. In this system, the media mirrors political divisions, further contributing to polarizing opinions on climate change among the public.

In such a polarized political landscape, where topics such as the defence of car use or meat consumption have become identified with conservatism, it is worth considering what climate communications specialist Maxwell Boykoff said about the need to find common ground—finding shared values and concerns—which he notes is essential to facing the climate crisis.[63] According to Boykoff, while there are no magic formulas for effectively communicating the climate crisis and reaching this common ground, adapting messages to different audiences and their needs will be vital. One potential way to reduce political polarization may be to find ways to show how closely aligned environmental concerns are with religious,[64] conservative, and centrist interests and values.[65] In short, we need not let the culture wars frame the climate problem as merely the concern of a 'green [leftist] ghetto', to paraphrase Boykoff. That approach may help us to avoid a situation in which, with each new political cycle, political polarization again threatens the little progress already made on climate.

As for industry, the late establishment of democracy in Spain may not be the only cause of this country's climate inaction, but it is a substantial one, in our view. A combination of late industrialization, which accelerated during the Franco dictatorship, and a subsequent restoration of democracy that merged a fascist legacy with neoliberalism resulted in a weak form of democracy guided by authoritarian liberalism. This political context has led to obstruction of climate action in the energy sector, with the practice of revolving doors between industry and government jobs serving as a main exponent of the vested interests between business and politics. The corruption of politics involves turning it into a lifetime career, as happens so often in Spain, thus creating an entrenched, privileged political class that makes decisions not for the public good but according to their career needs. Addressing this problem may require limiting the practice of allowing revolving doors.

Some measures to move us in this direction may include requiring more transparency in the management of polluting companies and their links to politicians as well as strict 'cooling off' periods before and after holding political posts. Reducing political corruption and the economic privileges of the elites in general may also result in a reduction of parallelism among political, social, and economic interests in Spain. That change may, in turn, promote a media system that is truly independent from the political system and the financial economy. There is a direct connection between the economic dependence of the media on polluting companies, the increasing precariousness of journalism jobs, and media coverage of climate change.

As mentioned above, however, polluting industries, policymakers, and media are not alone in perpetuating the climate status quo. Citizens, mostly in their role as consumers, are empowered to either sustain or reject social privileges. We believe the case of animal agriculture exemplifies this principle very well. Although this sector combats emerging concerns about the role of animal compassion and climate change in our dietary choices, it does so by tapping into an existing cultural substratum of Spanish society. Thus, although the meat business enjoys the privilege of being able to conduct communication and public relations campaigns, the reluctance of citizens to change their lifestyles underpins climate inaction.[66]

Reducing economic privileges (e.g. ending subsidies for polluting industries) encourages governance in the public interest rather than for the benefit of corporations. The resources gap between NGOs advocating for the ethical treatment of the environment and nonhuman life and powerful corporations lobbying for the right to exploit both is so vast that it has created a two-tiered democracy in Spain. Limiting resources devoted to corporate lobbying can be achieved partly by limiting the public relations expenditures of large corporate players, but perhaps the most feasible and urgent solutions involve transparency. In Spain, the regulation of lobbying activity is still pending as there has not yet been a parliamentary agreement on the matter. Political discussions on possible laws governing this issue have included options such as transparency registers, codes of conduct, transparency in the participation of interest groups in drafting legislation, and (once again) stricter control of revolving doors.[67] It would be desirable to advance such legislation to better control the lobbyists' activity, thus empowering the media and citizens with the information needed to better scrutinize them.

Finally, education is the best tool any society possesses to overcome the entanglements that produce environmental inaction and societal privileges, and this is particularly the case for Spain. In our view, an effective twenty-first-century education curriculum would include not only such basics as science, technology, and media literacy but also a foundation in the humanities, which have increasingly been devalued in today's schools and universities. In particular is the need for training in a form of universal ethics that produces empathy and respect for all types of life.

As for citizen action, Spain needs to move beyond recycling to promote real mindset changes in all lifestyle areas including diet, leisure, and mobility. The goals should be to avoid political polarization, promote understanding and cooperation, and foster respect for all beliefs while providing the skills to identify and cease rewarding vested interests. Such a complex scenario involves overcoming different types of entrenched inertia that feed off one another, revealing that climate obstruction is based not just in scientific scepticism but also in attempts to protect an institutionally and culturally embedded network of interests.

NOTES

1. Miguel Ángel Medina, Rodrigo Silva, and Antonio Alonso (2021, 14 June), 'Los Peores Desastres Ecológicos En España: Cuánto Costaron y Cuánto Han Dejado Sin Pagar Los Contaminadores', *El País*, https://elpais.com/clima-y-medio-ambiente/2021-06-14/los-peores-desastres-ecologicos-en-espana-cuanto-costaron-y-cuanto-han-dejado-sin-pagar-los-contaminadores.html.
2. Elena G. Sevillano (2014, 3 June), 'Europa Cumple La Reducción de Emisiones Del Protocolo de Kioto', *El País*, https://elpais.com/sociedad/2014/06/03/actualidad/1401798742_648544.html.
3. Henrique Morgado Simões, and Gema Andreo Victoria (2021), 'Climate action in Spain. Latest state of play', European Parliamentary Research Service, https://www.europarl.europa.eu/RegData/etudes/BRIE/2021/690579/EPRS_BRI(2021)690579_EN.pdf.
4. Observatorio de la Sostenibilidad (2020), 'Evolución de Las Emisiones de Gases de Efecto Invernadero En España (1990-2019)', https://www.observatoriosos

tenibilidad.com/documents/EVOLUCI%C3%93N%20EMISIONES%20GEI%20ESPA%C3%91A%20%281990-2019%29%20v03.pdf.
5. Ministerio para la Transición Ecológica y el Reto Demográfico (MITECO) (2023), 'Sectores difusos. Situación actual y objetivos', https://www.miteco.gob.es/es/cambio-climatico/temas/mitigacion-politicas-y-medidas/definicion-difusos.aspx.
6. Arend Lijphart (1999), *Patterns of Democracy: Government Forms and Performance in Thirty-Six Countries*. New Haven, CT: Yale University Press.
7. Daniel C. Hallin and Paolo Mancini (2004), *Comparing Media Systems: Three Models of Media and Politics*. Cambridge: Cambridge University Press.
8. Pablo Corral-Broto (2018), 'Historia de La Corrupción Ambiental En España, 1939–1979. ¿Franquismo o Industrialización?', *Hispania Nova*, 646–684, doi:10.20318/hn.2018.4051.
9. Naomi Oreskes and Erik M. Conway (2011), *Merchants of Doubt*. New York: Bloomsbury Publishing.
10. António Costa Pinto (2010), 'The Authoritarian Past and South European Democracies: An Introduction', *South European Society and Politics*, 15, 3: 339–358, doi:10.1080/13608746.2010.513598.
11. Paul Preston (1995), *Franco: A Biography*. London: Harper Press.
12. Jose Antonio González (2008), *La Derecha Contra el Estado: El Liberalismo Autoritario en España (1833–2008)*. Lleida: Milenio.
13. José Luis Velasco (2015), *Crónicas Eléctricas. Breve y Trágica Historia Del Sector Eléctrico Español*. Madrid: Akal.
14. Konrad Gürtler, Rafael Postpischil, and Rainer Quitzow (2019), 'The Dismantling of Renewable Energy Policies: The Cases of Spain and the Czech Republic', *Energy Policy*, 133: 110881, doi:10.1016/j.enpol.2019.110881.
15. See more at Cinco Días (2004, 13 March), 'El Gobierno aprueba in extremis el nuevo marco económico de las renovables', https://cincodias.elpais.com/cincodias/2004/03/13/empresas/1079188782_850215.html; Santiago Carcar (2019, 19 July), 'De Aznar a Zapatero: memorias de una burbuja', *La Información*, https://www.lainformacion.com/opinion/santiago-carcar/aznar-zapatero-memorias-burbuja/6506973/.
16. The cabinet of President Zapatero saw a period of more progressive politics since the 1970s, with measures such as the withdrawal of Spanish troops from Iraq and the approval of same-sex marriage. Examinging this progressive time is beyond the scope of our chapter, thus we focus on the short pioneering period in renewable energies that took place in this context.
17. El País (2007, 22 July), 'Rajoy cuestiona el cambio climático y afirma que no puede convertirse en el "gran problema mundial"', https://elpais.com/sociedad/2007/10/22/actualidad/1193004007_850215.html.
18. Santiago Carcar (2012, 28 January), 'El Gobierno decreta un parón en las renovables para taponar el déficit', *El País*, https://elpais.com/diario/2012/01/28/economia/1327705210_850215.html.
19. José María Llopis (2012, 12 January), '¿Moratoria o destrucción de la fotovoltaica?', *Cinco Días*, https://cincodias.elpais.com/cincodias/2012/01/12/economia/1326484548_850215.html.
20. Jose A. Moreno and Noelia Ruiz-Alba (2021), 'Journalism or Greenwashing? Sponsors of COP25 Chile-Madrid in the Spanish Press', *Mediterranean Journal of Communication*, 12, 2: 285–300, doi:10.14198/MEDCOM.19089.

21. Jose A. Moreno and Gina Thornton (2022), 'Climate Action Obstruction in the Spanish Far Right: The Vox's Amendment to the Climate Change Law and Its Press Representation', *Ámbitos: Revista Internacional de Comunicación*, 55: 25–40, doi:10.12795/Ambitos.2022.i55.02.
22. Miguel Ángel Medina (2023, 23 June), 'El PP y Vox eliminan carriles bici en varias ciudades: ¿son también una cuestión ideológica?', *El País*, https://elpais.com/clima-y-medio-ambiente/2023-06-23/pp-y-vox-eliminan-carriles-bici-en-varias-ciudades-son-tambien-una-cuestion-ideologica.html.
23. Enrique Naranjo (2023, 8 March), 'Isabel Díaz Ayuso, contra la prohibición de los coches de combustión en 2035', *Marca*, https://www.marca.com/coches-y-motos/trafico/2023/03/08/6408476ae2704e953a8b4632.html.
24. Juan José Mateo (2022, 10 November), 'Ayuso dice que la lucha contra el cambio climático favorece el comunismo y que la izquierda va "contra la evidencia científica"', *El País*, https://elpais.com/espana/madrid/2022-11-10/ayuso-dice-que-la-lucha-contra-el-cambio-climatico-favorece-la-pobreza-y-el-comunismo-y-que-la-izquierda-va-contra-la-evidencia-cientifica.html.
25. Although Vox showed a clearly obstructionist stance in its amendment to the Climate Change Law or the 2023 elections programme, its communications on climate issues showed also diverse positions that are linked to nationalism and the reterritorialization of climate action. See Johanna Hanson (2023), 'Looking beyond Climate Contrarianism: Nationalism and the Reterritorialization of Climate Discourse in Spain's Vox Party', *Nordia Geographical Publications*, doi:10.30671/nordia.121511
26. Ecologist NGOs have criticized the socialist government's lack of climate ambition: Público (2021, 13 May), 'Las ONG califican la Ley del clima de "imprescindible", pero "poco ambiciosa"', https://www.publico.es/sociedad/ong-califican-ley-del-clima-imprescindible-ambiciosa.html.
27. Puri Caro (2022, 18 March), 'Collboni se alía con el empresariado para relanzar la ampliación del Aeropuerto de Barcelona a la que se opone Colau', *20 Minutos*, https://www.20minutos.es/noticia/4972855/0/collboni-se-alia-con-el-empresariado-para-relanzar-la-ampliacion-del-aeropuerto-de-barcelona-a-la-que-se-opone-colau/.
28. Eduardo Robaina (2023, 23 July), 'El Tribunal Supremo desestima la demanda contra el Gobierno de España por inacción climática', *Climática*, https://www.climatica.lamarea.com/ts-desestima-demanda-litigio-climatico/.
29. Spain is the main producer of pork in the European Union, with revenues of €18.5 billion in 2021 (62% of all revenues of the Spanish meat industry that year), 58.3 millions of slaughtered pigs, and 41.2% of revenues coming from export (see https://www.eleconomista.es/retail/noticias/11727627/04/22/La-industria-del-cerdo-alcanzo-facturo-un-record-de-18500-millones-en-2021.html and https://www.mapa.gob.es/es/ganaderia/temas/produccion-y-mercados-ganaderos/indicadoressectorporcino2021_tcm30-564427.pdf). The main sources of pollutants from this industry are feed production, transportation, waste treatment, and energy used. The agriculture sector in Spain accounts for 19% of the diffuse GHG emissions, making the pork industry one of the main businesses responsible for them (i.e. https://www.lavanguardia.com/natural/cambio-climatico/20221129/8626533/ecologistas-denuncian-desorbitadas-emisiones-macrogranjas-porcinas-pmv.html).
30. Corporate Europe Observatory (CEO) (2020), 'Internal Documents Reveal Agribusiness Lobby to Keep Status Quo CAP and Derail "Farm to Fork"', https://

corporateeurope.org/en/2020/10/internal-documents-reveal-agribusiness-lobby-keep-status-quo-cap-and-derail-farm-fork.

31. Isidro Jiménez Gómez and Samuel Martín Sosa (2022), 'Análisis discursivo del escepticismo climático en los medios impresos y digitales españoles entre 2015 y 2021', *Estudios Sobre El Mensaje Periodístico*, 28, 3: 525–536, doi:10.5209/esmp.80779.
32. SciencePolicy (2023), 'Media and Climate Change Observatory', MeCCO, http://sciencepolicy.colorado.edu/icecaps/research/media_coverage/index.html
33. Melanie Joy (2010), *Why We Love Dogs, Eat Pigs, and Wear Cows: An Introduction to Carnism*. Newburyport, CT: Red Wheel/Conari Press.
34. Jose A. Moreno and Núria Almiron (2021), 'Representación en la prensa española del papel de la agricultura animal en la crisis climática', *Estudios Sobre El Mensaje Periodístico*, 27, 1: 349–364, doi:10.5209/esmp.73745.
35. European Commission (2021), 'Special Eurobarometer 513: Climate Change', https://europa.eu/eurobarometer/surveys/detail/2273.
36. Mònica Pons-Hernández (2022), 'Power(Ful) Connections: Exploring the Revolving Doors Phenomenon as a Form of State-Corporate Crime', *Critical Criminology*, 30, 2: 305–320, doi:10.1007/s10612-022-09626-z.
37. Rec. 683/2022 Condena al cese de la realización de manifestaciones publicitarias denigratorias de la carne [2022] Audiencia Provincial de Barcelona, Juzgado Mercantil no. 12 Barcelona, Sección 15ª.
38. Jose A. Moreno (2018), 'Estructura informativa y cambio climático: El caso de El País'. In: Rogelio Fernández-Reyes and Daniel Rodrigo-Cano (eds.), *La comunicación de la mitigación y la adaptación al cambio climático*, pp. 77–98. Sevilla: Egregius, https://hdl.handle.net/11441/81112.
39. Fundación Repsol (2023), 'Motor verde', https://motorverde.fundacionrepsol.com/en/homepage.html
40. Joel Calero (2022, 14 September), 'El lado oscuro de Ryanair: millones en subvenciones y permiso para contaminar', *Merca2*, https://www.merca2.es/2022/09/14/lado-oscuro-ryanair-millones-subvenciones-permiso-contaminar-1074272/
41. Insights informing these cases, as well as different ideas in this chapter, come from research projects funded by the Spanish State Research Agency (AEI) and the European Regional Development Fund (ERDF) under grants CSO2016-78421-R and PID2020-118926RB-100, and by the Spanish Ministry of Universities, under grant FPU18/04207. More info at https://www.upf.edu/web/thinkclima/ and https://www.upf.edu/web/compass.
42. Katrin Buchmann (2022), *European Climate Diplomacy in the USA and China Embassy Narratives and Coalitions*. Leiden, The Netherlands: Brill.
43. For this analysis we have built on our previous comparative analysis of the web publications of eight European think tanks against climate action, completing the JMI sample and using simpler inductive analytical categories; see Núria Almiron, Maxwell Boykoff, Marta Narberhaus, and Francisco Heras (2020), 'Dominant Counter-Frames in Influential Climate Contrarian European Think Tanks', *Climatic Change*, 162, 4: 2003–2020, https://doi.org/10.1007/s10584-020-02820-4.
44. Adolfo Lozano (2015, 17 November), 'Cinco mitos sobre el cambio climático', Instituto Juan de Mariana, https://juandemariana.org/ijm-actualidad/analisis-diario/cinco-mitos-sobre-el-cambio-climatico/

45. José Carlos Rodríguez (2019, 8 December), 'La ciencia del cambio climático y otras religiones', Instituto Juan de Mariana, https://juandemariana.org/ijm-actualidad/articulos-en-prensa/la-ciencia-del-cambio-climatico-y-otras-religiones/
46. Gabriel Calzada (2007, 22 July), 'La lucha contra el cambio climático cuesta una pasta', Instituto Juan de Mariana, https://juandemariana.org/ijm-actualidad/articulos-en-prensa/la-lucha-contra-el-cambio-climatico-cuesta-una-pasta/.
47. José Carlos Rodríguez (2019, 29 December), 'Los impuestos sobre el pecado y el cambio climático', Instituto Juan de Mariana, https://juandemariana.org/ijm-actualidad/articulos-en-prensa/los-impuestos-sobre-el-pecado-y-el-cambio-climatico/
48. For example, in: Francisco Capella (2007, 7 February), 'Películas Gore de vísceras climáticas', Instituto Juan de Mariana, https://juandemariana.org/ijm-actualidad/articulos-en-prensa/peliculas-gore-de-visceras-climaticas/.
49. Original excerpt: 'En España hemos estado solos durante muchos días informando a los lectores de esta canallada seudocientífica subvencionada por los políticos'. From Pablo Molina (2009, 4 December), 'Con razón quieren acabar con internet', Instituto Juan de Mariana, https://juandemariana.org/ijm-actualidad/articulos-en-prensa/con-razon-quieren-acabar-con-internet/.
50. Eva M. Ramírez (2023, 16 March), 'La cifra de negocio de la industria cárnica supone el 29% del sector agroalimentario español', *FoodRetail*, https://www.foodretail.es/especiales/sector-primario/negocio-industria-carnica-representa-agroalimentario_0_1748225171.html
51. Ourworldindata (2023), 'Meat and Dairy Production', https://ourworldindata.org/meat-production
52. Dani Cabezas (2016, 2 March), 'El lobby español de la carne pagó estudios para contradecir a la OMS', *La Marea*, https://www.lamarea.com/2016/03/02/oms-lobby-carne
53. Ministerio para la Transición Ecológica y el Reto Demográfico (MITECO) (2019), 'Las emisiones de CO2 disminuyen en España un 2,2% en 2018 con respecto al año anterior', https://www.miteco.gob.es/es/prensa/ultimas-noticias/las-emisiones-de-co2-disminuyen-en-españa-un-22-en-2018-con-respecto-al-año-anterior/tcm:30-497589
54. David Vigario (2021, 8 July), 'Alberto Garzón Enfurece al Sector Ganadero al Incitar a Comer Menos Carne Para Proteger al Planeta', *El Mundo*, https://www.elmundo.es/economia/2021/07/07/60e5e176e4d4d815248b4596.html.
55. European Commission (2019), 'Proud of EU Beef', https://ec.europa.eu/chafea/agri/en/campaigns/proud-eu-beef. Also see the campaign website, at: https://haztevaquero.eu/.
56. Manuela Ripa Homepage (2020, 9 December), 'EU Agricultural Promotional Programme Contradicting Sustainability Goals by Financing 'Proud of EU Beef' Campaign', Members of the European Parliament, https://manuela-ripa.eu/wp-content/uploads/2020/12/Letter-EC_Campaign-Beefatarian.pdf.
57. Carol Adams (1990/2010), *The Sexual Politics of Meat. A Feminist-Vegetarian Critical Theory*. New York: Continuum International.
58. A. Tena (2020, 12 November), 'La UE financia con 3,6 millones de euros una polémica campaña para "inciter" al consumo de carnes rojas', *Público*, https://www.publico.es/sociedad/carne-vacuno-ue-financia-3-6-millones-euros-polemica-campana-incitar-consumo-carnes-rojas.html.
59. For a more detailed commentary, which also includes these images as examples, on this and other similar campaigns in other European countries, see Sini Eräjää

(2021, 'Marketing Meat: How EU Promotional Funds Favour Meat and Dairy', Greenpeace European Unit, https://www.greenpeace.org/static/planet4-eu-unit-stateless/2021/04/20210408-Greenpeace-report-Marketing-Meat.pdf.

60. European Commission (2019), 'Let's Talk about Pork from Europe', https://ec.europa.eu/chafea/agri/en/campaigns/letstalkaboutpork. Also see the campaign website, at https://letstalkabouteupork.com.
61. Javier Moreno (2019, 13 June), 'Así controla lo que comemos el poderoso lobby español de la carne', *Vice*, https://www.vice.com/es/article/597jaa/lobby-carne-espana-isabel-garcia-tejerina-gobierno-poder-igualdad-animal.
62. Ignacio Fariza (2023, 28 June), 'El Gobierno eleva el objetivo de renovables para que generen más del 80% de la electricidad consumida en 2030', *El País*, https://elpais.com/economia/2023-06-28/el-gobierno-eleva-los-objetivos-de-renovables-que-generaran-mas-del-80-de-la-electricidad-consumida-en-2030.html.
63. Maxwell Boykoff (2019), *Creative (Climate) Communications: Productive Pathways for Science, Policy and Society*. New York: Cambridge University Press.
64. Rogelio Fernández-Reyes (2023), 'Aproximación a la mitigación climática en comunidades con fe en España. Comunicación Climática Con Audiencias Con Fe', ECODES, https://ecodes.org/images/que-hacemos/MITERD-2022/cambio_climatico/Aproximaci%C3%B3n_a_la_mitigaci%C3%B3n_en_las_comunidades_con_fe_en_Espa%C3%B1a.pdf.
65. Rogelio Fernández-Reyes (2019), 'Aproximación a la comunicación climática con audiencia conservadora en España', ECODES, https://ecodes.org/images/que-hacemos/01.Cambio_Climatico/Movilizacion_accion/Medios_Comunicacion_CC/Aproximacion_comunicacion_climatica_audiencia_centro_y_derecha.pdf.pdf.
66. We elaborate more on the nuances of climate obstructionism and its ideological roots in Núria Almiron and Jose A. Moreno (2022), 'Beyond Climate Change Denialism: Conceptual Challenges in Communicating Climate Action Obstruction', *Ámbitos: Revista Internacional de Comunicación*, 55: 9–23, https://repositori.upf.edu/handle/10230/52309.
67. During the XIV legislature, the coalition government of PSOE and Podemos presided over by Pedro Sánchez promoted a Draft Bill on Transparency and Integrity in the Activities of Interest Groups. It did not reach the end of its parliamentary route due to differences between parties and the early elections in 2023. More information about this proposal at Marta Gutiérrez San Blas (2022, 8 November), 'El Gobierno hará público qué 'lobbies' intervienen en las decisiones políticas en España', *Newtral*, https://www.newtral.es/anteproyecto-ley-lobbies/20221108/; and Civio (2022), 'De nada sirve regular la transparencia de los Lobbies sin un control independiente', https://civio.es/novedades/2022/12/12/de-nada-sirve-regular-la-transparencia-de-los-lobbies-sin-un-control-independiente/.

13

Climate Obstruction in the European Union

Business Coalitions and the Technocracy of Delay

DIETER PLEHWE, MORITZ NEUJEFFSKI, AND TOBIAS HAAS

INTRODUCTION: BUSINESS EFFORTS TO PREVENT EUROPEAN CLIMATE PROTECTION POLICIES

'This is Europe's man on the moon moment'.[1] With much fanfare, European Commission President Ursula von der Leyen announced the European Green Deal (EGD) on 11 December 2019. The EGD is a major political project that aims to achieve climate neutrality by 2050. Although the analogy to the moon landing is hyperbolic, the EGD does represent an attempt to significantly upgrade climate and environmental policy within the European Union. These are highly contested policy arenas in the European Union, and the achievement of the goals stipulated in the EGD, such as a 55% reduction in greenhouse gases (GHG) by 2030 compared with 1990, is by no means a foregone conclusion.

The European Union has gained increasing influence in setting climate policy. Its power to determine the general orientation of the integrated body and its weight in multilevel negotiations and decision-making gives the European Union a privileged role in the climate policies of the individual member states—and these processes, too, are contested. Unlike in the United States or Australia, the opposition to ambitious climate policy in the EU arena does not feature influential climate denial networks openly opposed to any form of climate policy regulation. Rather, the main policy obstructors in the European Union are large companies and business associations with a strong vested interest in the fossil economy and neoliberal or conservative think tanks. As such, climate obstruction efforts in the European Union are shaped mainly by efforts to sow doubt about specific climate policy measures and by strategies to weaken their level of ambition. This is what Ekberg and colleagues refer to as 'secondary obstruction', which 'includes all those calls which do not deny the human-induced nature of the climate crisis (science), but nevertheless delay or forestall meaningful climate action'.[2] We count it as obstruction when actors attempt to delay or otherwise shape climate policy in ways that undermine efforts to keep global warming below 1.5°C. Such efforts include, among other things, shirking responsibility, advocating non-transformative solutions, defending the fossil fuel status quo, and climate policy nihilism.[3]

Interest groups logically pursue their own interests and those of the members they represent. However, groups that are interested in long-term, good-faith relationships with politicians also need to accept defeat if competing interests prevail and be ready to compromise and adjust. Climate obstruction, however, frequently involves bad-faith efforts to push back climate policy goals in favour of other goals, such as profit or otherwise narrow objectives, by claiming adherence to the goals of the Paris Agreement but nevertheless engaging in a range of frequently difficult to recognize efforts to block ambitious policy.

The limited appeal of denialism at the EU level is due mostly to the fact that the European integration process is quite technocratic and elite driven, geared primarily toward the establishment or completion of a single European market and Economic and Monetary Union (EMU). European environmental and climate policy approaches, in turn, have supported regulatory cohesion and are strongly oriented toward the guiding principle of ecological modernization (i.e. a depoliticized approach that puts technological innovation and market-based approaches at its centre[4]), albeit not exclusively. The limited European public sphere also does not lend itself easily to the type of tabloid press style campaigns known best in the Anglosphere, and Brexit (the United Kingdom's disengagement from the European Union) considerably weakened the political forces in the European Union that do promote climate change denial.

Regardless of the low profile of radical denialism in EU climate politics, climate policy obstruction remains a serious obstacle on the European 'road to Paris'. The European Union depends on its member states to secure and police implementation of treaties. While the European Union can put pressure on member states in the case of non-compliance, this process takes time and thus becomes yet another delay mechanism.

A SHORT HISTORY OF THE EUROPEAN INTEGRATION CONTEXT OF CLIMATE POLICY

The history of European integration was strongly influenced by both the ideas of German economists from the 'ordoliberal' school, such as Walter Eucken, who opposes state interventionism, and by more state planning-oriented perspectives prevalent in France in the 1950s (such as those articulated by Jean Monnet). While economic liberalization steps were pursued rather cautiously in Europe until the 1980s, the economic liberalization agenda gained momentum with the introduction of the European Monetary System (EMS) in 1979. In the 1980s and 1990s, the European community took numerous steps to facilitate a European EMU, culminating in in the plan to complete the single market (via the Single European Act of 1986) and the introduction of the euro as a common currency in 1999. Over time, the balance of power between the neoliberal and regulatory-minded forces shifted in favour of the former.[5]

Until today, the aim to complete the single market is a common narrative used to enforce liberalization and market-based policies in the European Union. Within the European multilevel governance system, the distribution of regulatory responsibilities and the geographical scale of action are always contested. While the European Commission (EC) and the majorities in the European Parliament (EP) usually try to shift responsibilities to the European level, the European Council (EUCO), composed of the heads of government of the member states and the president of the EC, frequently try to preserve national sovereignty (Box 13.1).

As a result, climate policy in the European Union is characterized by a tension between the scalar dimension (i.e. the question of which responsibilities and obligations are located at which governance level) and the substance of proposals. The level of climate policy ambition and orientation are typically contested within the EP as well as the EUCO and EC. Accordingly, lobbying activities are directed at both the European institutions and national governments, the latter of which can exert significant influence through the EUCO.[6] This approach was crucial in, for example, the negotiation of the European Green Taxonomy for Sustainable Investments (Green

> **Box 13.1 KEY INSTITUTIONS OF THE EUROPEAN UNION**
>
> **THE INSTITUTIONS OF THE EUROPEAN UNION**
>
> **The European Commission**
> is the EU's executive branch, responsible for proposing and implementing EU laws and policies. It holds considerable powers, including the authority to initiate legislation, negotiate international agreements on behalf of the EU, and enforce EU laws.
>
> **European Parliament**
> is the EU's directly elected legislative body. It has the ability to pass laws, approve the EU budget, and exercise democratic oversight over other EU institutions. However, it cannot initiate its own legislative proposals.
>
> **The Council of the European Union**
> represents the Member States' governments. It shares legislative power with the European Parliament, with the ability to adopt EU laws, coordinate policies, and make important decisions on various EU matters, including foreign policy and economic coordination.

Taxonomy, discussed later) and in resolving disputes over the ban on internal combustion engines after 2035, which the German government refused to accept in 2023.

To create the least distorted competition possible in the single European market, the idea goes, European competition law needs to be developed in such a way that market interventions on a national level may only be made in case of an overriding European interest. An example of such an interest could be meeting binding targets in the fields of energy and climate policy. Nevertheless, climate policy approaches within the EU member states constantly exist in a state of tension with the single market, the EMU, and the corresponding competition and state aid law, which governs the provision of governmental financial support by member states to individual companies or industry sectors.[7]

In the meantime, however, we argue that environmental and climate policies have become central to the legitimacy of the European integration project. While the total GHG emissions within the European Union decreased overall between 1990 and 2020 (Figure 13.1), EU policies to deal with CO_2 emissions have risen in both number and substance. For example, environmental policy was advanced with the Amsterdam Treaty of 1997 (effective 1999), which established a duty to integrate environmental protection into all EU sectoral policies to help promote sustainable development.[8] The European Union has since established the precautionary

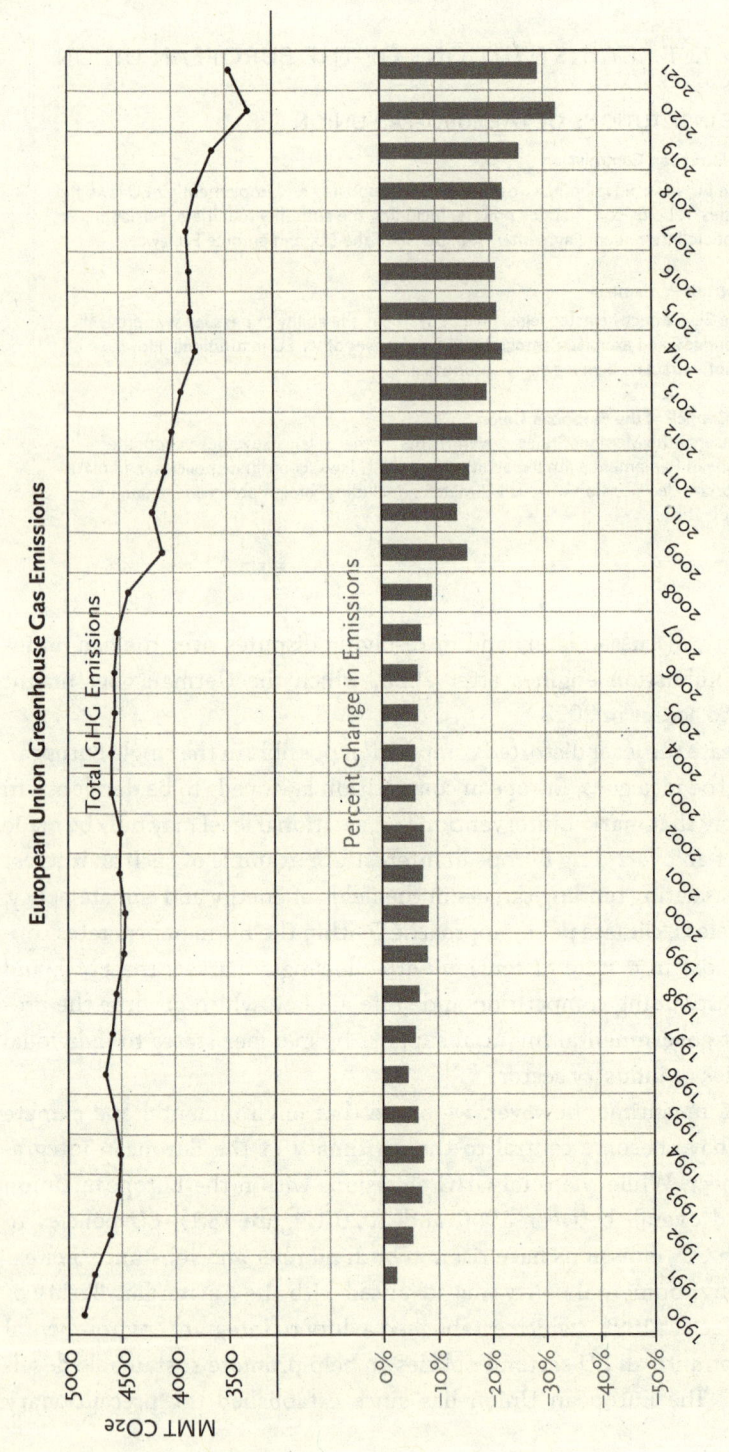

Figure 13.1 Total greenhouse gas (GHG) emissions (in MMT CO_2e) and percentage change in emissions in the European Union between 1990 and 2021, inclusive.

Source: Total GHG emissions based on data provided by Gütschow and Pflüger (2023) for Kyoto Six Greenhouse Gas Totals.

principle (now Article 191 of the Treaty on the Functioning of the European Union [TFEU]) and required the EC to engage in preventative action to achieve a high level of environmental protection. The precautionary principle requires policymakers to delay innovations with a potential for causing harm if there is not yet sufficient scientific evidence available on its safety. Preventative action also requires the integration of environmental considerations into all European policies.

The Amsterdam Treaty also stipulated majority voting in the EC and co-decision of the EP. This so-called community method of European lawmaking thereby eliminates the veto power of any single EU member state, which had been a major obstacle to ambitious policymaking. Despite persistent limitations, various environmental regulations have thereby been the result of EU environmental policymaking since the late 1990s.[9] Over time, green parties, environmental think tanks, and social movements also became more integrated into European policymaking. Today the Commission consults nongovernmental organizations (NGOs) on a regular basis, and Green MEPs are leading members of European Parliament Committees and working groups.

Lenschow and Sprungk have argued that the goal and 'myth of a Green Europe' has recently been important in driving the European integration process as such. In the academic literature, the European Union is often classified as a leader in global environmental and climate policy. Indeed, in the field of climate policy, numerous goals have been defined and instruments have been developed since 1990[10] (Table 13.1). For example, under the Kyoto Protocol of 1997, the European Union committed itself to reducing its emissions by 8% over the period 2008–2012 compared with 1990 levels, and, as part of this goal, the EU Emission Trading System (ETS) was introduced in 2005. By 2007, the European Union had already adopted three major climate and energy policy targets: By 2020, (1) GHG emissions were to be reduced by 20%, (2) the share of renewable energy in total energy consumption was to increase to 20%, and (3) energy efficiency was to be increased by 20%. Then, in 2014, the (preliminary) climate and energy policy framework for 2030 was adopted: by then, emissions were to fall by 40%, renewables were to be expanded to 27%, and energy efficiency was to increase to 27%.

At the time, this target horizon was already being criticized by environmentalists as extremely unambitious; indeed, it was also not backed up by nationally or sectorally binding targets. The situation changed with the incoming Commission headed by von der Leyen, a German Christian Democrat, in 2019. The Commission president is elected by the EP. Due to the weakening of the conservative faction in the EP following Brexit and

Table 13.1 EU CLIMATE TARGETS BY YEAR AND STATUS

Year adopted (and related UN climate development)	GHG emission reduction target (base year 1990)	Status
1990 (1992 UNFCCC)	Stabilization by 2000 (CO_2 only)	Achieved
1997–1998 (1997 Kyoto Protocol)	8% by 2008-2012	Achieved
2007 (2009 Copenhagen COP 15)	20% by 2020	32% in 2020
2009 (2009 Copenhagen COP 15)	80–95% by 2050	With the European Green Deal (EGD), by 2050 climate neutrality is the new target
2014 (2015 Paris Agreement)	40% by 2030	With the EGD and the Fit-for 55 package, the target was tightened
2019 (2021 Glasgow COP 26)	Climate neutrality by 2050	Achievement very uncertain
2020 (2021 Glasgow COP 26)	55% by 2030	Achievement very uncertain

Source: Based on von Homeyer et al. 2021, p. 961, reprinted by permission of the publisher (Taylor & Francis Ltd, http://www.tandfonline.com).

the split of the conservative parliamentary group, von der Leyen had to woo greens and liberals to support her candidacy.[11] The strategy of the far right-wing groups in the EP to end centrist cooperation between conservatives and social democrats ultimately achieved the opposite: the EC moved to the left, which allowed for more ambitious European climate policy. (The upcoming European elections are expected to strengthen the right and far right instead, which might lead to a reversal of ambitious European climate policy.)

Accordingly, much more ambitious targets were formulated as part of the EGD in 2019, namely, reaching an emissions reduction goal of between 50% and 55% by 2030 for the European Union as a whole and achieving climate neutrality by 2050.[12] Torres and Bongard have argued that the EGD substantially enhanced the sustainability dimension within EU economic governance: 'the EGD may thus be regarded as a third building block in the making of the European economic model, alongside the single market and EMU, and that any crisis would therefore need to be addressed through its framework'.[13] In the meantime, agreement was reached on a firm CO_2 emissions reduction target of 55% by 2030. In July 2021, the EC then presented the 'Fit for 55' package, which was adopted in October 2022.[14] This policy package includes a variety of initiatives, directives, and reforms. Among other things, it called for the establishment of the additional EU

ETS for buildings, road transport, and fuels; the sharpening of reduction targets within the framework of fleet limits for car manufacturers; the introduction of a carbon border adjustment mechanism (CBAM) to protect European producers of energy-intensive materials (such as concrete, aluminium, steel, and fertilizers) from imports that are not subject to carbon pricing; and more ambitious regulations in the areas of energy efficiency, the circular economy, and carbon dioxide removal, all of which are necessary for achieving climate neutrality. The individual member states are also required to make a greater contribution to climate protection.

But the climate policy push has also been controversial. Opposition to many of the Fit for 55 package measures was strong among the conservative political parties and almost unanimous in the far-right parties. Jacob also reported strong support for the inclusion of nuclear energy and fossil gas in the Green Taxonomy (which classifies economic activities according to their alignment with the net zero strategy to direct investments) from several governments, a position that runs counter to the scientific recommendations of the expert group installed to develop the taxonomy.[15]

Based on these insights, obstruction to ambitious climate policy can thus be located in the centre-right wing of the political party spectrum on the one hand and, on the other, in specific countries dependent on fossil and nuclear paths. To characterize the opposition to ambitious climate policy more completely, we next turn the spotlight on private sector lobbies and their allies.

CLIMATE OBSTRUCTION ACTORS IN THE EUROPEAN UNION

With each expansion of supranational negotiating and decision-making powers in Brussels, the lobbyist landscape has grown, seeking to expand the policy influence capacities of actors previously focused on relevant national jurisdictions. Scholars speak about the co-evolution of the EU political system, 'private interest governments',[16] and a new type of network governance in Europe.[17] And, indeed, by 2021, more than 13,300 organizations had been recorded in the European Transparency Register, an official database of organizations that try to influence the policy process of EU institutions.[18] Of these groups, 7,000 are business organizations and professional service firms, compared with about 1,000 trade union organizations, about 3,500 NGOs, and nearly 1,000 think tanks and academic institutions.[19] Observers and watchdog organizations estimate the number of lobbyists in Brussels to be around 25,000 people, working with a total budget of about €1.5 billion.[20] Contrary to efforts to characterize the European Union as a bureaucratic monster, corporate and other lobbyists target only about 32,000 staff members[21] of the commission, the 705 EP members, and their staffs. Due to the very high number of lobbyists, Brussels has become known as the second lobbying capital of the world after Washington, DC.[22] Although the European political system is considered somewhat more balanced than the United States's, the business-favouring asymmetries are formidable, and counterforces from labour and environmental organizations constitute a limited counterweight even as the number of NGOs registered in Brussels continues to grow.[23] Besides lobbying struggles that involve competing business forces, attention also needs to be paid to competing business–civil society alliances.[24] In the energy field, for example, renewable energy coalitions include solar and wind energy firms and green NGOs, which confront traditional energy firms and their allies in regions that stand to lose jobs.

While the European Union claims climate leadership, individual EU countries continue to emit a high share of global GHGs. This situation is due to strong social forces committed to defending the fossil fuel status quo and to slow down the transition to carbon neutrality. To portray the power asymmetries between supporting and opposing forces in the field of European climate policy, Tobias Haas has suggested distinguishing between 'green' and 'grey' energy coalitions (see Chapter 6).[25] Firms and their business associations invested in fossil industries as producers or (large) energy consumers comprise the grey energy faction, as opposed to the green energy coalition, which comprises supporters of renewables and their allies and represents industry groups and social forces advocating an energy transition. In the first group, we find not only the major European (and non-European) oil, gas, and coal companies as well as utilities and related services but also car manufacturers and providers of the infrastructure services on which they rely. In the fields of agriculture and food production, there are also many corporate interests aligned with traditional energy production and the European subsidy regime associated with it. In the second group, the green coalition extends to firms strongly involved in public transport and in the ecological and social transition to a sustainable ecological and social economy.

This grey–green division represents two poles. Firms that belong to the grey coalition can, of course, move in the direction of the green coalition if they decide to divest from fossil energy sources; similarly, fossil firms can also invest strategically in renewable energy, in part to co-opt the opposition, as has happened with gas companies investing in the renewables sector.[26] Haas has also reported on significant differences between manufacturing firms (the end consumers of energy) and utilities regarding support for specific policy instruments.[27] For example, utilities have come out in support of emissions trading because they can pass higher prices on to consumers, whereas industrial-sector firms remain more sceptical toward carbon pricing. While utilities score better in terms of alignment with Paris goals due to their support for the EU ETS, they have also defended the centralized system of energy production and distribution along with fossil path dependencies in Germany, for example (see Chapter 6).

The think tank InfluenceMap has examined how corporations and industry associations in Europe engage in climate policy debates and classifies them from A (supporting the Paris Agreement) to F (not supporting the Paris Agreement). Their analysis of advocacy at the European level showed that there is a strong preponderance of organizations that do not fully support the objectives of the Paris Agreement. Only four business associations (12.1%), including WindEurope and SolarPower Europe, were classified as fully supporting the Paris climate goals, versus twenty-nine associations (87.9%) classified as C or D, which do not support many of the policies that aim to achieve the Paris goals. Among them are powerful business associations such as BusinessEurope (the umbrella organization of European industry), the European Steel Association (Eurofer), and Germany's VDA.[28]

In terms of corporations that engage in EU climate policy processes, the picture looks only slightly more optimistic, according to InfluenceMap. Only one company (Vestas Wind Systems) falls into its A category, and twenty-three companies rank under the B category. These firms face 60 companies (71.4%) that are not fully in support of the Paris Agreement (rated C or D). This latter group consists of corporations such as Uniper, Volkswagen, and Airbus. In fact, all companies from the automotive and mining and metal sector fall into this category, as do the bulk of energy companies. The Russian oil and gas companies Rosneft and Gazprom bring up the rear with ratings of E and F, respectively.

Adding together the lobbying staff and financial resources available (according to the incomplete information provided in the EU Transparency Register), the corporations and associations not fully aligned with the Paris goals outnumber and outspend the green coalition by far. We obtained lobbying expenditure data for most of these companies (seventy-seven out of eighty-three) and for all the associations in the green and grey coalitions. Due to data availability, we used the year 2019 as a reference point for our comparison.[29] We found that when it comes to climate policies at the EU level, the small number of Paris-aligned associations spent a total of €2.4 million to lobby EU institutions. This figure contrasts with the €43.5 million spent by twenty-nine associations that favour weaker

Associations   Companies

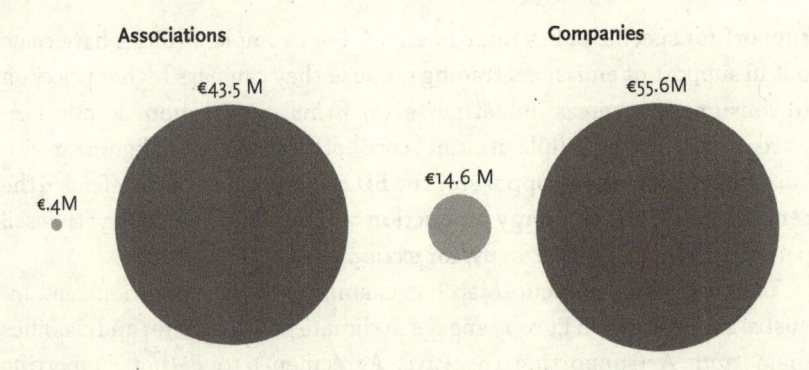

€43.5 M   €55.6M

€.4M   €14.6 M

Figure 13.2 Comparison of total lobby expenditures in 2019 for the 'green' and 'grey' alliances.
Source: Data retrieved from lobbyfacts.eu, which archives information from the EU Transparency Register. Authors' own calculation and presentation.

EU climate protection measures (Figure 13.2). The latter associations also spent three times more, on average, for lobbying at the EU level compared with the twenty-four Paris-aligned associations.

Regarding companies, the group of less-supportive corporations spent almost four times as much on lobbying in 2019 (€55.6 million) as the more supportive firms (€14.6 million). Again, these companies also spent more on average compared with the groups in the A or B category of Paris alignment. By and large, the group of economic actors who favour a less-ambitious EU climate policy also possess substantially greater financial resources. This additional financial power translates into a superior ability to organize policy events, commission policy briefs, lobby MEPs or the commission directly, participate in public consultations, and hire contract lobbying service firms. During the Fit for 55 negotiations, for example, Mohammed Chahim, member of the EP and in charge of drafting the EP's positions, remarked in an interview, 'I'm spending a lot of my time countering the information from the lobby, . . . That's what's happened the last weeks—every element in the deal was attacked, and in my group meetings and individual meetings, I needed to counter old arguments'.[30]

Superior financial resources allow companies and associations to cultivate specialists to lobby on specific issues, whereas green coalition lobbyists are forced to work as generalists, covering many different topics. These resources also allow the grey coalition to pursue simultaneous advocacy across multiple channels and venues, while the green coalition must concentrate on select venues. Multinational corporations, for example, can and do pursue many different strategies to get attention for their policy priorities: they operate public affairs departments at headquarters in their home country and at subsidiaries in other countries[31]; they also belong to sectoral and cross-sectoral national business associations and their federations at the European level. They have founded separate, large-company lobby groups and specialized lobbies for specific objectives such as 'green growth' and business-friendly sustainability perspectives. They invest in think tanks and hire consulting and law firms to conceive and run campaigns.

And they dominate the work in many of the European expert groups,[32] as in the case of the EU Energy Platform Industry Advisory Group. They organize events at specific venues to observe the positions and perspectives of the many actors in the climate arena and, to some extent, coordinate public affairs in various policy fields. The European Roundtable on Climate Change and Sustainable Transition, for example, was founded in 2016 to provide this function in the climate policy field. While NGOs also participate in lobbying and frequently become visible actors in the European Union, most of the time their power is limited to scandalizing the negative results of European policies, in contrast with industry's capacity to set, advocate, and shape agendas to their advantage.[33] Altogether, the EU lobbying scene is far from a level playing field.[34]

The data compiled by InfluenceMap cover only the very large emitters and their most important associations. These include the automotive industry, which is leading the fight against higher emission standards for cars and trucks and slowing the transportation transition in general[35]; the airline companies and aircraft manufacturers and their trade associations, which are fighting against the taxation of kerosine[36]; and the gas industry and its manufacturing allies, which are leading the fight for the perpetuation of gas in the name of energy security and to serve as a transition technology until so-called green hydrogen is ready to replace fossil gas. The strategy of the oil and gas majors and many smaller groups in the sector succeeded in locking in gas dependency for a long time and thus threatens to undermine the Paris goals.

Here, the industry cleverly concentrated on infrastructure investment and was able to successfully tap into European public programmes. The European Energy Union of 2015 (an effort to further the integration of the EU energy markets), the Trans-European Networks for Energy (a policy to link EU energy infrastructures), the legislation Save Gas for a Safe Winter (passed in July 2022), and the Third Energy Package (a bundle of laws to open up national gas and electricity markets) all were designed to secure the supply and commercial viability of the fossil gas sector in Europe. The Trans-European Networks are earmarked for funding from the European regional and structural fund, the European Union's second-largest budget after agriculture. In a major success for the European Network of Transmission System Operators for Gas (ENTSOG-G), there were no less than seventy-seven gas projects included in the Trans-European Networks.[37]

Notably, a report by Corporate Europe Observatory published in 2017 documented the limits of the Transparency Register, finding that most of the companies involved in Trans-European Gas projects were not even registered.[38] In addition to their work across the gas production and supply chain and direct lobbying by energy corporations and their business associations, the industry relied heavily on PR firms FTI Consulting, Fleishman-Hillard, Weber Shandwick, Gazprom's PR firm Gplus, and others to promote their position and public image.

To more closely observe some of the lobbying strategies employed in the effort to weaken, delay, and otherwise shape European climate policy according to business priorities, we briefly examine two key pieces of climate policy legislation more closely: the European ETS and the Green Taxonomy.

STRATEGIES OF DELAY AND TO MODERATE AMBITION
EU ETS and CBAM

The EU ETS was introduced in 2005. The emissions-trading system employs market mechanisms[39] to enable the reduction of GHG emissions where this is most feasible in terms of cost efficiency. Advocates for carbon pricing via emission trading hold it to be superior to carbon taxes set directly by public authorities. However, the concept has been widely criticized for being oriented primarily toward narrow economic calculations, representing a process of commodification,[40] and for becoming very complex to implement as it requires detailed regulations which, in turn, require specific expert knowledge to track and enforce. The EU ETS, which was anchored via agreements under the Kyoto Protocol, was preceded in the 1990s by an attempt to introduce a CO_2 tax. However, the attempt failed due to the resistance of the European industrial lobby,[41] which initially also opposed the EU ETS, claiming industry self-regulation was a superior mechanism.

While the defensive action of European industry did not suffice to block the new policy instrument altogether, the design of the EU ETS remained under massive pressure from large parts of industry, which explains its long gestation period. Until 2021, the price for CO_2 was so low that it did not have any significant environmental incentive effect before the fourth phase of ETS development (since 2021). In the first two phases (2005–2012), the allocation of allowances took place in their respective national contexts. The flexible mechanisms of the Kyoto Protocol, such as the Clean Development Mechanism (a carbon offset scheme which allows countries to finance projects that reduce GHG emissions in the Global South), combined with a substantial over-allocation of allowances, led to CO_2 prices falling and remaining in the single digits for years. This result was related to the fact that the Alliance of Energy Intensive Industries (AEII), which includes about fifteen associations including CEFIC (the European Chemical Industry), CEMBUREAU (the European Cement Association), EUROFER (the European Steel Industry), and Fertilizers Europe had developed and successfully invoked the concept of 'carbon leakage' at every phase of the negotiations.[42] According to Ehrenstein and Neyland, the concept refers to the risk that 'in a globalised world, climate policy in Europe might lead to domestic producers losing market share to foreign competitors and imports from regions where emissions are not regulated'.[43]

In addition, the EU ETS almost exclusively covered large industrial plants, and the majority of allowances have been distributed free of charge to date. Due to the free allocation, various industrial sectors have been able to benefit from the certificate trade, generating windfall profits (i.e. non-performing income), which they would not have achieved without the EU ETS.[44]

Notwithstanding the industry-friendly reality of the ETS, the danger of carbon leakage has been summoned in each new round of negotiations. In some cases, studies have substantiated carbon leakage claims. In the run-up to the negotiation of the third round of the EU ETS in 2008, the cement industry commissioned the Boston Consulting Group to prepare a report, which claimed,

> Based on the expected cost of production in the EU assuming the carbon cost of CO_2 versus the cost of producing in non-ETS countries, clinker and cement production in the EU is not competitive without free allowances allocation. As a result, the 'wise businessman' will prefer to relocate production to more competitive countries, this leading to production offshoring. At CO_2 prices above €35/t (expected for the 2013–2020 period 3) the current proposal of the Directive will lead to the complete offshoring of the cement industry. At CO_2 price of €25/t, more than 80% of EU clinker production will be at risk of offshoring by 2020.[45]

Based on such alarmist reports, the 'carbon leakage' myth'[46] has been constantly renewed. Although it is difficult to prove success of individual business studies and lobbying strategies, the CO_2 price only cracked the €30 mark for the first time in December 2020 (€30 was targeted when the EU ETS was introduced in 2005). Overall, it is obvious that the EU ETS so far has served mainly to protect the competitiveness of European industry. As a result, the climate policy record of the flagship of European climate policy is unimpressive: 'Climate policy appears condemned to suffer from having the precautionary principle applied for the benefit of the economy instead of the environment'.[47]

Only toward the end of the third phase of the EU ETS in 2020 did prices pick up somewhat due to the introduction of the Market Stability Reserve (MSR). And with the new ambition level in the wake of the EGD (a reduction of carbon emissions by 55% by 2030 compared with 1990), the linear reduction factor (LRF) has also been increased, meaning that up to 4.2% fewer allowances will be issued each year until 2030. In addition to the increase in the LRF, the introduction of the CBAM in 2022 (scheduled to take full effect in 2026) was a second key reform project of the ETS. Addressing carbon leakage concerns, the CBAM represents a levy on imported goods produced under less stringent climate regulations outside the European Union. This levy would remove the basis for the free allocation of emission allowances.

However, large parts of European industry and neoliberal think tanks such as the Freiburg-based Centre for European Policy (CEP) see this situation differently. An alliance of thirteen associations from energy-intensive industries warned in May 2021: 'The carbon leakage risk is more pressing than ever given the recent evolution of the carbon price reaching unprecedented values in spite of the economic impact of the COVID-19 pandemic, and considering the further increase expected in the fourth trading period'.[48] BusinessEurope, the umbrella organization of European industry, took the following position in November 2021:

> The Carbon Border Adjustment Mechanism (CBAM) could prove to be a tool to fight carbon leakage and level the playing field. Ensuring WTO compatibility and avoiding retaliation from trading partners is key not only for CBAM sectors, but also for the EU industry as a whole. The CBAM *should not be considered as an alternative to free allowances*, but should complement them, until the mechanism has proven its ability to effectively prevent carbon leakage and level the playing field.[49]

The CEP also criticized the plan to abolish free allowances through the introduction of a CBAM. Referring to the concept of carbon leakage, the CEP warned: 'The provisions on reduction of free allowances are misguided because they increase the risk to EU industry of production and emissions being relocated to third countries (carbon leakage)'.[50] In addition, the think tank raised the issue of the growing international trade controversies: 'In view of current geopolitical tensions, the EU should base its climate policy on global cooperation instead of conflict-prone unilateral initiatives like the CBAM'.[51] This position resonates with one feature of discourses of delay, namely redirecting responsibility.[52]

In this respect, the negotiations of the EU ETS and the introduction of the CBAM reveal a long-standing pattern: conjuring up the danger of carbon leakage to maintain existing business models as profitably as possible for as long as possible and thus prevent effective climate protection. Lamb and colleagues have described this as the delay strategy of pursuing non-transformative solutions.[53] Various lobbying forces are jointly working on advancing this strategy in policy disputes and thus help the major CO_2 emitters to continue to operate in large and diversified business coalitions.[54]

The EU's Green Taxonomy for sustainable activities

A wide range of corporate actors and think tanks has also attempted to influence the Green Taxonomy. To meet the European Union's 2030 climate goals of reducing CO_2 emissions by 55% compared with 1990 levels, the Brussels-based think tank Bruegel estimated an annual investment gap of €300 billion.[55] The Green Taxonomy was created to steer additional private investments into sustainable assets in addition to the €1.8 trillion earmarked in the EGD.

Six objectives are central to this effort: (1) climate change mitigation, (2) climate change adaptation, (3) promoting a circular economy (an economic system minimizing waste and promoting recycling), (4) fighting pollution, (5) protecting water, and (6) protecting biodiversity. From the outset, the question of what types of investments can be classified as sustainable was contested. A technical expert group consisting of thirty-five members from civil society, academia, business, and the finance sector[56] was established in 2018 to assist in the development of the Green Taxonomy. The expert group was committed to a science-based approach, which led it to propose the exclusion of fossil gas and nuclear energy from the scope of the Green Taxonomy.[57]

Contrary to the recommendations derived from the science-based findings of the technical expert group, small hydropower projects have been included in the taxonomy due to Scandinavian lobbying efforts. Bioenergy groups also succeeded in removing the requirement to limit bioenergy to advanced feedstocks, and various agricultural lobby groups managed to remove declining GHG emission standards.[58] However, the biggest success was secured by fossil and nuclear interests, which managed to obtain the inclusion of gas and atomic projects in the Green Taxonomy under certain conditions. Nuclear energy had been omitted by the technical expert group because it violates the precautionary principle because of the unresolved issue of securing long-term storage of nuclear waste. Gas does not meet the requirement in principle, either, but has been included for projects that keep CO_2 emission low due to plant efficiency and/or the additional deployment of carbon capture and storage (CCS) technologies, among other exceptions. Yet in contrast to such relatively stringent conditions, fossil gas-powered projects already permitted until 2030 and designed to switch to hydrogen by 2035 can still be funded despite their higher GHG emissions.[59]

Based on the entries in the Transparency Register, a study by Reclaim Finance identified no fewer than 189 nuclear and gas lobby actors employing 825 lobbyists. The groups were found to spend between €71.4 million and €86.6 million a year to influence European policymaking. The study found a very high frequency of meetings between the lobby groups and the EC, totalling 310 such meetings between 2018 and 2020. The fossil gas lobby accounted for the bulk of the lobbying force involved in the effort.[60] The effort to maintain fossil gas dependency in the European Union thus involved many lobby groups working in coalition, including gas companies, the oil majors, turbine producers such as Rolls Royce, and associations of large energy customers such as the European chemical industry.[61] Earlier work from InfluenceMap also found that the shadow banking industry had lobbied for less rigid classification criteria for green investments and for soft methods to assess the sustainability of funded projects.[62]

In addition to the industry lobby groups, think tank allies objected to the Green Taxonomy. For example, the EPICENTER think tank network circulated a policy brief by Carlo Stagnaro and Stefano Verde, the network's energy policy experts, 'Only a Sith deals in absolutes: how to nudge the Taxonomy towards the Light Side'. Stagnaro and Verde, who have close relationships with the Italian gas industry, accuse the Commission of

> actively picking technological winners among the existing clean(er) technologies, to the detriment of other technologies that may well be as clean and even more so to the detriment of technologies that are not yet available. The Taxonomy deals in absolutes: it is founded upon the claim that a bureaucratic document can draw a line between Good and Bad, by attaching a label of Absolute Good

to technologies that have the capability to create an environmental Eden in an imperfect, dirty world.[63]

The authors thus attempted to include fossil gas in the list of clean industries, arguing that fossil gas is bridging technology—an argument the Commission ultimately adopted.

The CEP, mentioned earlier, also attacked the Green Taxonomy in six policy papers published between 2018 and 2020.[64] The CEP objected to a single view of sustainability (thus rejecting the science-based precautionary approach of the technical expert group), suggesting that the Green Taxonomy stands in contrast to the 'risk-based' approach of financial market regulation and declaring the whole effort useless: 'Detailed measures on how corporate actors should consider sustainability aspects and long-termism are unnecessary. They risk being inefficient and may run counter to the interests of owners, customers and other stakeholders'.[65]

Think tanks generalize the arguments of lobby groups, which in turn are used to legitimate demands submitted by firms and business associations in the Commission's policy consultation processes. Regarding the Green Taxonomy, diverse business associations such as BusinessEurope, the European cement association Cembureau, and the European Steel Association Eurofer have insisted on the inadequacy of binding sustainability criteria and demanded a more comprehensive assessment of products such as steel beyond their CO_2 emissions records. Meanwhile, Germany's BDI—the voice of German Industry—demanded Europe wait for a science-based method of life-cycle analysis.[66] Until such a unified method can be applied, they argued, only direct emissions should be considered in calculations of sustainability. When it comes to delaying ambitious efforts to achieve climate policy objectives, much creativity in the defence of fossil interests can be observed in the lobbying against the Green Taxonomy, ranging from insistence upon technology neutrality (remaining unbiased toward which specific technologies are to be used) and promoting the need for multiple perspectives on sustainability to demanding new and allegedly superior methods of assessing products and advocating the continued use of fossil sources until better technology is available.

KEY FRAMES OF EU CLIMATE POLICY OBSTRUCTION

The two examples just discussed illustrate that discursive framings are important tools in climate policy disputes. Climate obstructionists typically advance numerous arguments against ambitious climate policy.

Since European integration and its climate policy are strongly shaped by the logic of competitiveness and the concept of ecological modernization, respectively, the frames utilized in EU arguments need to resonate with these two points of reference. Three main narratives can be found in the obstructionist's toolbox: first, that market-based instruments are preferable to regulatory 'command and control' mechanisms (such as bans and binding commitments) and allow companies maximum flexibility. Second, that climate policy approaches should be technology neutral. Third, that policy instruments should always be designed in such a way that the competitiveness of European companies does not suffer.

In terms of promoting market-based solutions, the CEP claimed as early as 2008 that 'fixed quota systems for reducing greenhouse gas emissions are inferior to emissions trading on the grounds of efficiency'.[67] Similar narratives can be identified in the CEP's evaluation of the commission's recent offshore renewable energy strategy, which aims to significantly increase the capacity for offshore wind production by 2030. On the one hand, the legal think tank stated that 'the Commission is right to pursue the goal of bringing offshore renewable energy into competition and exposing it to market risks'.[68] On the other hand, the CEP opposed plans for concrete sector targets. The organization declared that these targets 'should be rejected as a dirigiste presumption of knowledge '.[69] By associating the commission proposal with a planned economy measure in this way, they tarred it with a socialist stigma. When, in the 2000s, some EU member states, including Germany and Spain, triggered a boom in renewable energy via guaranteed feed-in tariffs, the grey spectrum of actors criticized these instruments as incompatible with the European single market. With the reform of state aid guidelines in 2014, the renewable expansion system based on feed-in tariff mechanisms was largely undermined.[70]

Regarding the second narrative, the call for a technology-neutral approach appears in many current efforts to obstruct a stronger EU climate policy. One example is the stance of the umbrella organization of European car manufacturers, ACEA. In January 2020, the association issued a position paper on EGD, with ten recommendations for action. In the first point, the demand for technology neutrality is intertwined with an appeal for undistorted competition. ACEA stated,

- *Technological neutrality* must be maintained in order to reflect the diverse requirements of different vehicle segments and the many use cases of customers.

- The Commission should *refrain from (directly or implicitly) mandating certain technologies* for specific vehicle segments.
- It should also maintain the integrity of the European single market by *discouraging national and local bans on specific technologies which can deliver further CO_2 improvements*'.[71]

In a similar vein, FuelsEurope has repeatedly called for a technologically neutral approach to achieve the EU climate goals in the transportation sector. In so doing, the industry association argued against a ban on combustion engines.[72] In addition, after German Minister of Transportation Volker Wissing managed to extend the time frame for an EU-wide ban on combustion engines that would run only on CO_2-neutral fuels, he also stated that this move would 'ensure the element of technology neutrality'.[73] The concept of technology neutrality, or openness, also plays a central role in the arguments advanced by climate obstruction actors such as Shell,[74] as well as by the electricity industry.[75]

These examples illustrate the way in which industry representatives have repeatedly used the narrative of technology neutrality to enable the preservation of climate damaging technologies. Coupled with technological optimism and empirically unsubstantiated and misleading references to the innovative power of humanity in general, the advancement of renewable energy sources and other effective climate solutions can thereby be slowed.[76]

Turning to the third narrative, climate policy approaches in the European Union are subsumed under the primacy of competitiveness. For example, BusinessEurope's March 2023 Reform Barometer was titled 'The EU's global competitiveness under threat'.[77] In this report, the organization concluded, among other things, that stricter climate protection measures are counterproductive because they could lead to a migration of industry to countries with much lower levels of climate protection. This carbon leakage narrative is particularly popular at the EU level because promoting the European Union's competitiveness has been a key political project of the European Union. As noted earlier, the concept of carbon leakage was particularly prominent in the construction of the EU ETS. However, as shown by the creation, in 2023, of the European Union's Net Zero Industry Act (in response to the US's Inflation Reduction Act), climate policies are closely intertwined with global competitiveness in grey as well as in green industries.

Apart from these three strands of discourse, which are central to climate obstruction in the European Union, other narratives also delay ambitious climate policies. For example, the belief in new breakthrough technologies is reflected in the EGD itself, without any certainty that such technological optimism will ever be realized. Such optimistic forecasts regarding the eventual availability of hydrogen, e-fuels, or carbon dioxide removal systems are often deployed to delegitimize regulatory measures that could contribute immediately to emissions reductions and thus also serve as excuses to shirk responsibility for more appropriate climate action. At the same time, industrial sectors know very well how to position themselves as part of the solution to the climate problem (Tilsted and colleagues have elaborated upon this strategy in for the petrochemical industry).[78] In this respect, discourses of delay in the European Union are based on structuring the discursive field in such a way that existing business models can be continued for as long as possible and remain protected from attack.

CONCLUSION

The European Union's climate policy is best understood as a contested field. At the European level, denialism is relatively uncommon; secondary obstruction is the dominant form of blocking ambitious climate policy. The grey spectrum of actors relies on different strands of discourse to implement their obstructive goals, emphasizing the European single market, the primacy of competitiveness for European business, and the technocratic concept of ecological modernization. Despite the material superiority of grey companies, their associations, and related think tanks, far-reaching climate policy mandates including the precautionary principle and ambitious CO_2 reduction goals, such as the EGD, have been set. Nevertheless, the disputes over the EU ETS and the Green Taxonomy show that the achievement of ambitious goals is highly uncertain. Strong forces of delay are at work promulgating discourses and lobbying practices in both the European arena and, via the EUCO, the nation-state arenas. Consequently, the European climate policy picture remains unclear, oscillating between the justifiable claims of global environmental leadership and strong impediments, resulting in a general orientation toward global competitiveness and hardened power relations that continue to protect fossil interests.

Although academic research on European lobbying has clarified the relevance of lobby coalitions in impeding climate action,[79] more work needs to be done on the ways in which green and grey alliances form and fare in key climate policy conflicts across the European Union. Beyond an assessment of lobbying success and failure with regard to specific legislation, a key research priority will be to learn more about if and how climate proposals (e.g. a revival of public transportation or a modification of state aid to support policy instruments like the feed-in tariff, etc.) have been blocked from entering the European Union's policy debate in the first place. All the work thus far supporting the pluralist perspective of lobby balance—which holds that asymmetry in lobby power relations is not a problem—fails to address the 'second face of power', or 'non-decision', namely agenda-setting ability.[80] Finally, it will be critical to examine the history and mobilization of the 'innovation principle' that industry lobbyists and neoliberal think tanks invoke in an effort to counter the impact of the precautionary principle.[81]

NOTES

1. European Commission (2019, December), 'President von Der Leyen on the European Green Deal', December 2019, https://ec.europa.eu/commission/prescorner/detail/en/SPEECH_19_6749.
2. Kristoffer Ekberg, et al. (2022), *Climate Obstruction: How Denial, Delay and Inaction Are Heating the Planet*, New York: Taylor & Francis, p. 12.
3. William F. Lamb, et al. (2020), 'Discourses of Climate Delay', *Global Sustainability*, 3: 1–5, doi:10.1017/SUS.2020.13.
4. Amanda Machin (2019), 'Changing the Story?The x of Ecological Modernisation in the European Union Machin, A.', *Environmental Politics*, 28, 2: 208–227.
5. Dieter Plehwe and Karin Fischer (2022), 'Tales of Passage from the North to the South and Back: Constitutionalizing (European) Neoliberalism'. In: Fernando López-Castellano, Carmen Lizárraga, Roser Manzanera-Ruiz (eds.) *Neoliberalism and Unequal Development: Alternatives and Transitions in Europe, Latin America and Sub-Saharan Africa*, pp. 123–140. New York: Taylor and Francis, doi:10.4324/9781003153306-11/TALES-PASSAGE-NORTH-SOUTH-BACK-DIETER-PLEHWE-KARIN-FISCHER.
6. Sylvain Laurens (2017), *Lobbyists and Bureaucrats in Brussels: Capitalism's Brokers*, *Lobbyists and Bureaucrats in Brussels: Capitalism's Brokers*. New York: Taylor and Francis, doi:10.4324/9781315267258/LOBBYISTS-BUREAUCRATS-BRUSSELS-SYLVAIN-LAURENS.
7. Tobias Haas (2019, February), 'Struggles in European Union Energy Politics: A Gramscian Perspective on Power in Energy Transitions', *Energy Research & Social Science*, 48: 66–74, doi:10.1016/J.ERSS.2018.09.011.
8. Hence the moonshot analogy; see Gengnagel, Vincent, Zimmermann, Katharina (2022), 'The European Green Deal as a Moonshot: Caring for a Climate-Neutral Yet Prospering Continent?' *Historical Social Research*, 47,4: 267–302, doi 10.12759/hsr.47.2022.47
9. Andrea Lenschow and Carina Sprungk (2010, January), 'The Myth of a Green Europe', *JCMS: Journal of Common Market Studies*, 48, 1: 133–154, doi:10.1111/J.1468-5965.2009.02045.X.
10. Ibid.
11. Alexandra Brzozowski and Claire Stam (2019), 'Greens Set Conditions to Approve Nominee von Der Leyen – EURACTIV.Com', *Euractiv*, https://www.

euractiv.com/section/future-eu/news/greens-set-conditions-to-approve-nominee-von-der-leyen/.
12. Ingmar von Homeyer, Sebastian Oberthür, and Andrew J. Jordan (2021), 'EU Climate and Energy Governance in Times of Crisis: Towards a New Agenda', *Journal of European Public Policy*, 28, 7: 959–979, doi:10.1080/13501763.2021.1918221.
13. Annette Bongardt and Francisco Torres (2022), 'The European Green Deal: More than an Exit Strategy to the Pandemic Crisis, a Building Block of a Sustainable European Economic Model', *JCMS: Journal of Common Market Studies*, 60, 1: 170–185.
14. European Commission (2021), 'Fit for 55 – The EU's Plan for a Green Transition', https://www.consilium.europa.eu/en/policies/green-deal/fit-for-55-the-eu-plan-for-a-green-transition/.
15. Jacob, 'Umwelt- Und Klimapolitik' (2022), in Werner Weidenfeld and Wolfgang Wessels (eds.), *Jahrbuch der Europäischen Integration 2022*, pp. 308–309. Baden Baden: Nomos.
16. Volker Eichener and Helmut Voelzkow (1994), *Europäische Integration Und Verbandliche Interessensvermittlung*. Maburg: Metropolis.
17. Beate Kohler-Koch and Rainer Eising (eds.), *The Transformation of Governance in the European Union*. London: Routledge.
18. Following the US Lobby Disclosure Act of 1995, which requires registered lobbyists to file quarterly activity reports with Congress, the European Union established its own lobby register in 2008 (first joined by the EC, then eventually the EP, although the Council of Ministers has never taken part). Unlike the US register, the European register is voluntary, though some restrictions apply if companies or other organizations involved in lobbying do not register.
19. Transparency Register Management Board (2022), 'Annual Report on the Functioning of the Transparency Register 2021'. https://www.europarl.europa.eu/at-your-service/files/transparency-and-ethics/lobby-groups/en-annual-report-on-the-operations-of-the-transparency-register-2022.pdf
20. Yearly reports by corporate watchdog organizations such as Corporate Europe Observatory or LobbyControl provide details of the shortcomings of the current register, which nevertheless provides a better overview of the lobbying landscape than had previously been available. See Will Dinan (2010), 'The Battle for Lobbying Transparency'. In: Alter EU (ed.), *Bursting the Brussels Bubble*, pp. 139–147. Alter EU, www.alter-eu.org/book/bursting-the-brussels-bubble.
21. European Union Employment Advisor (2022), 'How Many People Work for the European Commission?', https://euemployment.eu/european-commission-employee-number/.
22. Mahoney, 'Brussels versus the Beltway'; Cornelia Woll (2012), 'The Brash and the Soft-Spoken: Lobbying Styles in a Transatlantic Comparison'. *Interest Groups and Advocacy*, 1, 2: 193–214.
23. Beate Kohler-Koch and Christine Quittkat (2011), *Die Entzauberung Partizipativer Demokratie: Zur Rolle Der Zivilgesellschaft Bei Der Demokratisierung von EU-Governance*. Frankfurt am Main: Campus Verlag.
24. Heike Klüver (2013), *Lobbying in the European Union: Interest Groups, Lobbying Coalitions and Policy Change*. Oxford: Oxford University Press, 2013.
25. Haas, 'Struggles in European Union Energy Politics'.

26. BelénSabido, Pascoe Balanyá (2017, October), 'The Great Gas Lock-In: Industry Lobbying behind the EU Politics for New Gas Infrastructure', https://corporateeurope.org/sites/default/files/the_great_gas_lock_in_english_.pdf.
27. Haas, 'Struggles in European Union Energy Politics'.
28. InfluenceMap (2022, April), 'Europe: Does Corporate Europe Support Climate Policy?', https://europe.influencemap.org/report/Does-Corporate-Europe-Support-Climate-Policy#13.
29. In cases where we could not obtain information on lobby expenditure in 2019, we used the year 2018 or the year 2020 instead. The two figures do not include the group of companies that are completely opposed to the European Union's climate policy (groups E and F) for which we could not identify their expenditure for lobbying at the EU level.
30. Weise, Zia (2022, June). 'The Lobbying War over Cutting EU Emissions', Politico, https://www.politico.eu/article/europe-emissions-climate-change-lobbying-war/
31. See the overview article Puck, Jonas, Lawton, Thomas, Mohr, Alexander (2018) The Corporate Political Activity of MNCs: Taking Stock and Moving Forward. Manag Int Rev58, 663–673 (2018). https://doi.org/10.1007/s11575-018-0364-0 and the Swedish case study Fjelkman, Anton and Rolanda Fromholde (2021) Strategic Management of Public Affairs in Swedish Multinational Corporations—The Recipe for Success. Lund: Lund University (https://www.lunduniversity.lu.se/lup/publication/9050100)'.
32. Corporate Europe Observatory (2017, February), 'Corporate Interests Continue to Dominate Key Expert Groups', https://corporateeurope.org/en/expert-groups/2017/02/corporate-interests-continue-dominate-key-expert-groups.
33. Anne Therese Gullberg (2011), 'Equal Access, Unequal Voice. Business and NGO Lobbying on EU Climate Policy'. In: Transparency International, *Global Corruption Report: Climate Change*, pp. 39–44. Transparency International. https://issuu.com/transparencyinternational/docs/global_corruption_report_climate_change_english
34. Klemens Joos (2014), 'Erfolg Durch Prozesskompetenz. Paradigmenwechsel in Der Interessenvertretung Nach Dem Vertrag von Lissabon'. In: Doris Dialer and Margarete Richter (eds.), *Lobbying in Der Europäischen Union. Zwischen Professionalisierung Und Regulierung*, pp. 29–45. Wiesbaden: Springer VS; Andreas Dür and Gemma Mateo (2012, September), 'Who Lobbies the European Union? National Interest Groups in a Multilevel Polity', *Journal of European Public Policy*, 19, 7: 969–987, doi:10.1080/13501763.2012.672103; Christian Lindner (2014), 'Lobbyismus Und Interessenvertretung Auf Europäischer Ebene. Zwischen Professionalisierung Und Regulierung?'. In: Doris Dialer and Margarete Richter (eds.), *Lobbying in Der Europäischen Union. Zwischen Professionalisierung Und Regulierung*, pp. 47–59. Wiesbaden: Springer VS.
35. InfluenceMap (2021, November), 'German Automakers and Climate Policy', https://influencemap.org/EN/report/German-Automakers-And-Climate-Policy-a3edf15c64b2e258c29f83beb93337f6.
36. Susanne Götze and Annika Joeres (2020), *Die Klimaschmutzlobby*. München: Pieper,; Influence Map (2021), 'Aviation Industry Lobbying and European Climate Change Policy'. https://influencemap.org/report/Aviation-Industry-Lobbying-European-Climate-Policy-131378131d9503b4d32b365e54756351
37. Balanyá, 'The Great Gas Lock-In'.

38. Ibid.
39. A Canadian environmental economist developed the basic idea; see John Harkness Dales (1968), *Pollution, Property and Prices*. Toronto: University of Toronto Press.
40. Steffen Böhm, Maria Ceci Misoczky, and Sandra Moog (2012, November), 'Greening Capitalism? A Marxist Critique of Carbon Markets', *Organization Studies*, 33, 11: 1617–1638, doi:10.1177/0170840612463326.
41. Véra Ehrenstein and Daniel Neyland (2021, April), 'Economic Under-Determination: Industrial Competitiveness and Free Allowances in the European Carbon Market', *Journal of Cultural Economy*, 14, 5: 596–611, doi:10.1080/17530350.2021.1908397.
42. Carbon Market Watch (2021, July), 'Survival Guide to EU Carbon Market Lobby: Debunking Claims From Heavy Industry', https://carbonmarketwatch.org/wp-content/uploads/2021/06/Survival-guide-to-industry-lobbying_WEB.pdf.
43. Ehrenstein and Neyland, 'Economic Under-Determination', p. 597.
44. According to a study by TU Delft, such windfall profits of European industry totaled more than €50 billion between 2008 and 2019. These profits were generated because surplus allowances could be sold and because allowances themselves were generated by flexible mechanisms; thus, their cost was factored into the price of companies' products. See: Carbon Market Watch, 'Survival Guide to EU Carbon Market Lobby', pp. 4–6.
45. Boston Consulting Group (BCG) (2008), 'Assessment of the Impact of the 2013–2020 ETS Proposal on the European Cement Industry', p. 1. https://www.oficemen.com/wp-content/uploads/2017/05/BCG-Assessment_IMPACT-2013-2020.pdf
46. Carbon Market Watch, 'Survival Guide to EU Carbon Market Lobby', p. 8.
47. Ehrenstein and Neyland, 'Economic Under-Determination', p. 608.
48. IFIEC Europe (2021, May), 'Energy-Intensive Industries' Recommendations on the "Fit for 55 Package"', Brussels: IFIEC. https://www.businesseurope.eu/publications/fit-55-package-businesseurope-position-paper
49. Kohler-Koch and Quittkat, *Die Entzauberung Partizipativer Demokratie*; emphasis added.
50. Martin Menner and Götz Reichert (2022), 'Fit for 55: EU-Emission Trading Scheme (EU ETS I) for Industry and Energy', p. 1, www.cep.eu.
51. Ibid.
52. Lamb et al., 'Discourses of Climate Delay'.
53. Ibid.
54. According to the European Transparency Register, 178 organizations engage in CBAM policy discussions. Among them are eighty-two companies, thirty-six trade and business associations, thirteen professional consultancies, ten NGOs, nine in-house lobbyists and trade/business/professional associations, eight think tanks, seven trade unions (and a few self-employed, or from other organizations like the City of London), and two law firms. Furthermore, twenty organizations have submitted consultations, including seven companies and one in-house lobbyist, three NGOs, and nine trade associations. In terms of sectors, steel and materials account for the largest share of submissions.
55. Simone Tagliapietra and Reinhilde Veugelers (2020), 'A Green Industrial Policy For Europe', Bruegel, https://www.bruegel.org/sites/default/files/wp_attachments/Bruegel_Blueprint_31_Complete_151220.pdf.

56. European Commission (2020), 'Technical Expert Group on Sustainable Finance (TEG)', https://finance.ec.europa.eu/publications/technical-expert-group-sustainable-finance-teg_en.
57. Activities in support of the climate policy objectives of reducing CO_2 emissions could not be included if they would be found to violate another sustainability condition.
58. InfluenceMap (2020), 'Lobbying on the EU Taxonomy's Green Criteria: An InfluenceMap Report', www.politico.eu/wp-content/uploads/2020/12/14/IM_Taxonomy_Industry_Lobbying_Dec2020_final.pdf.
59. European Commission (2022), 'Commission Delegated Regulation (EU) 2022/1214', https://eur-lex.europa.eu/legal-content/EN/TXT/?uri=CELEX:32022R1214.
60. Reclaim Finance (2020), 'Insuring Our Future: 2020 Scorecard on Insurance, Fossil Fuels and Climate Change' https://u6p9s9c8.rocketcdn.me/site/wp-content/uploads/2020/12/IOF-REPORT-2020-FINAL-1-1.pdf.
61. A detailed study of the ten major corporations with fossil gas activities nevertheless shows the extent to which the majors and their associations need to be considered drivers in delay strategies, pushing back science-based solutions for an energy transition in line with the Paris goals. See Influencemap (2022), 'The International Gas Union's Climate Strategy'. https://influencemap.org/landing/-a794566767a94a5d71052b63a05e825f-20189
62. Including the European Fund and Asset Management Association, the Association for Financial Markets in Europe, the European Banking Federation, and European Issuers; see InfluenceMap (2019), 'Asset Managers and Climate Change'. https://influencemap.org/report/Asset-Managers-Climate-Change-2023-22976
63. Carlo Stagnaro and Stefano Verde (2022), 'Only a Sith Deals in Absolutes: How to Nudge the Taxonomy towards Light Side', https://ec.europa.eu/info/publications/220202-sustainable-finance-taxonomy-complementary-climate-.
64. Bert Van Roosebeke and Philipp Eckhardt (2018), 'Sustainable Finance', *CepPolicyBrief 32/2019*, https://www.cep.eu/fileadmin/user_upload/cep.eu/Analysen/COM_2018_97_Nachhaltige_Finanzwirtschaft/cepPolicyBrief_COM_2018__97_Sustainable_Finance.pdf; Martina Anzini and Bert Roosebeke (2019), 'Green Taxonomy', *CepPolicyBrief 2019/08*, https://www.cep.eu/fileadmin/user_upload/cep.eu/Analysen/COM_2018_353_Green_Taxonomy/cepPolicyBrief_COM_2018__353_Green_Taxonomy.pdf; Philipp Eckhardt, Anne-Carine Pierrat, and Bert Van Roosebeke (2019), 'The EU Green Bond Standard (GBS) An Analysis of the TEG's Draft', https://www.cep.eu/fileadmin/user_upload/cep.eu/Studien/cepInput_Green_Bonds/EU_Green_Bond_Standard__GBS_.pdf; Bert Van Roosebeke and Martina Anzini (2019, April), 'Green Taxonomy: The Position of the EU-Parliament', www.cep.eu; Bert Van Roosebeke (2020), 'EU Taxonomy for Sustainability: Summary and Assessment', *CepAdhoc*, www.cep.eu; and Philipp Eckhardt and Anne-Carine Pierrat (2020), 'The Renewed Sustainable Finance Strategy: Questionable Measures in the Regulatory Pipeline', *CEPInput*, 23: 3–14.
65. Eckhardt and Pierrat, 'The Renewed Sustainable Finance Strategy', p. 1.
66. EUROFER (2019, April), 'Position Paper Sustainable Finance Taxonomy Update', https://www.eurofer.eu/assets/Uploads/20190424-EUROFER-Position-Paper-Taxonomy-Regulation.pdf; 'Stellungnahme: Inception Impact Assessment EU-Taxonomie Zur Konsultation Zum Delegierten Rechtsakt Der EU-Kommission

über Die Einführung Technischer Kriterien Zur Eindämmung Und Anpassung an Den Klimawandel'.Bundesverband der Deutschen Industrie e.V. (2020), https://newsletter.vero-baustoffe.de/fileadmin/user_upload/NL-Maerz-2021/d-BDI-Konsultation_1.pdf
67. Bert Van Roosebeke (2008), 'Reduktion Der Treibhausgas-Emissionen Der Mitgliedstaaten', www.cep.eu.
68. Svenja Schwind and Götz Reichert (2021), 'Offshore Renewable Energy', *CepPolicyBrief No. 18/2021*, p. 1, www.cep.eu.
69. Ibid., p. 3.
70. Haas, 'Struggles in European Union Energy Politics'.
71. Ehrenstein and Neyland, 'Economic Under-Determination' (emphasis in original).
72. Referring to technology neutrality in a recent publication, the association asked the rhetorical question, 'Surely, it is about reducing, and stopping fossil combustion emissions, rather than stopping all combustion technology equipment?' FuelsEurope, 'Genuine Technology Neutrality in Transport Benefits 2050 Climate Targets', https://www.fuelseurope.eu/publications/publications/genuine-technology-neutrality-in-transport-benefits-2050-climate-targets. Accessed April 4, 2023.
73. Bundesministerium für Digitales und Verkehr (2023, March), 'Wissing: Weg Frei Für Verbrenner Mit CO2-Neutralen Kraftstoffen', https://bmdv.bund.de/SharedDocs/DE/Pressemitteilungen/2023/026-wissing-weg-frei-fuer-verbrenner-mit-co2-neutralen-kraftstoffen.html.
74. Shell (2022, July), 'Sectoral Decarbonization under the EU "Fit For 55"', https://www.shell.com/sustainability/transparency-and-sustainability-reporting/advocacy-and-political-activity/advocacy-releases/_jcr_content/root/main/section/simple/promo_copy_954155441_1493642008/links/item0.stream/1659686999280/2b84ed678d80f8055e0db7b9de7b7e9ad5bf5556/ff55-statement-design-july-2022.pdf.
75. The umbrella organization Eurelectric wrote in its position paper on the EGD: 'it is important to remain technology open'. Eurelectric (2019), 'Powering the Green Deal Eurelectric Position Paper', https://cdn.eurelectric.org/media/4127/eurelectric_2030_high_level_paper-2019-030-0736-01-e-h-C36F5F4E.pdf.
76. Lamb et al., 'Discourses of Climate Delay', p. 3.
77. BusinessEurope (2023, March), 'Reform Barometer 2023 – The EU's Global Competitiveness under Threat', https://www.businesseurope.eu/sites/buseur/files/media/reports_and_studies/reform_barometer_2023/2023-03-21_reform_barometer_2023.pdf.
78. Joachim Peter Tilsted, et al. (2022, December), 'Petrochemical Transition Narratives: Selling Fossil Fuel Solutions in a Decarbonizing World', *Energy Research & Social Science*, 94: 2–13, doi:10.1016/J.ERSS.2022.102880.
79. See, for example, Klüver, *Lobbying in the European Union*.
80. See, for example, Peter Bachrach and Morton S. Baratz (1970), *Power and Poverty*. New York: Oxford University Press.
81. For different views, see 'The 'innovation principle' trap' by corporate Europe Observatory (https://corporateeurope.org/en/environment/2018/12/innovation-principle-trap) and the European Commission's article on research and innovation: 'Innovation Principle Makes EU Laws Smarter and Future-Oriented, Experts Say' (https://research-and-innovation.ec.europa.eu/news/all-research-and-innovation-news/innovation-principle-makes-eu-laws-smarter-and-future-oriented-experts-say-2019-11-25_en).

14
Conclusion

Ten Lessons about Climate Obstruction in Europe

J. TIMMONS ROBERTS AND ROBERT J. BRULLE

INTRODUCTION: OVERVIEW OF THE EUROPEAN CASE STUDIES

Europe has, since the 1990s, sought to become a global leader on climate change. But if we have learned anything from this set of eleven national cases and the situation in the European Union, it is that building consensus on this goal has been extraordinarily difficult. The reasons lie in the way the region's national economies have developed, the core problem being the deep interdependencies and interlinkages between national governments and fossil fuel and other heavy industries. This situation can be seen in the heavily coal- and oil-dependent Eastern European nations examined here: Poland, the Czech Republic, and Russia. But it is also visible in the Netherlands, with its virtual marriage to Shell, and Scotland's rising dependence upon North Sea drilling.

But it is not just the oil, gas, and coal industries whence comes the obstruction of ambitious climate action. Germany's economy is centred on its automobile industry; Ireland's, on agriculture (which is high-carbon). Sweden's forestry sector resists realistic carbon rules, and oil refining and distribution monopolies are intertwined with state agencies in Spain and Italy. In a wide variety of national contexts, history and the structure of the economy play formative roles in shaping the actors, discourses, and tactics of climate obstruction across Europe. Civil society and others seeking to act on the climate emergency have sought to tailor their responses to these contexts, with widely varying degrees of success.

This conclusion synthesizes the major findings of the volume, highlighting common characteristics of climate obstruction across Europe, and discusses the impact of particular historical circumstances and governance structures in shaping the unique dynamics of obstruction efforts in various European countries. Here, we draw out ten lessons that apply across these cases. These lessons include the roles of national histories and economies, but also some common factors such as the efforts of international right-wing think tanks, PR firms, lobbyists, and the 'legacy' and social media. We also discuss this volume's limitations, including the absence of information on major countries such as France and Norway, and then outline a research agenda for further developing our scholarly understanding of climate obstruction across Europe and the rest of the world. We end with a discussion of areas that should be a special focus of efforts toward change and some policy recommendations that arose from the national cases and that we think might be more widely applied.

Our goal is to expand our understanding of climate obstruction in Europe and is both scholarly and practical. First, we seek to expand scholarly knowledge beyond the most-studied case: the United States. In our introductory chapter, we presented the current state of knowledge on the 'structure of obstruction' and identified the actors and organizations blocking climate action, revealing how they are connected and how they manage to achieve their objectives so often. The main take-home was that industries, wealthy individuals, and their foundations coordinate their efforts to fund right-wing media, think tanks, university programmes and centres, PR and law firms, and front groups, all of which do their parts in developing and advancing the ideas that government action on climate change infringes on individual rights and is ineffective. Our practical goal in developing this volume is to provide key information on how these actors have succeeded in blocking climate action in a way that may prove useful to those seeking to advance the ambitious climate action that science tells us is needed: rapid decarbonization achieved by ceasing to burn fossil fuels and by addressing other sources of emissions, such as agriculture.

The network of obstruction organizations (sometimes referred to as the 'denial machine'[1]) seeks to advance three main ideas: that climate change is not a major problem, that current industry efforts are taking care of the problem, and that governments should stay out of the way. Climate solutions, such as transition to a renewable energy system, have often been attacked as unreliable and expensive, but as the reality of climate change has become harder to deny, 'discourses of climate delay'[2] and attacking climate solutions have grown more common. These contrarian ideas are advocated directly to decision-makers and the public via the media. The task of the denial machine is not to definitively prove that the climate isn't changing or that solutions never work but to instill just enough doubt to slow or stop regulatory efforts and keep the status quo systems in place.[3] Voluntary initiatives have been particularly effective in avoiding binding regulation of industry. A complex set of organizations have been set up and networked to accomplish their goal: think tanks, right-wing media outlets, PR firms, university programmes, political campaign groups, 'astroturf' local groups, lobbying firms, trade organizations, and coalitions.[4]

Turning to Europe, these case studies show a great range of discourses and strategies to slow the region's efforts to take ambitious steps to address the climate crisis. These narratives range from outright climate denial to dismissing and resisting solutions. Even nations that fancy themselves climate leaders (like Germany, Sweden, and the United Kingdom) face a range of positions opposing climate action. While the United States provides a useful starting place to consider actors and methods of resisting climate action in Europe, our twelve cases reveal substantial differences from the US case. In some countries, such as those of Eastern Europe, the situation is almost completely different, with climate change a second- or third-tier issue, ignored by political leaders and simply not on the public agenda. In others, right-wing think tanks have imported the extreme language and tactics of US organizations such as the Atlas Network, which began in the United States and exported its tactics around the world.[5] Though not prompted, these studies also confirm findings from the States[6] that public opinion only plays a marginal role in climate policy development. In no case did the authors in this collection find that public opinion was a powerful force in developing public policy on climate change in Europe. Rather, the essays show that it is the political elites and the media that drive action on climate.

In the remainder of this chapter, we briefly summarize the country studies and then look across them thematically. Many new findings emerge from the studies, which we have distilled into the these ten main lessons. These ten lessons are followed by a discussion of future research needs and suggestions for best policy and action directions.

THE NATIONAL CASES: A BRIEF RECAP

The UK chapter showed how incumbent interests utilize their structural, institutional, and discursive power to shape climate policy and obstruct ambitious climate action. Gas and oil from the North Sea are locking in direct and indirect greenhouse gas (GHG) emissions in Scotland (as is the case for Norway). The gusher in oil revenues allowed Scotland to consider being a viable independent state, so protecting oil and gas jobs continues to be prioritized over climate action. Despite Ireland's recent industrialization now that it is part of the European Union, the country continues to have a primarily agricultural economy and identity, and major agricultural interests act to obstruct climate regulations there. Sweden's national identity since the Cold War, centred on its 'middle way' between communism and capitalism, has had a formative impact on policy approaches to climate change there. Climate scientists and especially activists are called 'extremists' and 'kooks' by the country's right wing, and only moderate and incremental changes have been adopted. Germany, despite having long-term support from Angela Merkel and the centre-right parties, faces neoliberal opposition to climate action, further advanced by think tanks and campaign organizations. In the Netherlands, Royal Dutch Shell, along with other actors, has used PR influence techniques including co-opting public education, cultural institutions, and scientific inquiry to slow regulatory action there.

Poland, Russia, and the Czech Republic were exempted when it came to early international treaties on reducing emissions; the 1990 baseline adopted by the UN for gauging action on climate came, conveniently enough, just before the unintended plunge in their emissions after the collapse of the Soviet economy. In Poland, the state-controlled coal monopolies are tightly woven with governmental agencies and utilities; together, they work to perpetuate fossil fuel use with little contest. In Russia, the call for climate action is often expressed as a plot by Europe and the United States to keep the nation poor, and a transition away from fossil fuels is rarely discussed in public. That same low 'issue saliency' (including in the media, which in some cases is controlled by fossil fuel firms and is always patrolled by the state) relieves any pressure on politicians to reduce carbon emissions. Italy, meanwhile, has a notably strong climate *counter*movement, led by conservative think tanks linked to the United States and supported by Italian oil and gas companies and their allies. Spain, with its history of authoritarian government, obstructed action on climate change by allowing vested interests including large corporations, oil and gas monopolies, and large-scale agriculture to dominate the country politically and culturally and to control policy levers. In Brussels, the European Union faces a conflict between its core mandate to develop a centralized and competitive European economy and a secondary effort to reduce the region's overall carbon emissions. Lobbying efforts in the capital have sought to reduce the ambition of climate policy initiatives and make them more market friendly.

The map that is emerging from these national cases is of a difficult but not impossible to master terrain on which to advance climate action in Europe. Together, the cases in this volume point to a series of important, overarching insights, which we have distilled into the following lessons. We hope they prove useful for scholars, activists, and civic and global leaders.

TEN LESSONS ON CLIMATE OBSTRUCTION
Lesson 1: The importance of structural economic conditions

How a country's economy has been structured both creates and constrains the possibilities for climate action there. This factor can vastly favour those seeking to slow and block climate action. This seemingly obvious point is worth remembering as it implies that strategies to move forward in each nation will need to keep that history and structure in mind. Countries with powerful fossil fuel industries prove much more difficult terrain on which to mount climate action than those whose economies are dominated by service jobs. At the extreme end of this scale, we saw how Russia is so dependent upon oil and gas revenues that climate issues are simply not considered a primary or even a secondary concern. State-owned companies including Gazprom and Rosneft blur the line between government and corporations—it is not clear who has captured whom. The state itself has a vested interest in perpetuating national and global dependence on their product—fossil fuels—so it asserts cultural and political dominance in a number of subtler and more heavy-handed ways.

In Poland and the Czech Republic, we saw the real and symbolic importance of coal as the bedrock of important regional economies and as a core part of the national culture and economy. In the Polish case, authors Szulecka, Maltby, and Szulecki describe the 'soft impossibilism' expressed in discussions of why any national shutdown of coal and transition to renewables must be undertaken slowly. Calls for faster climate action have been portrayed as 'irrational, ideologically driven', while slow-boating is considered 'realism'. Civil servants cannot imagine anything else. Even oppositional parties in the nation have agreed that the European Union is 'imposing' climate requirements on Poland. Poland thus sought to, and succeeded in, weakening EU climate policies.

Eastern Europe is not alone in this dependency, of course—similar patterns are also evident in the Netherlands, with the profound structural and cultural influence of Royal Dutch Shell there, and in the United Kingdom, home of British Petroleum. Though Shell is not producing oil locally, both companies are woven into the colonial histories of those nations, building their empires on the extraction of oil wealth from distant, low-income, developing countries. Similarly, Italy's oil monopoly Eni dominates national culture and politics, and revolving doors between such huge firms and the government ministries assigned to regulate them makes the system unable to effectively control their actions. Clearly, national economies predetermine obstruction arenas by creating and empowering major incumbent interests based on how the country was integrated into the regional and global economy.

Lesson 2: History counts

We learned in these eleven single-country cases how different national histories lead to different forms of climate obstructionism. Some very important patterns can be discerned, as well as some of what sociologists call 'middle-range generalizations',[7] but each case varied significantly from the others and quite substantially from the US case framework we outlined in Chapter 1. As social scientists, we find this diversity more interesting than if we had discovered one homogenous reality across the region. For those seeking to advance climate action, this variability represents a challenge: a need to know both the history and economic structure of a country well enough to be able to anticipate which strategies for combating climate obstruction might be worth trying nationally and whether they might be 'exportable'.

The case of Spain is striking, with the authoritarian legacy of Franco setting in place a sort of flywheel driving decades of climate denial and delay, as the national economy and the interests of the elites were held to be paramount over any social or environmental concerns. As authors Moreno and Almiron write, 'Franco's fascist administration (1939–75) applied tactics to ignore, deny, hide and absolve the industries of the environmental impacts they produced'. They argue that it took until 2021 for this legacy to be overcome with the passage of Spain's Climate Change and Energy Transit Bill, albeit the legislation still includes an insufficient emissions reduction target of just 23% of 1990 levels by 2030.

Many other examples in this volume show just how important historical forces, events, and trends have been in shaping the nature of climate obstruction in European countries. Cold War dichotomies of communism versus liberalism have led to rejection or at least suspicion of state involvement in climate action in both Western and Eastern Europe (as described in Chapters 5 and 10). The centuries-long tensions between Great Britain and its Commonwealth members in the regions of Scotland, Ireland, and Wales have driven an especially strong interest in developing gas and oil in Scotland so the country could break away from England. The drifting of a radioactive cloud from Chernobyl in Ukraine to Germany led to an anti-nuclear movement that led to the formation of the first Green Party. Brunnengräber, Neujeffski, and Plehwe describe how, although it was initially marginal, the party in Germany gained the ability to drive climate policy by joining centre-left and centre-right coalitions needing a parliamentary majority. The idea of a Green Party (which originated in New Zealand in 1972[8]) spread from Germany to other European nations with parliamentary systems, but the model has had less impact elsewhere.[9]

In Chapter 13, Haas, Neujeffski, and Plehwe argue that the founding of the European Union has proven decisive in the way it handles the need to act on climate and the way it shapes its initiatives around broader goals of the union and its bureaucracy. As a way to diffuse political differences between member nations, the Union has promoted principles of technocracy, attempted to insulate the bureaucracy from their influence, and allowed science to drive policy in areas such as environmental protection. Unifying environmental standards was always the European Union's priority, in part to expedite trade between members and thus make multinational supply chains and production more efficient. The technocracy and unification focus was elite-led, the authors argue, seeking regulatory cohesion and economic cooperation. At its core, they write, technological and market-based solutions became the dominant (and the only adequately acceptable) ways for the Union to address the climate problem.

Denial of the reality of climate change has never existed at the level of the European Union, but there has still been plenty of obstruction there. We learned of the vast lobbying effort that diverges from the technocratic vision associated with Brussels: 25,000 lobbyists ply the long avenues of European Community (EC) and EU office towers, and twenty times more of them represent 'grey' (polluter) industries than 'green' (environmental) groups. One interesting twist in the EU story is that the departure of the United Kingdom from the Union has liberated the trade zone to *increase* its ambition on climate, setting a goal of 55% reductions in emissions by 2030. This goal leads the world but is constrained by the European Union's overriding desire to fit all regional plans into the framework of protecting or improving 'competitiveness' and not constrain companies by requiring them to choose certain technologies over others. The EU Emission Trading System (ETS) has been a herculean effort but has had little success in reducing emissions due to a resistance to regulation in favour of 'market-based' approaches and pressure from entrenched interests such as utilities and major industries.

Then, in February 2022, we saw the invasion of Ukraine by Russia, which created profound pressure on the energy systems of Western Europe. The Continent is now faced with a decision—whether to scale up renewable energy rapidly to reduce dependence on Russian oil and gas or to quickly gather fossil fuel sources from abroad.[10] As these chapters were being written, this history was still unfolding. Fossil fuel corporations and industry organizations used this moment of fear and uncertainty to lock in another generation of infrastructure to deliver their product. They also blatantly exploited the price surge triggered by the war to charge customers billions of additional euros. A step change has occurred, but the fossil fuel dependence will likely endure.

Lesson 3: A wide range of actors is engaged

The beginnings of our understanding of climate obstruction by corporations came from a series of reports about the oil giant ExxonMobil, conducted in 2015 by reporters from a small independent journalism outlet, *Inside Climate News*.[11] Today, many portrayals of climate denial and obstruction still place the blame on just a few actors, often Exxon and the billionaire brothers Charles and David Koch, who made some of their fortune on natural gas pipelines. These are two of the most culpable actors in the story, but this volume's twelve detailed cases from Europe show definitively that a much wider range of actors is engaged in climate obstruction efforts. Who they are varies in each country, from agricultural interests in Spain and Ireland and electrical utilities and coal mines in the former Soviet republics to major oil companies and heavy industries in virtually all of the other countries. This reality shows the limitations of a research effort focused exclusively on the major oil companies and the value of these chapters' findings.

The United Kingdom chapter provides a useful typology of institutions involved in obstructionism in that nation. These include organized sceptic groups and think tanks, business lobby groups and trade organizations, government actors and institutions, and 'floating' organizations that 'bolster incumbent interests . . . from time to time'. Understanding the complexity of the networks between these different types of organizations is fundamental to advancing any effective strategy to advance action on climate change. It also must be at the core of any social science enterprise in this area. For example, the Global Warming Policy Foundation (GWPF) in Britain has for decades fought relentlessly against any regulatory efforts, most recently launching Net Zero Watch, a project that capitalized on Russia's invasion of Ukraine to argue that governments must 'recommit to fossil fuels'. International links, especially to libertarian think tanks in the United States, bolster these efforts with talking points, strategies, and funding. Business associations infiltrate and lobby government agencies and boards and penetrate government agencies through private meetings and 'revolving doors' through which businesses and governments hire each other's employees. Trade unions and the media often 'float' into discussions of, or efforts to block, climate initiatives, claiming dire job losses if they are adopted.

Lesson 4: International networks of think tanks are highly influential

Extreme right-wing think tanks are actively creating and promoting misinformation about climate change and solutions to the problem, and these organizations are active across the region. As the chapters on climate obstruction in the United Kingdom, Italy, Spain, Poland, Germany, and the Czech Republic show, conservative think tanks play an important role in promulgating misinformation about climate change. More specifically, they serve as supposedly independent 'third-party' spokespersons who provide allegedly objective and unbiased information about possible government policies. Across Europe, they continue to question the veracity of climate change, and, when that is not plausible, they question the need to decarbonize, criticize its expense, and argue for fossil fuels' unique ability to provide 'energy security'. Local chapters of these think tanks build and develop ties with political parties through which they 'disseminate talking points' and studies to politicians and the media. These efforts provide political parties opposed to climate action with data, ideas, and documents that can be utilized to advance their position in policy debates on climate change. As the extreme ideas behind these talking points get repeated across borders, they become mainstream.[12]

The role of conservative think tanks in promulgating scientific misinformation about climate change originated in the United States, where the groups have had an enormous influence, especially in informing the policies of successive Republican presidential administrations. Given its success in the United States, the wider conservative movement expanded its efforts to develop conservative think tanks across Europe. Central to this effort has been the Atlas Network, established in the 1980s to establish think tanks advocating for free market and neoliberal policies throughout the world. To realize this goal, the network identified and enlisted individuals in numerous countries to found these conservative think tanks. It then worked with them to secure a steady funding stream for the new organizations and to institutionalize these relationships in a global network. The Atlas Network builds this community through a series of regular meetings and residential training programmes. It now claims more than 500 affiliated groups[13] in nearly every nation on Earth.

Thus this worldwide network has enabled US conservative think tanks to export their approach to obstructing climate action to Europe and other countries across the globe. Prominent US climate sceptic think tank figures have visited their sister European think tanks, attended conferences with their staffs, and assisted them in developing and promulgating materials designed to obstruct climate action. In half of the countries examined in this volume, conservative think tanks were found to play a major role in the development of the nations' climate change discourse. For example, in Italy, the Instituto Brono Leoni was heavily influenced by the Competitive Enterprise Institute and utilized the 'Climategate' incident to undermine public confidence in climate science. In Germany, the think tank EIKE (Euroäisches Institut für Klima und Energie), which is connected to the conservative US think tanks Committee for a Constructive Tomorrow (CFACT) and the Heartland Institute, serves as the country's intellectual hub for climate obstruction. This pattern is repeated throughout Europe. This cross-national network of conservative think tanks has served as a crucial link in facilitating the promulgation of climate science misinformation and the export of climate obstruction efforts beyond the United States.

Lesson 5: Industrial lobbying is powerful

The cases in Europe support the pattern we have observed in the United States by which industrial groups mobilize former politicians and professional influencers to lobby for the policies they prefer.[14] Though lobbying rules vary significantly across the Continent, industries manage to leverage their power over politicians and bureaucrats systematically to distort government decision-making. We see examples of this process in the chapters on the Netherlands, Germany, Sweden, and the European Union. Industries essentially get five bites at the apple to stop or slow climate regulations in Europe. First, they can use back channels to express their displeasure with decarbonization rules in their national parliaments, and then, if they cannot sway the politicians, they undermine the rules' implementation by their national ministries and even local/district authorities, as we saw in several cases. Then they can lobby their MEP and, if that tactic fails, they can go straight to the EU/EC ministries or attempt to intervene in enforcement of the rules by administrative agencies. As mentioned earlier, 25,000 lobbyists ply the halls of Brussels ministries on behalf of grey industries, about twenty times more than those working for 'green' interest groups.

Effective lobbying can preclude the role of the public in climate policy development. In the Scotland case, we saw how industries increasingly seem to be skipping the step of appealing to the public on the merits of their product or the processes by which they produce it. Rather, firms are engaging directly with policymakers. For example, the UK chapter tells a grim story of how the nation's energy firms lobbied against the original 1992 EU-wide carbon tax and won. Lobbying works, and the field is vastly unequal.

Lesson 6: Denial has given way to 'delay'

A clear pattern emerges in these studies: hard denial of the reality of climate change has been taken up in only a few countries, and such claims have become both rare and confined to fringe right-wing groups. However, a well-worn discourse has been honed and developed to become the core of resistance to climate action: scepticism about climate solutions and, more generally, 'discourses of delay' that make the continued use of fossil fuels seem inevitable. An influential article led by William Lamb identified twelve such discourses, grouped into four categories: redirecting responsibility, pushing non-transformative solutions, emphasizing the downsides, and surrender.[15] This framework has proven useful for observers to understand the tactics used by today's climate obstruction actors: greenwashing, technological optimism, appeals to social justice and wellbeing, 'whataboutism' (e.g. 'what about larger nations like China and the US?'), and policy perfectionism are reported all across the Continent. Together, these are all flavours of 'secondary obstruction', a useful term advanced by Eckberg and colleagues in their important book on the subject to distinguish delay and solution scepticism from outright denial of scientific evidence.[16]

We can see variations on climate delay in several of the chapters. In Germany, repeated arguments that renewables are too expensive, and that adopting them aggressively will raise costs and damage the country's competitiveness on global markets, get raised again and again. We see the same arguments in several other countries and in EU and EC discussions of climate policy. There, the argument that market-based approaches were the preferred or only acceptable ones for regional policymaking dominated for decades until the latest round of 'European Green Deal' approaches adopted in 2023, which focused on incentives and investments. The Poland chapter uncovered many discourses of delay, including one heard elsewhere: that only a 'gradual transition' away from fossil fuels is possible, given the country's deep dependence on coal. Facing a recent surge in heat waves and other climate impacts, in Italy a 'We will adapt' discourse is increasingly being advanced, making reductions in GHG emissions seem futile. These examples suggest a rise in 'surrender' discourses of doom, which imply that change is impossible.

Lesson 7: Public opinion is not that important: Climate politics is done in private

Though not prompted to discuss it, in no chapters did the author teams describe any substantial influence of public opinion on climate action in European national arenas despite that their citizens clearly care about the issue. EuroBarometer and other public opinion studies are frequently cited in the media and by environmental campaigners. A recent one, for example, showed that those surveyed felt widespread alarm about climate change, attributed responsibility for climate action to corporations and governments, and desired stronger action from their governments and the European Union.[17] And thirteen times as many said they thought their government is not doing enough to tackle climate change as thought it is doing too much.

More than three-quarters (77%) of EU citizens said they think climate change is a very serious problem at this moment. A majority of Europeans believe that the European Union (56%), national governments (56%), or business and industry (53%), should be responsible for tackling climate change. Thirty-five percent hold themselves personally responsible. More than eight in ten respondents said they think that it is important for their national government (86%) and the European Union (85%) to take action to improve energy efficiency by 2030 (e.g. by encouraging people to insulate their home, install solar panels, or buy electric cars).

However, in our twelve case studies, the diverse authors did not describe *any* mechanism by which public opinion could routinely influence policy outcomes. In the Sweden chapter, public opinion merited only a passing mention. This seems to be because major national climate policies, such as carbon taxes or feed-in tariffs, are negotiated in parliamentary corridors and chambers, not in plebiscites. It is not that people's opinions about the need for climate action or the approaches their country's government are taking aren't important, but that their opinions can only influence policy when people organize themselves into social movements, labour unions, and political parties.

Corporate interests in Europe often manage to influence politics and policy without bothering to influence the public. In Scotland Dinan, Esteves and Harkens report that 'the political influencing strategies of the oil and gas industry in Scotland appears to largely avoid engaging in media and public debate and seeks instead to build relationships and understanding with key political advisors and decision-makers'. Those routes of influence include lobbying and political campaign support, working through industry organizations to influence parties' candidate choices, and assisting their chosen advisors in gaining access to key positions inside and outside government. Sometimes these advisors are climate sceptics and deniers, scientists and economists, as we saw in the case of Sweden. Corporatist negotiations not only with politicians but also with unions, industries, civil society groups, etc. go on in back rooms all around the Continent. We now see why, despite the overwhelming support for stronger climate policy, European nations are moving slowly.

Lesson 8: Public relations techniques and discourses are universally used

Public relations firms and their climate obstruction strategies are ubiquitous in the studies presented in these chapters. These efforts focus on influencing elites and the media to advance the preferred policies of industries, think tanks, and even environmental campaigners. Studying these efforts is difficult: the work of public relations firms is designed intentionally to remain hidden such that clients' discourses and media appearances seem to have happened by themselves. Effective PR campaigns include many other efforts to legitimate an industry and its activities: the UK chapter reminds us that 'discursive power also extends to cultural legitimation through arts and sports sponsorship', reflecting Antonio Gramsci's theories on how power is created and maintained. Industries come to this part of the struggle for influence with vast advantages over their opponents.

The language used in Europe's professional PR campaigns on climate falls into the category of 'secondary obstruction'. For example, in the Swedish and UK cases, we heard not that climate change isn't happening, but that 'transition fuels' and the potential of carbon capture and storage are a viable solution to climate change that lessen the need for emissions reductions right now. In Poland, we heard about the merits of techno-fixes called 'carbon forestry' and 'clean coal'. The Italy chapter uncovered a series of such PR tactics: greenwashing, redirecting responsibility away from companies to individuals, fossil-fuel solutionism, appeals to energy security (especially given the Ukraine invasion), and attacks on climate scientists. These same tactics were employed across the Continent.

Lesson 9: Media coverage of climate often reinforces the status quo

While some major media outlets in Europe routinely cover the urgent need for climate action, in country after country we heard how, intentionally or not, the news media's way of handling climate change often helps sceptic and delayist arguments to influence the public's and decision-makers' thinking. In the Czech case, the media overwhelmingly support the status quo on energy and climate change. As mentioned, Russia's government-owned fossil gas company Gazprom controls more than 60% of media outlets in the country, and so the motive for denial and delay discourses there couldn't be clearer.

However, the reasons for media outlets' hesitancy to take strong positions on the need to rapidly transition off fossil fuels are not always clear. The Italy case showed how even 'progressive' media outlets have advanced climate denial, delay, and obstruction on the air. On Italian TV, contrarians are often treated as experts and elevated to prominent platforms on popular shows and channels. In the case of Ireland, coverage of climate change is often event-based, mentioned only during extreme weather. This approach forecloses deeper discussions and an understanding of how larger trends triggered by human GHG emissions are exacerbating those events.

With the global reach of its major media companies, the United Kingdom merits special attention. The right-wing media in the United Kingdom has achieved a large degree of influence through 'giving space to columnists offering different forms of obstructionism'. A right-wing think tank, the GWPF, for years had great success in placing its 'experts' on prominent programmes; only recently have these individuals been excluded from major outlets such as BBC1, ITV, and Channel 4. However, editorials supporting fracking and attacking climate campaigners and national and EU climate policies have soared once again in the wake of oil supplier Russia's invasion of Ukraine. Further study on how the media are influenced by external forces as well as their internal dynamics is a clear priority going forward for understanding climate obstruction across Europe.

Lesson 10: Greens and the European Union are targets

One final lesson from these cases is that individuals and groups actively advancing climate action are often portrayed as irresponsible, 'ideologically driven', 'irrational', or even dangerous to a nation's identity, economy, or security. In Ireland, Kelly, McNally and Stephens cite 'anti-environmentalists who deride those advocating climate action or attack environmentalist stances for being overly earnest or sanctimonious', where activists are called "environmental nutters', 'lunatic environmentalists', 'headbangers', and 'Luddites marching us back to the 18th Century'. A 'Church of Green' is said to be advancing 'green authoritarianism'. More troubling, many of these attacks often appear in the headlines of Irish newspapers.

The situation in Poland, the Czech Republic, and Russia is at another level entirely. Climate advocates in Russia have been officially listed as 'foreign agents', and their safety threatened. Journalists covering the climate story risk imprisonment or worse. In these countries, the European Union and its environmental policies are often characterized as seeking to destroy these nations' economies—before the Ukraine war, the Russian government was calling climate policy an 'undeclared war against Russia'. In this way, even science and scientists working on climate are seen as vulnerable to the dynamics of a new Cold War, according Poberezhskaya and Martus. In Poland, we heard how Prime Minister Beata Szydło argued that coal was 'a synonym of development and modernity' and that 'attacks on coal are cast as "an attack on sovereignty"'. Among right-wing politicians in that country, the attacks on the European Union and its climate policy are unrelenting: 'unfair and pathological', '[a] hypocrisy of people who usually belong to the elite'. They cause 'chaos', and, in trying to lead the world on climate by example, are undertaking 'kamikaze politics'.

As these lessons illustrate, in spite of some stark national differences in economic and political structures, cultural and media landscapes, and histories, a series of trends are emerging in climate obstruction across Europe. These lessons represent key elements to consider in future research on climate obstruction in the region and globally and key insights for developing strategies to combat it. We also see a set of central actors that appear across the continent and need to be carefully studied and understood: industry organizations, political parties, labour unions and environmental organizations, international think tanks, PR firms, and the news media. All of these lessons are relevant at the EU level as well, with an additional set of structural elements determining how battles over climate policy play out there. There is much more research to be done and more lessons to be drawn.

FUTURE RESEARCH NEEDS

Because of time constraints and a dearth of scholars of and information on climate obstruction in many countries, this volume represents an initial set of snapshots of a complex and shifting situation. As such, there are many directions in which future research can be launched. Our understanding of climate obstruction across Europe will improve as we conduct and gather extant and original research on industries, organizations, and strategies representing a wider range of countries.

We especially need studies on which types of discourses and tactics are most effective in overcoming climate obstruction in different national contexts. As Dinan, Esteves, and Harkins pointed out in the Scotland chapter, we can speculate, but we need concrete data on the actual policy *impact* of discourses of delay and other tactics. Which are working and in which contexts? What is working to counter obstruction? What are the conditions for progress? Which lobbying techniques might environmentalists employ to advance ambitious climate policy? Many other avenues call to be explored, informed by the needs of campaigners, policymakers, reporters, investigators, and litigators.

EMERGENT IDEAS FOR POLICY AND ACTION

The national case studies in this volume and the ten lessons recounted earlier hold extensive policy implications and suggest new possibilities for action by climate advocates and agencies. First and perhaps unfortunately, we learned that, at least in the countries covered in this volume, public opinion is not particularly salient to those who would obstruct climate action. Rather, political elites are the target of industry organizations seeking to slow or block steps to curb emissions quickly. This reality suggests the widespread emphasis on developing strategies seeking to sway the public may be misplaced and that resources could be much more closely focused on swaying key players in European institutions. The international Right understood this reality decades ago.

This volume's authors were not asked to develop detailed policy recommendations, but, based on the insights their chapters generated, there are several we can list for consideration. The Ireland and UK chapters, for example, suggested the formation of 'new coalitions' to fight obstructionism. Daley, Newell, McKie and Painter also suggest the creation of aggressive transparency laws (e.g. on lobbying, political contributions, and committee and parliamentary processes) and pointed out a need for rules to stop or regulate the 'revolving door' between industry and government agencies. They also discussed the need for regulatory oversight of the media (and social media), including rules on claims-making and disinformation, establishing a definition of greenwashing, and more controls over social media platforms.

There is much to be done. This volume has provided key information on the factors that shape the nature, extent, and form of climate obstruction efforts. All of the countries in Europe have various forms of climate obstruction, some blatant, others operating in the smoke-filled rooms of corporate lobbyists. This volume provides one small step in understanding the intentional barriers to climate action. We hope this work will inspire new strategies and tactics by those seeking to advance ambitious climate action.

NOTES

1. Riley E. Dunlap and Aaron M. McCright (2011), 'Organized climate change denial'. In: *The Oxford Handbook of Climate Change and Society*, pp. 144–160. New York: Oxford University Press.
2. William F. Lamb, et al. (2020), 'Discourses of Climate Delay', *Global Sustainability*, 3: 1–5. doi:10.1017/SUS.2020.13.
3. Oreskes, Naomi and Erik M. Conway (2011). *Merchants of Doubt: How a Handful of Scientists Obscured the Truth on Issues from Tobacco Smoke to Global Warming*. New York: Bloomsbury Publishing.
4. Robert J. Brulle (2013), 'Advocating Inaction: A Historical Analysis of the Global Climate Coalition', *Environmental Politics*, 32, 2: 185–206. Robert J. Brulle, Galen Hall, Loredana Loy, and Kennedy Schell-Smith (2021), 'Obstructing Action: Foundation Funding and US Climate Change Counter-Movement Organizations', *Climatic Change*, 166: 1–7. Kristoffer Ekberg, Bernhard Forchtner, Martin Hultman, and Kirsti M. Jylhä (2021), *Climate Obstruction: How Denial, Delay and Inaction Are Heating the Planet*. New York: Taylor & Francis; Robert Brulle (2021), 'The Structure of Climate Obstruction: Understanding Opposition to Climate Action in the United States', CSSN Primer, p. 1, https://cssn.org/wp-content/uploads/2021/04/CSSN-Briefing_-Obstruction-2.pdf.
5. See M. L. Djelic and R. Mousavi (2020), 'How the Neoliberal Think Tank Went Global: The Atlas Network, 1981 to the Present', *Nine Lives of Neoliberalism*, 257.

6. Robert J. Brulle, Jason Carmichael, and J. Craig Jenkins (2012), 'Shifting Public Opinion on Climate Change: An Empirical Assessment of Factors Influencing Concern over Climate Change in the US, 2002–2010', *Climatic Change*, 114, 2: 169–188; Jason T. Carmichael and Robert J. Brulle (2017), 'Elite Cues, Media Coverage, and Public Concern: An Integrated Path Analysis of Public Opinion on Climate Change, 2001–2013', *Environmental Politics*, 26, 2: 232–252.
7. Robert K. Merton (1968), *Social Theory and Social Structure* (enlarged ed.). New York: Free Press. ISBN 978-0-02-921130-4; Alejandro Portes (2010), *Economic Sociology: A Systematic Inquiry*. Princeton, NJ: Princeton University Press.
8. Wolfgang Rüdig (1991), 'Green Party Politics around the World', *Environment: Science and Policy for Sustainable Development*, 33, 8: 6–31.
9. See E. Van Haute (ed.) (2016), *Green Parties in Europe*. London: Routledge.
10. J. Korosteleva (2022), 'The Implications of Russia's Invasion of Ukraine for the EU Energy Market and Businesses', *British Journal of Management*, 33, 4: 1678–1682.
11. https://insideclimatenews.org/project/exxon-the-road-not-taken/.
12. Djelic and Mousavi, 'How the Neoliberal Think Tank Went Global'. Also see J. Smith, S. Thompson, and K. Lee (2017), 'The Atlas Network: A 'Strategic Ally' of the Tobacco Industry', *The International Journal of Health Planning and Management*, 32, 4: 433–448.
13. https://www.atlasnetwork.org/. Accessed 28 August 2023.
14. R. J. Brulle (2018), 'The Climate Lobby: A Sectoral Analysis of Lobbying Spending on Climate Change in the USA, 2000 to 2016', *Climatic Change*, 149, 3–4: 289–303.
15. Lamb, et al. 'Discourses of Climate Delay'.
16. Ekberg, et al. Climate Obstruction.
17. European Commission. (2023), Eurobarometer 2023: Climate Change. Online at https://europa.eu/eurobarometer/surveys/detail/2954

INDEX

For the benefit of digital users, indexed terms that span two pages (e.g., 52–53) may, on occasion, appear on only one of those pages.

Tables and figures are indicated by an italic *t* and *f* following the page/paragraph number.

ABDUP lobby group, 171–72
Abdusamatov, Khabibullo, 218
ACC. *See* anthropogenic climate change (ACC)
ACEA (European Automobile Manufacturers' Association), 338–39
Action of Dissatisfied Citizens 2011 (ANO 2011), Czech Republic, 249
Adams, Carol J., 311
adaptation discourse, Czech media, 258
Adenauer, Konrad, 137–38
advertising firms, climate change countermovement (CCCM), 14
advocacy coalitions, CCCM, 14
Ag Climatise, Ireland, 96
Agenda 21, 116–17
Agenzia Nazionale per la Protezione dell'Ambiente (ANPA), 278–79
Agri Aware, Ireland, 98
Agrofert, 249, 251–52
Åkesson, Jimmie, 125
Akzo, 167, 171–72
alarmism, 277
Alders, Hans, 166
All Europe Academies of Science (ALLEA), 18
Alliance of Energy Intensive Industries (AEII), 332–33
Alliance Party, Scotland, 72–73
All-Island Strategic Rail Review, Ireland, 95
All-Party Parliamentary Groups (APPG), 37
All-Poland Alliance of Trade Unions (OPZZ), 200–1
Alternative for Sweden, 123
Alternative für Deutschland (AfD), 139–40, 143, 149–50
American Clean Energy and Security Act (2009), 12
American Energy Alliance, 307–8
American Enterprise Institute, 307–8
American obstructionism, 2–3
American Petroleum Institute, 'Fueling It Forward' campaign, 11
Amsterdam Treaty of 1997, 323–25
An Bord Bia (The Food Board), 98–100
Anderson, Kevin, 111
anthropogenic climate change (ACC)
  doubts over, 219, 221–22
  as myth, 218
  theory of, 218
Apostolic Exhortation Laudate Deum (Francis), 277
Atlas Network, 14, 31, 199, 349, 355–56
attribution scepticism, Poland, 193
Aurora Energy Resources, 70
authoritarian liberalism, 297
Automobilclub Mobil Germany, 154–55

Ayuso, Isabel Díaz, 300
Aznar, José María, 298

Babiš, Andrej, 248, 249, 252–53, 258
Baerbock, Annalena, 156
Baker, Steve, 32–34
Balkenende, Jan Peter, 168
Bareiss, Thomas, 147–48
Barratt Developments, 42
Bates, Ray, 90
Battaglia, Franco, 278
Baxi, 37
BBC, 44–45
BCG Energy, 70
Bełchatów, 186–87, 190, 191
Bellamy, David, 89
Benedict XVI (Pope), 272, 277
Berkeley Group, 42
Berkhout, Guus, 170
Berlusconi, Silvio, 272, 275
Bern, Lars, 116–17, 118
Beyond Oil and Gas Alliance (BOGA), 80
'Beyond Petroleum' campaign, BP, 11
Blair, Tony, 30, 168
Blog, 154
Blue Planet in Green Shackles (Klaus), 260
Bolin, Bert, 109, 114–15
Bolkenstein, Frits, 176
Bosch, 37
Boston Consulting Group, 333
Böttcher, Frits, 119, 166–67, 170, 173
Boykoff, Maxwell, 312
BP, 37, 40, 41–42, 43, 62
    corporate strategy in Scotland, 64–65
    rebranding as 'Beyond Petroleum', 64, 65
brain drain, Soviet Union and, 219–20
Bratby, Philip, 39–40
Brexit Party, 45–46
'Brexit' referendum (2016), 31, 325–26
'Britain Means Business', 45
British American Tobacco, 98
British Coal, 31
British Energy, 62
British Gas Cooperation, 30, 62
British media, United Kingdom (UK), 35–36
British National Oil Corporation (BNOC), 62
British Petroleum (BP), 30, 62

Broadcasting Authority, 90
Brothers of Italy (Fratelli D'Italia), 275, 276
Browne, Lord, BP, 41–42
Brundtland Report, 166
Budyko, Mikhail, 219
Bundesverband der Energie und Wasserwirtschaft (BDEW), 147–48
Buns, Melina Antonio, 109–10
Burmatov, Vladimir, 222
business lobby groups, United Kingdom, 36–37
business opportunity, Czech media, 258–59

Calenda, Carlo, 276
Cameron, David, 30–31, 41–42
campaign funding, climate change countermovement (CCCM), 14
Campaign to Protect Rural England (CPRE), 32, 39–40
Canadian Friends of Science, 170
cap and trade schemes, 65
Captus, 120–21
carbon border adjustment mechanism (CBAM), European Union (EU), 326–27, 332–35
carbon capture, usage and storage (CCUS), 71
carbon capture, utilization, and storage (CCUS), 202
carbon capture and storage (CCS), 42–43, 65, 169, 335–36
carbon forestry, 359–60
carbon leakage
    concept of, 332–33
    flight and fears of, 40
    term, 98
carbon lock-in, national energy systems, 94–95
Carbon Plan, 30–31
Carbunión, 302
car culture, Ireland, 94–95
Carmena, Manuela, 300
Carta Bianca (television show), 280
Cassa Depositi e Prestiti, bank in Italy, 271
Catholic Church, 272–73, 287
CEFIC (European Chemical Industry), 332–33

Celtic Tiger, 82
CEMBUREAU (European Cement
    Association), 332–33, 337
Centre for Clean Coal Technologies,
    Poland, 202–3
Centre for Climate and Energy Analyses
    carbon capture, utilization, and
        storage (CCUS), 202
    Poland, 198
Centre for Economics and Politics (CEP)
    concept of science and ideology, 262
    Czech Republic, 250, 256, 257f
    global warming by, 261
    Ideology master frame, 261–62, 263t
    Science master frame, 261, 263t
Centre for European Policy (CEP),
    149, 334
Centre for Policy Studies (CPS), 34
Chahim, Mohammed, 330
Chambers of Commerce, 63
Charles Koch Foundation, 31
Chase, Howard, 43
Chatham House, 63
Chernobyl, 352–53
Chevron Corporation, 14
Chicago School theories, neoliberal
    economics, 10–11
Chill Out (Bern), 121
Christian Democratic Party (Democrazia
    Cristiana, DC), Italy, 270–71, 272
Christian Democratic Union (CDU),
    Germany, 137–38
Christian Democrats, Germany, 150–51,
    157, 158
Christian Social Union (CSU),
    Germany, 137–38
'Church of Green' discourse, 89
Cingolani, Roberto, 274
Citizens' Assembly on Climate Change,
    Ireland, 93–94
'Citizens Environmental
    Manifest', 116–17
Citizens Environmental Manifesto
    (Bern), 118
Civic Democratic Party (ODS), Czech
    Republic, 248–49
Civic Institute (OI), Czech Republic, 251,
    256, 257f
clean coal, 359–60
Clean Development Mechanism, 332–33

Clement, Wolfgang, 152
climate action, Europe's global role
    in, 1–3
Climate Action Act (2015), Ireland, 80
Climate Action Amendment Bill (2022),
    Ireland, 97
Climate Action Plan, Ireland, 85–87,
    95, 101
Climate Action Tracker, 58–60, 189
climate change, contentious politics of, 2
Climate Change (Scotland) Act (2009),
    60, 65–66
Climate Change Act (2008), 27, 30
climate change action, understanding
    opposition to, 7f
Climate Change Advisory Council,
    Ireland, 86
Climate Change Agreements (CCAs),
    33f, 42
Climate Change and Energy Transition
    Bill (2021), Spain, 299–300
Climate Change Committee (CCC),
    United Kingdom, 27–29
climate change countermovement
    (CCCM), 6–9, 13
Climate Change Law, Spain and, 301
Climate Change Levy (CCL), 42
Climate Change Performance Index,
    Czechia, 247
climate delay
    discourses on, 284–85
    Irish media, 87–88
Climate Delegation
    (Klimatdelegationen), 118
Climate Doctrine, Russia and, 218
'Climategate' affair (2009), 11–12, 121,
    173, 276–77, 309, 356
Climategate.nl, 168
Climate Intelligence (CLINTEL), 90–91
climate obstruction
    Czech Republic, 243–48, 263–64
    emergent ideas for policy and
        action, 362–63
    Europe, 15–18, 348
    European Union (EU), 327–
        32, 337–40
    future research needs, 362
    Ireland, 81
    Italy and, by period, 270–
        73, 285–88

Index [367]

climate obstruction (*cont.*)
   long-term activities, 10–11
   medium-term activities, 11–12
   objectives, activities, and players in, 8*t*
   Poland, 191–201
   practice of, 10–13
   recap of national cases, 349–51
   Russia, 214–16, 233–35
   Scotland, 72–76
   short-term activities, 12–13
   Spain, 296–300
   structure of, 7*f*, 13–15
   term, 6
   United Kingdom, 46–47
climate obstruction lessons
   climate politics in private, 358–59
   denial and delay, 357–58
   Greens and European Union, 361–62
   history and, 352–54
   importance of structural economic conditions, 351–52
   industrial lobbying, 356–57
   international networks of think tanks, 355–56
   media coverage of climate, 360–61
   public relations techniques, 359–60
   range of actors engaged, 354–55
climate policy, Russia on, as Western tool of dominance, 230–31
Climate Policy after Kyoto (Lindzen and Böttcher), 119
climate politics, public opinion and, 358–59
Climate Protection Plan (2050), Germany, 149
Climate Realists, 120–22
Climate Sceptism (magazine), 74
Climate Social Science Network (CSSN), 2–3, 18
Climate Target Plan (2030), 17
climate washing
   greenwashing and, 282–83
   term, 283
CLINTEL, 126, 169, 170, 173
Cluff National Resources, 70
CO2 Coalition, 121–22
coalbed methane (CBM) project, Scotland, 66–67
Coimisiún na Meán, 100
Colau, Ada, 300

Cold War, 349–50, 352–53, 361
Colijn, Hendrikus, 176
Commerzbank, 145–46
Committee for a Constructive Tomorrow (CFACT), 149–50, 356
Common Agricultural Policy (CAP), 302
Common Sense about Energy and Climate (Widding), 126
communication strategies in Czech media, 257–59
   adaptation, 258
   business opportunity, 258–59
   mitigation, 259
   open denialism, 257–58
Community Power, Ireland, 93–94
Competitive Enterprise Institute (CEI), 2, 120–21, 276–77, 356
Confederation of British Industry (CBI), 29, 36–37, 63
Confederation of Netherlands Industry and Employers (VNO-NCW), 171
Confederation of Swedish Enterprise, 113–14, 115, 120–21, 124
Confederation of Swedish Industry, 115
Confindustria
   Italian Chamber of Commerce, 275
   Radio 24, 280
conservative foundations, climate change countermovement (CCCM), 13
conservative media, climate change countermovement (CCCM), 15
Conservative Party, 34–35, 41
   Liberal Democrats and, 30–31
conservative think tanks, climate change countermovement (CCCM), 14
Conservative Woman, The (magazine), 74–75
consumption-based emissions accounting, 15
Cooler Heads Coalition, 121–22, 276–77
Corporate Europe Observatory, 332
corporate social responsibility (CSR), BP in Scotland, 64
corporations, climate change countermovement (CCCM), 13
Corral-Broto, Pablo, 297
Corriere della Sera (newspaper), 271–72
COVID-19 pandemic (2020), 30–31, 82, 206
   economic impact of, 334

emissions during, 162–64
Ireland, 82, 102
Italy's recovery funds, 274
virus, 123–24
CPRE. *See* Campaign to Protect Rural England (CPRE)
Cramer, Jacqueline, 168
Crimea, Russia's annexation of, 144
Crok, Marcel, 90–91, 170
Cross, Clark, 74–75
Crotty, Ted, 68–69
CSSN. *See* Climate Social Science Network (CSSN)
Cullinan, Tim, 99*t*
culture wars, climate, of Spain, 299–300
Czech Coal, 251
Czech media
  business opportunity, 258–59
  climate change adaptation discourse, 258
  mitigation discourse, 259
  open denialism, 257–58
Czech National Bank, 256–57
Czech Republic, 20, 347
  Action of Dissatisfied Citizens, 249
  activists' relationships and networks, 257*f*
  actors and types of institutions involved in, 248–53
  businesses, 251–52
  Centre for Economics and Politics (CEP), 250
  Civic Democratic Party (ODS), 248–49
  Civic Institute (OI), 251
  climate obstruction in, 243–47, 263–64
  coal economy of, 243, 351
  communication strategies, 257–59
  Czech climate story, 244–47
  discursive framings, 260–63, 263*t*
  discursive obstruction categories and actors involved, 260*t*
  far right, 249–50
  greenhouse gas (GHG) emissions, 243–44, 245*f*
  historical description of climate obstruction, 247–48
  Kyoto Protocol, 244
  Liberal Institute, 251
  media and internet-based platforms, 252–53

political parties, 248–50
political strategies and tactics, 253–57
reducing emissions, 350–51
think tanks, 250–51, 255*t*, 355
Václav Klaus Institute (IVK), 250
winning by default, 247
Czyzewski, Adam, 200

Dabrowski, Wojciech, 200
Daily Mail, The (newspaper), 32, 35–36
Daily Sceptic, The (magazine), 74
Daily Telegraph, The (newspaper), 32, 35–36
Dart Energy, 66, 70
defeatism, discourses on, 285
defensive obstruction, Italy and, 286–87
delayism, Poland, 193
Delingpole, James, 36
denial bloggers, climate change countermovement (CCCM), 15
denial machine, 10
Department of Business, Energy and Industrial Strategy (BEIS), 38
  (now Department of Energy Security and Net Zero), 32
Department of Energy and Climate Change (DECC), North Sea, 70–71
Desiderius Erasmus Foundation, 149
DeSmog Climate Disinformation Database, 199
Deutsche Bank, 145–46
Die Familienunternehmer e.V., 148
discursive power, obstructing climate change, 42–46
dismissal obstruction, Italy and, 287
doomism, discourses on, 285
Draghi, Mario, 274
DSM, 167, 171–72
Duda, Andrzej, 186, 193
Dutch-Indian Oil Company (NIAM), 175–76
Dutch Oil Company (NAM), 175–76
Dutch Research Council (NWO), 175
Dvorkovich, Arkady, 231–32
DWS, subsidiary of Deutsche Bank, 145–46

Ecodefence, 223
Economic and Monetary Union (EMU), 321

Index [369]

Ecos (magazine), 274–75
Edelman, 174
eFuel Alliance, 148
EIKE (Euroäisches Institut für Klima und Energie) think tank, 149–50, 153, 154, 156, 356
Ekberg, Kristoffer, 244
Ekho Moskvy (Echo of Moscow), 221
Electricity Feed-In Act (1990), 150
Electricity Supply Board (ESB), Ireland, 93
Emerald Isle, Ireland as, 80
Emissions Trading Scheme (ETS), 42
Enagás, 305
Ende Gelände, 154–55
Endesa, 297–99, 302–3, 305
Enerclub, 302–3
Energetic and Industrial Holding company (EPH), 251–52
Energias de Portugal (EDP), 298–99
Energiewende (energy transition), 136, 158
Energiewirtschaftliche Institut, University of Köln, 148–49
Energy Security Doctrine (ESD), 224–25
energy xenophobia, Poland, 197
Eni, 281
  oil company in Italy, 270–71, 273–75
  promotional campaigns, 284
environmentalism, 201
environmental obstructionism, Spain (1939–1996), 297–98
Environmental Pillar, Ireland, 97
EPICENTER think tank, 336
Equinor, 37
Eucken, Walter, 322
Eudoxa, 120–21
EU Energy Platform Industry Advisory Group, 331
EU Renewable Energy Directive, 201
Eurobarometer, 304, 358
EUROFER. *See* European Steel Association (Eurofer)
Europe
  climate obstruction in, 15–18
  consumption-based emissions accounting, 15
  global role in climate action, 1–3
  overview of case studies, 347–49
  subregions of, 2–3

  total quantity of greenhouse gases (GHGs) by country, 16*f*
European Climate Realist Network, 170
European Commission (EC), 2–3, 322, 323*f*
European Council (EUCO), 322–23, 323*f*, 340
European Court of Justice, 194
European Economic Community (EEC), 82, 166
European Energy Union, 331–32
European Green Deal (EGD), 320, 326–27, 339–40, 357
European Green Taxonomy for Sustainable Investments, 322–23, 327, 335–37
European Monetary System (EMS), 322
European Network of Transmission System Operators for Gas (ENTSOG-G), 331–32
European Parliament (EP), 322, 323*f*
European Roundtable on Climate Change and Sustainable Transition, 331
European Science and Environment Forum (ESEF), 116
European Severe Weather Database, 268
European Steel Association (Eurofer), 329, 332–33, 337
European Transparency Register, 327–28
European Union (EU), 1–2, 20–21
  business preventing climate protection policies, 320–22
  carbon border adjustment mechanism (CBAM), 326–27, 332–35
  climate obstruction actors in, 327–32
  climate policy, 340
  climate targets by year and status, 326*t*
  comparing lobby expenditures in 2019 for green and grey alliances, 330*f*
  denial of reality of climate change, 353–54
  eleven national cases and, 18–21
  EU Emission Trading System (ETS), 325, 328–29, 332–35
  'Fit for 55' climate package, 244
  founding of, 353
  greenhouse gas (GHG) emissions, 323–25, 324*f*
  Greens and EU as targets, 361–62

Green Taxonomy for sustainable
    activities, 335–37
industry, 328–29, 330, 331–32
institutions of, 323f
key frames of climate policy
    obstruction, 337–40
lobbying, 327–28, 329–31, 332, 340–41
National Energy and Climate Plans
    (NECPs), 294–96
Net Zero Industry Act, 339
Poland vetoing 2050 net zero
    emissions, 186–87
public opinion and climate
    politics, 358–59
reducing emissions, 350–51
Russia and, 234
short history of climate policy
    integration context, 322–27
strategies of delay and to moderate
    ambition, 332–37
think tanks, 329, 331
European Union Emissions Trading
    System (EU ETS), 17, 117, 119, 191–
    92, 325, 328–29, 332–35, 353–54
European Union Nature Restoration Law
    (2023), 101–2
EU Transparency Register, 329–30
Ewing, Fergus, 64–65, 69–70
Extinction Rebellion (XR), Germany, 139
Exxon, carbon tax proposal (2017), 11–12
ExxonMobil, 14, 34–35, 37, 43, 277, 354
Exxon Mobile, 2
Ezra, Lord, British Coal, 31

Farage, Nigel, 45
Farm to Fork, 310–11
Federal Environment Agency (UBA),
    Germany, 137
Federal Law N31, Russia, 220
Federal Law N-121, Russia, 220
Federal Law N139, Russia, 220
Federal Ministry of the Environment
    (BMU), Germany, 137
Feicht, Andreas, 148
Fertilizers Europe, 332–33
Fiala, Petr, 262
Fingleton, John, 88
*First Global Assessment of Climate
    Obstruction*, 18
Fitzmaurice, Michael, 99t

FLICC framework (fake experts, logical
    fallacies, impossible expectations,
    cherry picking, and conspiracy
    theories), 287
'floating organizations'
    term, 38–39
    United Kingdom, 38–40
Food Harvest 2020, Ireland and, 85, 96
Foods for Tomorrow, Spain, 305–6
Food Vision 2030, Ireland, 97
Food Wise 2025, Ireland and, 85, 96
Forest Carbon Farms, 203
Formica, Rino, 273
Forum for Civic Development (FOR), 199
Forum for Democracy (FvD), 173
forum shopping, 60–61
Forza Italia, 275
fossil fuel companies, Italy, 273–75
fossil fuel industry, Ministry of
    Economic Affairs and, 19–20
fossil fuel reserves, North Sea, 57–58
fossil fuel saviorism, 288
Fotuyn, Pim, 167
Foundation for Analysis and Social
    Studies (FAES), 298, 302–3
fracking, advocating for unconventional
    gas, 66–70
France, 348
Francis (Pope), 277
Franco, Francisco, 296–97
Francoism, 297
Frankfurter Allgemeine Zeitung
    (newspaper), 156
Fransson, Josef, 122
Free Democratic Party (FDP), Germany,
    139, 158
Freedom and Dignity of Science, 278–79
Freedom and Direct Democracy Party
    (SPD), 249–50, 257–58
    Czech Republic, 249–50
free market, 10–11
'free rider' problem, 98, 99t
Freie Welt (newspaper), 150
Fria Tider, 124
Fridays for Future (FfF), Germany, 139,
    157, 187
Friends of the Earth Ireland, 101–2
Frondel, Manuel, 152–53
'Fueling It Forward' campaign, American
    Petroleum Institute, 11

FuelsEurope, 339
Fundacíon Civismo, 302–3

Gabriel, Sigmar, 150–51
Gas Act (1986), 30
Gas Natural (Naturgy), 305
Gazprom, 225
GB News, 73–74
General German Automobile Club, 154–55
Generation Discover Festival, 174, 175
George Marshall Institute, 286–87
Gerholm, Tor Ragnar, 115, 116, 118, 121
German Automotive Industry (VDA), 144, 146–47, 147t
German Council of Economic Advisors (SVR), 152
German Environment Agency (Umweltbundesamt), 140
Germany, 5, 19–20, 349
   academic and partisan think tanks, 148–49
   anti-wind power campaigns, 153–54
   automobile industry, 347–48
   business associations, 146–48, 147t
   climate policy at a crossroads, 139–40
   at a crossroads, 140–43
   green growth and limited mitigation coalition, 136–43
   greenhouse gas (GHG) emissions, 141f, 142f
   industry headwind push against climate ambitions, 157–58
   INSM campaigns, 155–57
   major grey (nuclear/fossil firms) companies, 144–46, 146t
   opponents of climate action, 143–50
   realm of climate change policy denial, 149–50
   right-wing extremist mobilization, 154–55
   scientific studies, lobbying, and media campaigns, 151–53
   short history of energy transition, 137–39
   strategies and tactics, 150–55
   think tanks, 148–49, 355
GIG Institute, Poland, 198
GIG Research Institute, Poland, 202–3
Giorgetti, Giancarlo, 275–76
Global Climate Coalition, 62–63

global warming, 121–22
Global Warming Policy Foundation (GWPF), 32–34, 35, 36, 74, 90, 91, 354–55
Global Warming Sceptic Organizations, 199
GMB trade union, 71–72
GMB Union, 39
González, Felipe, 297–98
Gore, Al, 119–20, 164, 168, 218
Gorodnitskiy, Aleksandr, 218
government actors and institutions, United Kingdom, 38
governmental-industrial complex (GIC), discourse coalition, 194–95
government obstruction, Russia, 222–25, 229t
Gramsci, Antonio, 29, 359
green authoritarianism, 89, 361
'green economy', Ireland, 98–100
Green Engine, 306
green growth, 36–37
greenhouse gas (GHG) emissions, 1
   Czech Republic, 243–44, 245f
   Europe, total by country, 16f
   European Union (EU), 323–25, 324f
   Germany, 141f, 142f
   Ireland, 80, 83f, 94–97
   Italy, 269f, 270
   Netherlands, 162–64, 163f
   Poland, 187–89, 188f
   Russia, 214–16, 215f, 231
   Scotland, 59f, 349–50
   Spain, 294–96, 295f
   Sweden, 111, 112f
   United Kingdom (UK), 26, 28f
Green New Deal, 118–19, 310–11
Green Party
   European nations, 352–53
   Germany, 17, 137–38, 139–40, 143, 150–51, 156
   Ireland, 96–97
   Poland, 189–90
   Scotland, 64–65
   Spain and, 301–2
   Sweden, 115–16
Greenpeace, 200
Greenpeace Italy, 274–75
Greenpeace Russia, 223
Greens/European Free Alliance, 310–11

greenwash-and-delay, 288
greenwashing
　climate washing and, 282–83
　political tool in Italy, 283
　Spain, 306–7
　strategies, 288
Group of Seven (G7), United
　　Kingdom of, 27
Growth and Environment (Radetzki), 116
Grupo PRISA, 306
Guardian, The (newspaper), 43
Guerreiro, Francisco, 310–11

Haas, Tobias, 328
Habeck, Robert, 143
Hartley-Brewer, Julia, 36
Heartland Institute, 2, 121–22, 149–50,
　　170, 276–77, 356
Hedegaard, Connie, 43
Hendry, Charles, 41–42
Herdan, Thorsten, 148
Heritage Foundation, 307–8
Heydon, Martin, 99*t*
HIF EMEA, 148
Hildingsson, Roger, 117
Hinde, Dominic, 109–10
Hoechst, 167
Holyrood (Scottish Parliament), 57–60,
　　66, 69–70, 75
Hoofnagle, Mark, 287
Hoogovens, 167
Horgan, David, 90–91
Hosking, Jeremy, 31, 41
Hosking Partners, 31
HYBRIT-project, 124

Iberdrola, 298–99, 302–3, 305
ideological obstruction, Italy and, 288
IGas, 70
Il Corriere della Sera (newspaper), 281
Il Foglio (newspaper), 276–77, 279–80
Il Giornale (newspaper), 279–80
Illarionov, Andrei, 217, 230
Il Mattino (newspaper), 279–80
impossibilism, 201
Inconvenient Truth, An (film), 119–20,
　　164, 168, 309
individuals, climate change
　　countermovement (CCCM), 13–14
Industriförbundet, 115

industry obstruction, Russia, 225–28, 229*t*
INEOS, 65, 67–69
InfluenceMap think tank, EU, 329,
　　331, 336
Initiative for a New Social Market
　　Economy (INSM), 149, 155–57
　campaigns, 155–57
　redirection of German climate
　　policy, 156–57
　'Stop the RESA-do the energy
　　transition', 155
InLac, 302–3
Inside Climate News (magazine), 354
Institute for Advanced Sustainability
　　Studies (IASS), 138
Institute for Applied Ecology, 137
Institute for Fuel Technology and
　　Energy, Poland, 198
Institute for Water and Air
　　Quality, 114–15
Institute of Directors, 63
Institute of Economic Affairs (IEA), 34
Institute of Energy Research
　　(IER), 152–53
Institut für Unternehmerische Freiheit
　　(IUF), 149–50
institutionalized pillage,
　　dictatorship, 297
institutional power, obstructing climate
　　change, 41–42
Instituto Brono Leoni, 356
Instituto Juan de Mariana (IJM), 307–9
instrumental realism, obstruction, 271–72
Intergovernmental Panel on Climate
　　Change (IPCC), 3, 270
　Czech Republic and, 256–57
　Dutch merchants of doubt, 173
　Fifth Assessment Report, 44–45
　formation of, 3–4
　Fourth Assessment Report, 119–20
　historical failure of, 5
　improving assessments, 3–5
　Irish media, 87
　Italy and, 270, 278–79, 286–87
　Sixth Assessment Report, 44–45, 281
　Summary for Policymakers (SPM), 4
　Sweden's Bolin and, 109
　Thatcher and, 30
　Third Assessment Report, 278–79
　Working Group One (WG I), 44–45

Index [373]

International Conference on Climate Change, 276–77
International Development Authority (IDA), 98–100
International Energy Agency (IEA), 186
International Futures Forum (IFF), 64–65
International Republican Institute, 248
International Society for Individual Liberty, 248
Interporc
  animal agriculture, 302
  meat culture in Europe, 309–12
  meat lobby, 305–6
Interprofessional Dairy Organization, 302–3
Intesa Sanpaolo, Italy, 281
IPCC. *See* Intergovernmental Panel on Climate Change (IPCC)
Ireland, 19
  actors driving climate obstruction in, 89–100
  agricultural sector and discursive delay tactics, 98–100, 99t
  agri-food sector, 95–96
  background of Irish climate context, 82–87
  climate delay tactics, 87–88
  energy sector, 93–94
  evolving climate policy in, 85–87
  fringe academics and think tanks, 90–91
  greenhouse gas (GHG) emissions, 83f
  historical economic context, 82–84
  lobbying activities on 'climate' by organizations (2015–2022), 92t
  mainstream media and climate obstruction, 87–89
  nuanced landscape of, 80–81
  platforming climate contrarians and sceptics, 88–89
  policymaking in, 84–87
  political influence of agri-food sector, 96–98
  sectoral lobbyists, 91–93
  transport sector reinforcing car culture, 94–95
Irish Business and Employers Confederation (IBEC), 91–93, 92t
Irish Climate Science Forum (ICSF), 90–91

Irish Daily Mail (newspaper), 87, 88
Irish Environmental Protection Agency, 82
Irish Examiner, The (newspaper), 87
Irish Farmers' Association (IFA), 91–93
  delay discourse, 99t
  lobbying activities, 92t
Irish Farmers Journal (IFJ), 98
Irish Independent, The (newspaper), 87
Irish Sun (newspaper), 87
Irish Times, The (newspaper), 87
Istituto Bruno Leoni (IBL), 276–77
Italian Antitrust Authority, Eni fine and, 274
Italy, 20–21
  delaying action, 284–85
  discursive framings, 285–88
  financial institutions and banks, 281
  fossil fuel companies and industry groups, 273–75
  greenhouse gas (GHG) emissions, 269f, 270
  greenwashing and climate washing, 282–83
  historical overview of, 270–73
  individual climate deniers, 278–79
  major actors and types of institutions, 273–81
  media, 279–81
  outright climate change denial, 282
  period 1 (1990–2000), 271–72
  period 1 (1990–2000) and defensive obstruction, 286
  period 2 (2001–2007), 272
  period 2 (2001–2007) and oppositional obstruction, 286–87
  period 3 (2008–2013), 272
  period 3 (2008–2013) and dismissal obstruction, 287
  period 4 (2014–2018), 272–73
  period 4 (2014–2018) and ideological obstruction, 288
  period 5 (2019-present), 273
  period 5 (2019-present) and greenwash-and-delay obstruction, 288
  politicians and political parties, 275–76
  redirecting responsibility, 283–84
  tactics and strategies utilized, 282–85
  think tanks, 276–78, 355

IVL, 114–15, 116–17
Izrael, Yuri, 217, 218
Izvestiya (newspaper), 220–21

Jagiellonian Club, 199, 200–1
Jagiellonian Institute, 199
Jaroš, Petr, 261
Joy, Melanie, 303
Juan de Mariana Institute, 302–3
Jungk, Robert, 137
Just Russia party, Mironov, 230–31
Just Stop Oil, 34–35, 36
just transition
　Poland, 190–91
　Silesia Declaration, 205
Just Transition Fund, European Union, 191

kamikaze politics, 192
Kan, Alexander Rinnooy, 167
Kellner, Petr, 250
Kemerovo Oblast (Siberia), 225–26
Kenny, Pat, 89
Klaus, Václav, 244–46, 247, 298
Klimatsans (Climate Sense), 121–22
KLM Royal Dutch Airlines, 173–74, 175–76
Knaggård, Åsa, 117
Koch, Charles, 2, 13–14, 354
Koch, David, 2, 13–14, 354
Kohl, Helmut, 138
Kondratyev, Kirill, 218
Kowalsky, Janusz, 196
Kreab, 115, 120–21
Křetínský, Daniel, 248, 251–52, 257–58
Kretschmer, Michael, 147–48
Kristersson, Ulf, 111
Kroonenberg, Salomon, 167
Kurtyka, Michał, 197
Kyoto Protocol, 17, 119
　Czech Republic and, 244
　European Union and, 325, 332–33
　Ireland and, 85
　Netherlands and, 168
　Poland and, 187–89, 205
　Russia and, 216–17, 228
　Spain and, 294, 298, 308

Labohm, Hans, 167, 170
Labour Party, 39

Lamb, William, 357
Last Generation, Germany, 139
Latynina, Yulia, 221
Laudato Si (Francis), 277
La Verita (newspaper), 279–80
Law and Justice (PiS), 191, 192, 194
Lawson, Nigel, 32–34, 298
La Zanzara (radio show), 280
Leag in the Lausitz, 154–55
League (Lega), 275–76
Leibniz-Institut für Wirtschaftsforschung (RWI), 148–49
Liberal Democrats, Conservative Party and, 30–31
Liberal Institute (LI), Czech Republic, 251, 254–56, 257f
Libertad Digital (newspaper), 309
license to operate, CSR initiatives, 75
Limits to Growth (Club of Rome), 109
Lindzen, Richard, 119
LobbyControl, 147–48
lobbying firms
　climate change countermovement (CCCM), 15
　European Union (EU), 327–28, 329–31, 332, 340–41
　Germany, 151–53
　Netherlands, 171–72, 174–75
　power of industrial, 356–57
　Spain, 304–5
Lobbying.ie, 91–92
Lomborg, Bjørn, 64–65, 251, 256–57
Loneskie, William, 74–75
Looney, Bernard, 43
Lovins, Amory, 137
Lovins, Hunter, 137
Lukoil, Russia, 281
Lyon, Richard, 74

McAllister, Charles, 34–35
Mackinlay, Craig, 45
MAFRA, 251–52
Mail, 45–46
Makichyan, Arshak, 224
Mann, Michael E., 276–77
Market Stability Reserve (MSR), 334
material and structural power, obstructing climate change, 40–41
Mather, Peter, 41–42

Mattei, Enrico, 270–71
May, Theresa, 33f, 41
Mayer, Andy, 34–35
Mazzarella, Adriano, 278
Medborgarnas Offentliga Utredningar (MOU), 116–17
media
   coverage of climate, 360–61
   Czech Republic, 246–47, 252–53
   Germany, 151–53
   Ireland, 87–89
   Italy, 279–81
   Russia, 220–22, 229t
   Spain, 303, 306
   United Kingdom, 35–36
Mediazoo, 68
Medvedev, Dmitry, 218, 231
Melnichenko, Andrey, 228–30
Meloni, Giorgia, 276
Merchants of Doubt (Oreskes and Conway), 230–31, 278–79, 297
Merkel, Angela, 138, 150–51, 349–50
Michie, Deirdre, 72–73
middle-way policies, Sweden, 110–11
Miersch, Michael, 154
MinEnergo, Russia, 224–25
Ministry of Economic Affairs, fossil fuel industry and, 19–20
Ministry of Economic Affairs and Climate Policy, Netherlands, 170–71
Mironov, Sergey, 230–31
mitigation discourse, Czech media, 259
Mobil Oil, Masterpiece Theatre, 11
Monckton, Christopher, 90–91
Monnet, Jean, 322
Monsanto, 98
Mörner, Nils-Axel, 121–22, 123–24
Müller, Werner, 148
Munch, Edvard, 155–56
Musiał, Kasimierz, 109–10
Myers, Kevin, 88

Naimski, Piotr, 193
National Adaptation Plan (2018), Piano Nazionale di Adattamento al Cambiamento Climatico (PNAAC), 268
National Agency for the Protection of the Environment (Agenzia Nazionale per la Protezione dell'Ambiente, ANPA), 278–79
National Association of Meat Industries, 302
National Climate Adaptation Plan, 223
National Climate Agreement, 168–69
National Council for Science, Technology and Innovation, Ireland, 90
National Economic Chamber (KIG), Poland, 198
National Integrated Plan for Energy and Climate (2023), Piano Nazionale Integrato per l' Energia e il Clima (PNIEC), 270
nationally determined contributions (NDC), 27–29, 58
Natural Step (Det naturliga steget), 118–19
Naturgy, 298–99, 302–3
Nelligan, Maurice, 88
Netherlands, 19–20
   brief history of climate change, 166–69, 176–77
   climate action and inaction, 162–66
   denial and doubt tactics, 173
   discursive framings, 173–74
   Dutch merchants of doubt, 170
   first climate change wave (1987–1989), 166–67
   greenhouse gas (GHG) emissions, 162–64, 163f
   industry lobby groups, 171–72
   key Dutch climate obstructionists, 170–72
   lobbying and networking, 174–75
   Ministry of Economic Affairs, 170–71
   oil and gas in, 164–66, 165f
   oil wealth, 351–52
   polder model, 175
   Royal Dutch Shell, 349–50
   second climate change wave (2006–2011), 168
   Shell, 172
   strategic obstruction, 177
   strategies and tactics, 172–75
   systemic obstruction, 175–76
   third climate change wave (2015–2019), 168–69
Net Zero, 45–46
Net Zero Banking Alliance, 145–46
Net Zero Industry Act, European Union, 339
Net Zero Scrutiny Group (NZSG), 32–34

'Net Zero Strategy', 27
Net Zero Watch (NZW), 32–34, 354–55
New Labour Party, Blair and, 30
New Welfare (Den nya välfärden), 116–17
New York Times (newspaper), 11, 221, 271–72
Next Generation EU National Recovery and Resilience Plan, Italy, 270, 276
Nijpels, Ed, 164
Nongovernmental International Panel on Climate Change (NIPCC), 308
Nord Stream 2, 148
Norilsk Nickel, 217
Northern Sea Route (NSR), 219
North Sea
  fossil fuel reserves in, 57–58
  new climate obstructionism, 70–72
  oil and gas reserves, 61–62
North Sea Forties, 65
North Sea Transition Authority (NSTA), 70–71
Norway, 348
Novatek, 225
Novokuznetsk, 223
nuclear power, Sweden, 114–16
Nuclear Waste Management Fund (KENFO), 145–46

Obajtek, Daniel, 200
O'Brien, Jim, 90–91
Observatory of Sustainability, 294–96
obstruction, structure of, 7f
Office for Energy Regulation (URE), 195
Offshore Energies UK (OEUK), 71
Oil and Gas Authority (OGA), UK's, 70–71
Oil Change International, 281
O'Leary, Michael, 89
Oliver, Neil, 73–74
open denialism, Czech media, 257–58
Organisation for Economic Cooperation and Development (OECD), 186
organized sceptic groups, United Kingdom, 32–35
Osborne, George, 30–31
Otto e Mezzo (television show), 280

Palme, Olof, 109–10, 115
Paris Agreement, 27, 111
  European Union and, 321, 329–30
  Germany and, 140, 156
  Ireland and, 81
  Italy and, 275–77
  Netherlands and, 164, 168–69
  Poland and, 199
  Russia and, 226–27, 228
  Spain and, 301
  Sweden's commitments to, 127, 128
Paris Climate Accord, 86
Party for Freedom (PVV), Netherlands and, 168, 169
Pearson, Allison, 36
Peiser, Benny, 39
People's Party for Freedom and Democracy (VVD), 169
Persson, Göran, 118–19
Petrel Resources, 90–91
PHAUSD, lobby group, 172, 174–75
Philips, 171–72
Phillips Petroleum, 37
Piano Nazionale di Adattamento al Cambiamento Climatico (PNACC), 268
Planeta Azul (No Verde) (Blue Planet in Green Shackles) (Klaus), 298
Plan for the Ecological Transition (Piano per la Transizione Ecologica, PTE), 270
'Plenitude', Eni rebrand, 274
PNACC. See Piano Nazionale di Adattamento al Cambiamento Climatico (PNACC)
Poland, 20, 347
  challenges in phasing out coal, 205–6
  climate debate of, 192–93
  climate imposter, 205
  climate obstruction, 191–94
  coal-dependent economy of, 186–87
  coal economy, 351
  as coal land, 189–91
  from denialism to skepticism, 191–94
  discursive framings, 199–201
  greenhouse gas (GHG) emissions, 187–89, 188f
  key actors in climate obstruction, 194–201
  ministries for climate, energy and environment, 195t
  ownership of energy companies, 196t
  reducing emissions, 350–51

Poland (*cont.*)
   strategies and tactics of obstruction, 202–5
   techno-fixes, 202–4, 359–60
   think tanks, 355
polder model, Netherlands, 175
Policy Exchange, 34–35
Polish Economic Institute, 198, 201, 204
Polish Electric Energy Committee, 199
Polish Energy Policy by 2040 strategy, 203–4
political action committees (PACs), climate change countermovement (CCCM), 14
Pons-Hernández, Mònica, 305
Pontifical Council for Justice and Peace (Pontificio Consiglio della Giustizia e della Pace), 277
Popular Party (PP), Spain, 297–98, 300, 301, 302–3, 311–12
Potsdam Institute for Climate Impact Research (PIK), 138
Powering Past Coal Alliance, 186–87
Prague University of Economics and Business, 256
Preston, Paul, 297
Preuß enElektra (E.ON), 150–51
primary obstruction
   Poland, 193
   term, 113
Prodi, Franco, 279
Prodi, Romano, 279
Pro-Lausitzer Braunkohle e.V., 154–55
Provacuno
   meat culture in Europe, 309–12
   meat lobby, 305–6
Public Broadcasting Service, 11
Putin, Vladimir, 219, 221–22, 223

Radetzki, Marian, 116
Rafael del Pino Foundation, 308
Raidió Teilifís Éireann (RTE), 87
Rajoy, Mariano, 298–99
Ramella, Francesco, 277
Ratcliffe, Jim, 68
Reagan, Ronald, 6
Reclaim Party, 31
ReCommon, Italy, 274–75
Red Flag, 98
Rees-Mogg, Jacob, 32–34, 41
Reform UK, 31, 45–46
Regulation of Lobbying Act (2015), Ireland, 91–92
Reinfeldt, Fredrik, 119–20
Renewable Energy Sources Act (RESA), 136, 138, 139, 143, 144, 150, 151–53, 155, 157–58
renewables, Spain and pioneering, 298–99
Renzi, Matteo, 273, 276
Repsol, 298–99, 302–3, 305, 306
Repubblica (newspaper), 271–72
RESA. *See* Renewable Energy Sources Act (RESA)
Research Institute for Sustainability (RIFS), 138
resource nationalism, Poland, 197
response scepticism, Poland, 193
Ricci, Renato Angelo, 278–79
'Right 4 Coal' campaign, 227–28
right libertarianism, Instituto Juan de Mariana (IJM), 307–9
Roman Catholic Church, 272
Rörsch, Arthur, 167
Rosneft, 148, 225
Rotterdam School of Management, 174
Rowe, Wilson, 218
Royal Dutch Airlines (KLM), 173–74, 175–76
Royal Dutch Shell, 172, 349–50
Royal Irish Academy, 90
Royal Society Edinburgh (RSE), 73–74
Russia, 20, 347
   climate advocates, 361
   climate change as opportunity, 231–32
   climate change as 'second-order' problem, 232–33
   climate obstruction, 233–35
   climate policy as Western tool of dominance, 230–31
   discursive framings, 227–28
   foundations of climate obstruction in, 214–16
   government obstruction, 222–25, 229*t*
   as great ecological power, 228–30
   greenhouse gas (GHG) emissions, 214–16, 215*f*
   historical examples of climate obstruction, 216–17

industry obstruction, 225–28, 229t
Kyoto Protocol and, 216–17
lobbying, 226–27
media obstruction, 220–22, 229t
reducing emissions, 350–51
science obstruction, 218–20, 229t
Russian Union of Industrialists and Entrepreneurs (RUIE), 225, 226–27
Rutte, Mark, 171–72
Ryanair (airline), 89

SACE, Italian export credit agency, 281
Samhällsnytt, 124, 125
Sánchez, Pedro, 299–300, 310
Sandvik, 120–21, 123–24
Save Gas for a Safe Winter, 331–32
Sceptical Environmentalist, The (Lomborg), 251
Schäffler, Frank, 139
School Strike for Climate, 187
Schröder, Gerhard, 147–48
Science and Environmental Policy Project (SEPP), 116, 308
science obstruction, Russia, 218–20, 229t
Scotland, 19
  acts on climate, 65–66
  advocating for unconventional gas, 66–70
  BP's corporate strategy in, 64–65
  brief history of politics of gas and oil in, 61–72
  climate obstruction and delay in, 75–76
  climate obstructionism in, 72–75
  contextualizing corporate climate obstructionism, 60–61
  fracking the transition, 66–70
  greenhouse gas (GHG) emissions, 59f
  North Sea for net zero, 70–72
  obstruction of climate action, 57
  online discourses of delay and denial, 73–75
  Scottish Parliament (Holyrood), 57–60
  Westminster, Holyrood and the black, black oil, 57–61
Scottish and Southern Energy, 62
Scottish Gas (Centrica), 62
Scottish National Party (SNP), 64–65
Scottish Parliament (Holyrood), 57–60, 66, 69–70, 75
Scottish Power, 62

Scottish Sceptic, The (blog), 74
Scottish Utilities Forum, 63
Scream, The (Munch painting), 155–56
Sechin, Igor, 230–31
secondary obstruction, 359–60
  Poland, 193
  term, 113
Sedigás, 302
Seeheimer Kreis, 139
Seibt, Naomi, 149–50
Seitz, Fred, 286–87
self-generated electricity, penalization in Spain, 298–99
Ševčík, Miroslav, 256
Sexual Politics of Meat, The (Adams), 311
Shall We Be Destroyed by Climate or by Our Fighting the Climate? (Klaus), 257–58
Shell, 14, 37, 38, 40, 41, 43, 116–17, 167, 171–72, 175
Siberian Generating Company, 227–28
Silesia Declaration, just transition, 205
Silvia, Eni and, 284
Singer, Fred, 116, 286–87
Skidmore, Chris, 27
Skovgaard, Jakob, 17
Smith, Gary, 39, 71–72
Smith, Iain Duncan, 32–34
Snam, 273–74
SNS Energy, 116
Social Democratic Party, Sweden, 109–10
Social Democrats, Germany, 137–38, 157
Socialist Party (PSOE), Spain, 297–98, 299, 301–2
Solidarity trade union, 190–91, 202–3
Solidarna Polska, 194, 204
Soviet republics, 20
Soviet Union, climatology, 219–20
Spain, 20–21
  citizens, 304, 314
  climate culture wars (2018–2023), 299–300
  climate denial and delay, 352
  climate obstruction, 312
  climate obstructionism in, 296–300
  discursive framings distorting climate debate, 307–12
  education, 314
  genealogy of environmental obstructionism (1939–1996), 297–98

Index [379]

Spain (*cont.*)
  greenhouse gas (GHG) emissions, 294–96, 295*f*
  greenwashing, 306–7
  ideology, 312
  industry, 302, 312–13
  Interporc, 309–12
  legal action, 305–6
  lobbying, 304–5
  media, 303
  media co-optation, 306
  obstructionist stakeholders, 301–4
  pioneering renewables and penalization of self-generated electricity (1996–2018), 298–99
  politicians, 301–2
  privilege, 313–14
  Provacuno, 309–12
  public funds fuelling meat culture in Europe, 309–12
  right libertarianism and climate obstruction, 307–9
  strategies to obstruct climate policies, 304–7
  think tanks, 302–3, 355
Spanish Association of Oil Product Operators, 302
Spanish Electricity Industry Association (AELEC), 302
SPD. *See* Freedom and Direct Democracy Party (SPD)
Spectator, The (magazine), 36, 121–22
Stagnaro, Carlo, 276–77, 336
Starmer, Sir Keir, 35–36
State of Britain, The (blog), 74
Stay on the Ground movement, 128
Stern Review, 119–20
Stevens, Paul, 62
Stockholm Initiative, 120–21
Stockholm University, 121–22
Stoiber, Edmund, 147–48
Stone, Diane, 253–54
Stroukal, Dominik, 262–63
Stuart, Graham, 41
Summary for Policymakers (SPM), IPCC, 4
Sun, The (newspaper), 35, 45–46
Sunak, Rishi, 34–35
Svantesson, Elisabeth, 126
Svenska Arbetsgivarföreningen (SAF), 113–14, 115, 118

Svobodní, 249–50, 257–58
SwebbTV, 123–24, 126
Sweden, 19–20, 347–48, 349
  carbon budget, 111
  climate debates in UNFCCC era, 118–19
  deploying a far-right media ecosystem, 123–24
  early contestations of environmental policies, 114–16
  far right as countermovement ally, 122–23
  far-right nationalism defending industrial modernity, 125–26
  greenhouse gas (GHG) emissions, 111, 112*f*
  mediocre environmental middle way, 109–11
  middle way as secondary obstruction, 113–17
  middle-way policies, 110–11
  national protectionism and allure of middle way, 127–28
  Paris Agreement and, 127, 128
  primary obstruction, 113
  public climate change countermovement, 119–24
  secondary obstruction, 113
  sustaining the unsustainable, 116–17
  Svenska Arbetsgivarföreningen (SAF, later Confederation of Swedish Enterprise), 113–14
  tertiary obstruction, 113
Sweden Democrats, 122–24, 125–26
Swedish Climate Act (2017), 111
Swedish Climate Council, 111, 126
Swedish Environmental Research Institute, 114–15
Swedish Trade Union Confederation, 119–20
Szydło, Beata, 189–90, 361
Szyszko, Jan, 203, 205

Tata Steel, 168–69
Teagasc, Ireland, 96
Telegraph, The (newspaper), 45–46
tertiary obstruction, term, 113
Thatcher, Margaret, 6, 30, 33*f*
think tanks
  Czech Republic, 246, 250–51, 255*t*
  European Union, 329, 331, 336

Germany, 148–49, 355
  international networks of, 355–56
  Ireland, 90–91
  Italy, 276–78
  Poland, 355
  Spain, 298, 302–3
  United Kingdom, 32–35
Third Energy Package, 331–32
Third Pole (Terzo Polo), 276
Thunberg, Greta, 111, 123, 128, 149–50
Timbro, 115, 116–17, 118–19, 120–21, 122, 124
Total, 37
trade associations
  climate change countermovement (CCCM), 14
  United Kingdom, 36–37
Trans-European Gas, 332
Trans-European Networks for Energy, 331–32
transition fuel, natural gas as, 19–20
transition fuels, 359–60
Transparency Register, 336
transport sector, Ireland, 94–95
trend scepticism, Poland, 193
Trikolóra, 249–50, 257–58
Trittin, Jürgen, 150–51
True State of the Planet, The, Timbro and, 118–19
Truss, Liz, 33f, 34–35, 41
Twitter (now X), 74
Tykač, Pavel, 251–52, 257–58
Tynkkynen, Nina, 228–30

UK. *See* United Kingdom (UK)
UK Committee on Climate Change (CCC), 60
UK continental shelf (UKCS), 58–60
UK Petroleum Industry Association (UKPIA), 32, 36, 37
Ukraine
  Chernobyl in, 352–53
  Russian invasion of, 32, 143, 206, 230, 354, 360–61
Una Mirada Fría al Calentamiento Global (An Appeal to Reason) (Lawson), 298
UNFCCC. *See* United Nations Framework Convention on Climate Change (UNFCCC)

Unicredit, Italy, 281
Unidas Podemos, Spain, 299–300
Unilever, 171–72
United Kingdom (UK), 19, 349
  actors and institutions of climate obstructionism, 32–40
  British media, 35–36
  business lobby groups and trade associations, 36–37
  climate obstructionism in, 46–47
  'floating' organizations, 38–40
  government actors and institutions, 38
  greenhouse gas (GHG) emissions, 26
  historical description of climate obstruction in, 30–31
  institutional strategies, 41–42
  institutions involved in obstructionism, 354–55
  making sense of climate obstruction in, 26–29
  material and structural strategies, 40–41
  oil wealth, 351–52
  organized skeptic groups and think tanks, 32–35
  timeline of obstruction activities in, 33f
  total greenhouse gas (GHG) emissions, 28f
United Kingdom Onshore Oil and Gas (UKOOG), 34–35, 67
United Nations (UN), 1–2, 27
United Nations Conference on the Human Environment, 109
United Nations Framework Convention on Climate Change (UNFCCC), 20, 71, 117, 118, 138, 205, 268, 306–7
United Nations General Assembly, 30
Universidad Rey Juan Carlos (Madrid), 307–8
universities, climate change countermovement (CCCM), 14
University College Dublin, 90
University Federico II, Naples, 278
University of Bologna, 278
University of East Anglia, 276–77, 309
University of Ferrara, 279
University of Limerick, 90
University of Modena, 278

Index [381]

University of Padua, 278–79
US Institute for Energy Research (IER), 307–8

Václav Klaus Institute (IVK), Czech Republic, 250, 257f
Vacca, Roberto, 278–79
Vahrenholt, Fritz, 145, 154
van Wijgaarden, William, 90–91
Vattenfall, 116
Vellinga, Pier, 173
Verde, Stefano, 336
Vernunftkraft, 153, 154
Vesa, Juho, 17
Vestas Wind Systems, 329
Visegrad Group, Polish presidents of, 191
Vitol, 41–42
Volvo, 116–17, 120–21
von der Leyen, Ursula, 320, 325–26
Vox, Spain and, 300, 301

Walsh, Edward, 90
Warsaw Enterprise Institute, 199, 201
Washington Post, The (newspaper), 271–72
Waxman-Markey Act, 12
Western Fuels Association, 277
whataboutism, 357
Whitehouse, Sheldon, 5
Widding, Elsa, 123–24, 126
Wiebes, Eric, 168–69

Wildtierstiftung, 153, 154
Wildtier-Webinar, 154
Wilkinson, Harry, 34–35
Wimpey, Taylor, 42
wind energy, anti-wind power campaigns in Germany, 153–54
Wind Energy Ireland (WEI), 91–93, 92t
Windwahn, 153–54
winning by default, Czech Republic, 247
Wissing, Volker, 339
Worker's Educational Foundation (ABF), 109
Working Group III (WG III), IPCC, 3–4
World Bank, 27
World Coal Institute, 31
World Travel and Tourism Council (WTTC), 306–7
World Wildlife Federation (WWF) Russia, 223, 224

Yale Program on Climate Communication, 82
Yara, 168–69
Youth Climate Strikes, 194
YouTube, 74
Yukos, 217

Zapatero, José Luis Rodríguez, 298, 307–8
Zeman, Miloš, 247, 252
Zichichi, Antonino, 278
Zukunft Gas (Future Gas), 147–48